Engineering in
Process Metallurgy

Engineering in Process Metallurgy

R. I. L. Guthrie

Macdonald Professor of Metallurgy
Department of Mining and Metallurgical Engineering
McGill University, Montreal

CLARENDON PRESS · OXFORD
1989

5-12-93

Oxford University Press, Walton Street, Oxford OX2 6DP

Oxford New York Toronto
Delhi Bombay Calcutta Madras Karachi
Petaling Jaya Singapore Hong Kong Tokyo
Nairobi Dar es Salaam Cape Town
Melbourne Auckland

and associated companies in
Berlin Ibadan

Oxford is a trade mark of Oxford University Press

Published in the United States
by Oxford University Press, New York

British Library Cataloguing in Publication Data
Guthrie, R. I. L.
Engineering in process metallurgy.
1. Chemical metallurgy
I. Title
669'.9
ISBN 0–19–856222–5

Library of Congress Cataloging in Publication Data
Guthrie, R. I. L.
Engineering in process metallurgy.
Includes bibliographies and index.
1. Metallurgy. 2. Heat—Transmission. 3. Mass
transfer. I. Title.
TN665.G95 1988 669 87–35026
ISBN 0–19–856222–5

Typeset by Macmillan India Ltd,
Printed and bound in
Great Britain by Biddles Ltd,
Guildford and King's Lynn

To   Margaret

PREFACE

The last twenty years have witnessed much progress aimed at improving and controlling the kinetics of metallurgical processing operations.

Thanks to continued advances, metal products of higher chemical and physical quality are now being produced more efficiently than ever before. In a highly competitive world, all metallurgical graduates working in the processing industries need to understand the central role that heat, mass, and momentum transport phenomena play in determining process kinetics. Similarly, they need to learn how thermodynamic relationships, or equilibria, must form an integral part of process kinetics, and how these can be incorporated into the subject.

The primary purpose of this text, therefore, is to provide a fairly comprehensive, but basic, treatment of process engineering metallurgy. While many books have been written on the subject of transport phenomena for chemical and mechanical engineers, and these are of value to the process metallurgist, most of the examples and illustrations cited are naturally external to the field (of process metallurgy). Further, many of the texts are too extensive, or specialized, for easy incorporation into a metallurgical/materials science curriculum.

On the other side, many practising mechanical/chemical engineers entering the metallurgical field will not normally have been exposed to the peculiar features of metallurgical operations; these include very high processing temperatures, and unusual fluids (e.g. liquid metals and slags), whose chemical and physical properties can result in phenomena unique to metallurgical, as opposed to chemical, engineering. Similarly, their background in chemical, rather than metallurgical, thermodynamics will tend to make their knowledge of mass transport phenomena more difficult to apply.

Since the amount of time devoted to process metallurgy varies markedly from one curriculum to another, the present text has been written to take this into account, and can be ingested piecewise, as it is divided into four major components.

Chapter 1 introduces the subject, illustrating the role of transport phenomena in typical metallurgical operations. It goes on to present the fundamental or atomistic, origin of those transport coefficients governing the flow of momentum, heat, and mass within a system. Radiation properties are also presented at this point.

Chapters 2 and 3 go on to treat fluid statics, and then fluid dynamics in metallurgy, followed by process modelling techniques, using a series of relevant examples.

Chapter 4 provides an integrated approach to the subject of conductive heat flow, and diffusion in solids (or other stationary systems), leading into

Chapter 5, where the effects of convection on heat and mass flows are incorporated.

Chapter 6 completes the text, introducing the reader to the numerical modelling of metallurgical processes, and demonstrating the various techniques whereby complex metallurgical operations can be both successfully analysed and synthesized.

In presenting the subject matter in this way, the book reflects three courses (heat and mass transfer, fluid dynamics, numerical modelling) presented to a generation of patient McGill students. Many worked examples are presented to illustrate the subject areas. A series of relevant exercises, often dealing with actual industrial examples the author has encountered over the past twenty years, are also included for study.

In conclusion, it is hoped that this book can provide a starting point for the metallurgical student or graduate entering the processing industry. Equally, the text can provide valuable orientation to engineers or scientists of a different background, but working, or about to enter, the metallurgical processing industries. Given the extensive data on physical properties, together with appendices providing useful thermodynamic data, and conversion factors, it is hoped that the text can also be used as a quick reference source and workbook.

McGill University   R. I. L. G.
February 1987

ACKNOWLEDGEMENTS

The author would like to acknowledge the influence of his various instructors at Imperial College some twenty years ago when Process Engineering Metallurgy was in its infancy: Professor F. D. Richardson, for his leadership in introducing the subject to the undergraduate curriculum, and showing its intrinsic coupling with metallurgical thermodynamics, Professor A. V. Bradshaw for his pragmatic and realistic approach to the subject, Professor J. Szekely for his theoretical insights, and Professor A. W. D. Hills for his superb teaching skills. The subject has now come of age and its demonstrated accomplishments such as fully three-dimensional, transient models of blast furnace behaviour, rank with the very best in equivalent disciplines.

In preparing this book, the author deeply appreciates the knowledge he has gained conducting research with his various graduate students, research associates, and visiting professors while at McGill (R. Anantharayanan, S. Argyropoulos, I. Cameron, F. M. Chiesa, F. Dallaire, D. Doutre, D. Frayce, L. Gourtsoyannis, J.-W. Han, M. Hasan, J. Herbertson, D. Holford, T. Iida, C. Jefferies, S. Joo, A. Kadoglou, J. Kamal, S.-H. Kim, B. Kulunk, S. Kuyucak, H.-C. Lee, D. Mazumdar, F. Mucciardi, H. Nakajima, E. Ozberk, R. Perrin, Y. Sahai, M. Salcudean, M. Y. Solar, A. Storey, M. Tanaka, S. Tanaka, Tian, R. Urquhart, D. Verhelst, and C. Xu).

Particular thanks go to Professor C. Bale of Ecole Polytechnique for help with the thermodynamic examples, to Dr M. Hasan for help with proof reading and his boundary layer program, and Dr J. Herbertson and Miss P. Khan, for helping assemble the appendices and correlations.

Finally, the author acknowledges the forbearance of his family, in the hope that it was a worthwhile enterprise, even though it took far too long to accomplish.

The author gratefully acknowledges permission to redraw, or reset, the following list of figures and tables from previous publications.

Fig. 1.5a	Pehlke et al (1974)	Iron and Steel Soc. of AIME, Warrendale, Pa., U.S.A.
Figs. 1.14, 1.15	Richardson (1974)	Academic Press, New York, U.S.A.
Figs. 1.16, 1.17	Schack (1965)	Academic Press, New York, U.S.A.
Fig. 1.18	Geiger & Poirier (1980)	Addison-Wesley, Philippines'
Fig. 1.19	Iida & Guthrie (1988)	Oxford Univ. Press, Oxford, U.K.

Fig. 1.22	Croxton (1975)	Wiley Int'l., London, U.K.
Figs. 1.23a,b,c,d	Iida & Guthrie (1988)	Oxford Univ. Press, Oxford, U.K.
Figs. 1.24–1.26	Croxton (1975)	Wiley Int'l., London, U.K.
Table 1.2	Iida & Guthrie (1988)	Oxford Univ. Press, Oxford, U.K.
Figs. 1.27–1.30	Iida & Guthrie (1988)	Oxford Univ. Press, Oxford, U.K.
Fig. 1.31	Iida & Morita (1975)	Iron & Steel Soc. of AIME, Warrendale, PA., U.S.A.
Figs. 1.32–1.34	Iida & Guthrie (1988)	Oxford Univ. Press, Oxford, U.K.
Figs. 1.35, 1.36	Iida & Morita (1975)	Iron & Steel Soc. of AIME, Warrendale, PA., U.S.A.
Tables 2.1, 2.2, 2.3	Bird, Stewart & Lightfoot (1966)	John Wiley and Sons, New York, U.S.A.
Fig. 2.9	Bird, Stewart & Lightfoot (1966)	John Wiley and Sons, New York, U.S.A.
Fig. 2.10	Prandtl and Tietjens (1957)	Dover Publications, Inc., New York, U.S.A.
Fig. 2.16	Davies (1972)	Academic Press, New York, U.S.A.
Fig. 2.18	Davies (1972)	Academic Press, New York, U.S.A.
Fig. 2.22	White (1979)	McGraw Hill, New York, U.S.A.
Fig. 2.27	Davenport (1979)	Pergamon Journals Ltd., Oxford, U.K.
Fig. 3.14	Levenspiel (1967)	Wiley & Sons Inc., New York, U.S.A.
Fig. 4.22a,b,c	Heisler (1947)	Amer. Soc. Mech. Eng., New York, U.S.A.
Fig. 5.11	Solar and Guthrie (1975)	Metallurgical Soc. Warrendale, U.S.A.
Fig. 5.20	Hills (1966)	Institute of Mining and Metallurgy, London, U.K.
Figs. A3-2, A3-3	Richardson (1974)	Academic Press, New York, U.S.A.

CONTENTS

1

AN INTRODUCTION TO TRANSPORT
PHENOMENA AND PROPERTIES IN
METALLURGICAL OPERATIONS

1.0. INTRODUCTION

In today's world of high-level competition, of technical competence, and of relatively low bulk-transportation costs, any company involved in chemical or metallurgical manufacturing must make maximum use of efficient processing operations. For engineers and researchers engaged in the development and operation of such practices, a knowledge of thermodynamics and process kinetics is essential.

The present text has been written with this in mind and is intended for those interested in the metallurgical industry. The various concepts of fluid, heat, and mass transfer are therefore introduced by way of a series of metallurgical examples and practical problems.

Since useful solutions to many industrial design problems are readily obtained without one having to become too deeply involved in the detailed complexities of transport phenomena, care has been taken to avoid unnecessary complications in the treatment. The reader is therefore referred to the specialist literature for more detailed and exhaustive treatments of the various topics and processes discussed in the present text.

1.1. EXAMPLES OF HEAT AND MASS TRANSFER IN METALLURGICAL ENGINEERING

Practically all metallurgical processing operations involve, in one way or another, fluid flow and the transfer of heat and/or mass. To illustrate this statement and to familiarize the reader with some typical metallurgical operations, it is worthwhile to look briefly at some of the heat, mass, and momentum transfer aspects involved in steel-making. Figure 1.1 illustrates the sequence of operations typically performed in going from raw iron oxides to steel.

(a) Coke-making

The first major operation (Fig. 1.2) involves the production of coke, which is required for the reduction of iron oxides to iron. The process involves the destructive distillation of metallurgical grade coals in coking chambers (sealed

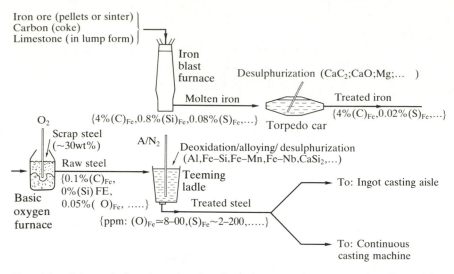

FIG. 1.1.   Schematic flowsheet showing the major operations conventionally used for converting iron into raw steel.

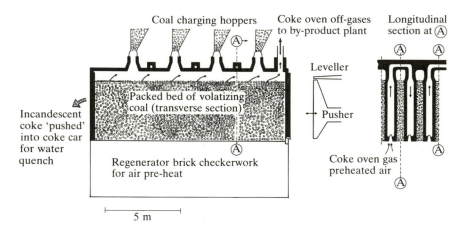

FIG. 1.2.   The vertical slot coke oven for the production of metallurgical grade coke for the iron blast furnace.

vertical slots 4 m high, 12 m wide, 0.5 m thick) of by-product coke ovens. The heat needed for distillation of the volatiles is transferred through the brickwork from adjacent vertical flues by combustion of enriched blast furnace off-gases. After about 17 hours' induction time, the incandescent coke is pushed out into transfer railway cars. During its fall, it breaks apart to form large lumps. The coke lumps are then transferred to the quenching tower, in which an intense and normally intermittent water spray quenches them for subsequent charging to the blast furnace.

This quenching procedure is an interesting example of heat transfer, in that the heat transferred from the inner portions of the coke lumps vaporizes the water running down their external surfaces. The overall operation involves the emission of vast plumes of steam (as well as certain undesirable vapours such as phenols and cyanides). The objectives here are to ensure the minimum amount of retained moisture in the coke, its endothermic effects contributing to poor fuel efficiency in the blast furnace.

(b)   Blast furnance

The iron-making blast furnace (Fig. 1.3), has evolved over many hundreds of years to become an efficient countercurrent exchanger of heat and counter-current exchanger of mass (i.e., of oxygen). Thus iron oxide (as pellets or sinter), together with coke and limestone, are successively charged through the top of the furnace. The charge slowly descends through the shaft (an eight-hour journey) and is gradually heated by hot ascending gases. As the gas lower down the furnace is richer in carbon monoxide, and therefore more reducing towards iron oxides, the pellets are gradually reduced as a result of mass transfer of carbon monoxide (and hydrogen) from the gas phase into the pellet, i.e.

$$3Fe_2O_3 + CO \rightarrow 2Fe_3O_4 + CO_2$$
$$Fe_3O_4 + CO \rightarrow 3FeO + CO_2$$

Final deoxidation is accomplished down in the cohesive zone where high temperatures and highly reducing conditions result in the reduction of wustite (FeO) to iron. Impurities such as silica, sulphur, alumina, and magnesia, present in the original pellets associate with the lime and are removed as a molten slag.

The final reduction from FeO takes place via mass transfer mechanisms:

1 $\qquad \begin{cases} FeO + CO \rightarrow Fe + CO_2 \\ CO_2 + C(coke) \rightarrow 2CO \end{cases}$ Indirect reduction
and

2 $\qquad\qquad FeO + C \rightarrow Fe + CO \qquad$ Direct reduction

The CO/CO_2 reaction is often termed the solution loss reaction (i.e., dissolution of coke by CO_2). Although the purpose of the coke is obviously to act as a reductant, it also plays another critical role in that it behaves as a supporting pillar for overlaying burden. In the region below the cohesive (or 'sticky') zone in Fig. 1.3, everything else is molten or melting (i.e., slag and pig iron).

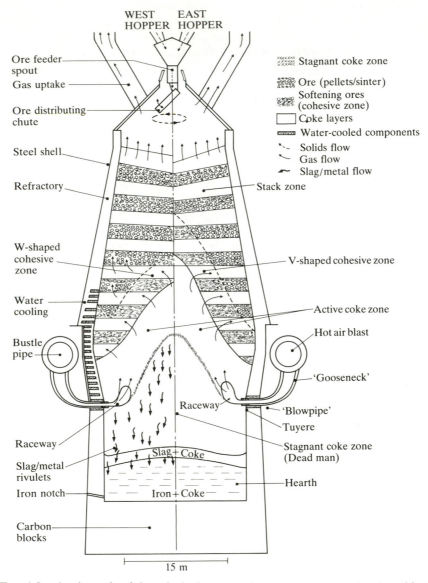

FIG. 1.3. A schematic of the principal zones and component parts of an iron blast furnace.

This brief description provides a glimpse of the complex heat and mass transport phenomena and chemical reactions involved in blast furnace technology.

(c) Stoves

To generate the high temperatures required for reduction to iron in the hearth region of the blast furnace, the incoming air is pre-heated prior to combustion with coke. This is done by passing the cold air blast over hot, pre-heated brickwork (checkerwork at about 1500 K) in the blast furnace stove shown in Fig. 1.4. Since the cold air gradually extracts the stored heat, a separate heating phase is also necessary, and this is effected by shutting off the cold air

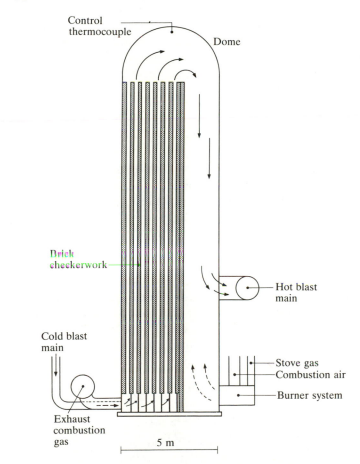

FIG. 1.4. A schematic of the principal zones and component parts of a stove for pre-heating the air blast to an iron blast furnace.

blast to the stove, opening up the gas valve, and combusting enriched blast furnace off-gas (cleaned via water scrubbers and electrostatic precipitators) to bring the cooled checkerwork back up to temperature. Once again, an interesting heat transfer problem is posed concerning the optimum recycling time for a batch of stoves (normally three per blast furnance, in parallel) so that a 'constant maximum' pre-heat temperature to the blast furnace can be achieved. This is of commercial significance, since higher pre-heat temperatures translate directly into lower coke rates per 'net tonne of hot metal' (NTHM).

(d)   Steel-making

Since the blast furnace produces hot metal that is saturated with carbon (from the coke), the next operation requires that this and other residual impurities (e.g., phosphorous, sulphur) be removed. Removal of impurities can be accomplished by blowing high-velocity (supersonic) jets of pure oxygen on to the hot metal, as shown in Fig. 1.5. Dissolved carbon is oxidized and escapes as carbon monoxide from the mouth of the vessel, while the other oxidized impurities are slagged off by addition of burnt lime. To compensate for the vast amounts of heat liberated during the oxidation of these and other impurities (e.g., silicon, manganese), about 30 per cent of the total charge comprises steel scrap as coolant. This scrap coolant is required to avoid the temperature of the molten steel exceeding 1923 K and causing unnecessary refractory erosion. Once again, highly complex heat, mass, and fluid transport mechanisms are involved. For example, mass transfer of bath carbon to the scrap metal surfaces effectively dissolves light section scrap, even though bath temperatures are below the melting point of the scrap during the major portion of a blow.

However, when the bath temperature exceeds the scrap melting range (1770–1810 K), normal thermal processes, involving turbulent heat transfer, melt the scrap, which finally becomes assimilated into the molten bath.

Typical mass transport reactions in the basic oxygen furnace (BOF)† involve oxygen transport to exposed metal surfaces. With simultaneous transport of dissolved carbon from inside the liquid to the gas–liquid interface, carbon is removed from the melt as carbon monoxide:

$$2(C)_{Fe} + O_2 \rightarrow 2CO$$

Similarly,

$$(Si)_{Fe} + O_2 \rightarrow (SiO_2)_{slag}$$
$$(Mn)_{Fe} + \tfrac{1}{2}O_2 \rightarrow (MnO)_{slag}$$
$$2(P)_{Fe} + \tfrac{5}{2}O_2 \rightarrow (P_2O_5)_{slag}$$

†BOF is alternatively known as BOP (Basic Oxygen Process) or LD (Linz–Donovitz) process.

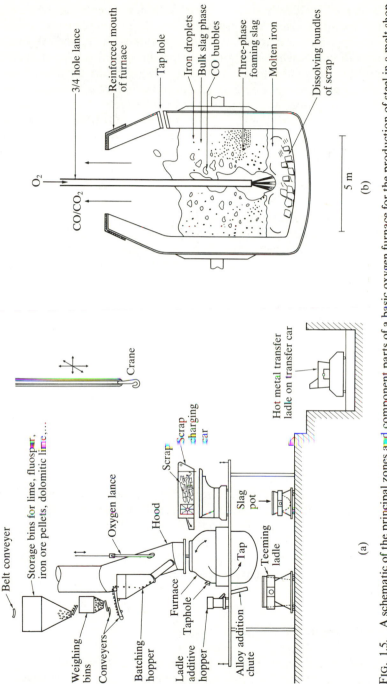

FIG. 1.5. A schematic of the principal zones and component parts of a basic oxygen furnace for the production of steel in a melt shop. (a) Typical plant layout; (b) BOF vessel.

for other impurities dissolved in the melt. It should be emphasized that the exact transfer mechanisms are obscure and tend to remain so, owing to the opacity of the system and to the experimental difficulties of and restrictions involved in direct measurements of important process variables at 1600 °C.

To emphasize the importance of hydrodynamics, one notes that it is now technically possible to inject pure oxygen into the bottom of the furnace via a series of submerged jets, without involving excessive refractory erosion at the nozzle tips. This process (known as the Q-BOP or OBM) produces better slag–metal contact and a more thorough mixing of the steel than does the 'conventional' BOF operation. The result is reduced processing times and a higher metallic yield. To meet this competition, however, many of today's basic oxygen furnaces have been retro-fitted with submerged gas-stirring facilities. It is appropriate to note that smaller non-integrated steel plants generally recycle steel scrap by remelting this in electric arc furnaces. The product is a raw steel, similar in composition to that produced by BOFs but often higher in residual elements such as copper and tin. In some countries (e.g. Russia), the open hearth process (described later) is practised, which again produces a molten product also requiring ladle treatment.

(e) Steel ladle additions

The raw steel poured from a steel-making furnace into a teeming ladle is usually too highly oxidized for immediate use, as it contains 0.04–0.1 wt% oxygen. This would cause blowholes in the steel if it were then solidified. Steel deoxidants (e.g., aluminium, ferro-silicon, or carbon) are therefore required to bring dissolved oxygen contents down to acceptable levels through precipit-ation of condensed oxides. At the same time, additions of other ferroalloys (e.g., Fe–Mn, Fe–Nb, Si–Mn, Fe–V) are made as appropriate, so as to meet the chemical specifications required for the variety of steel grades commonly produced by any integrated steel company.

These additions melt as a result of heat transfer processes and subsequently disperse into the bath. A knowledge of heat transfer rates and melting phenomena can help in deciding what maximum sizes of additions can be made, so that costs of crushing ferro-alloys and dust losses can be minimized. Similarly, a study of their hydrodynamic trajectories in the metal bath can lead to improved injection techniques (e.g., aluminium wire feeding), for better control and recoveries.

(f) Teeming ingots and continuous casting

Heat losses occur from the hot steel contained in the teeming ladle and tundish during continuous casting operations (Fig. 1.6(a)). These have to be

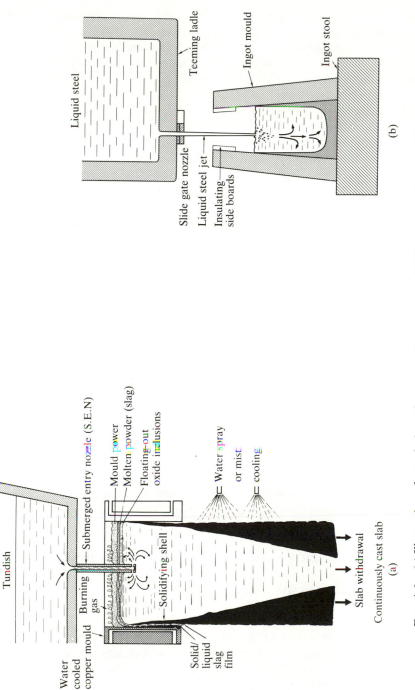

FIG. 1.6. (a) Illustration of a typical continuous casting operation. (b) Illustration of a typical ingot pouring operation.

minimized, since slight temperature differences can profoundly affect the quality of the slabs or billets that are produced.

Heat is lost partly by conduction through the side and bottom walls of a ladle, and partly as a result of convection and radiation from the top surface. These latter losses are minimized by the use of a slag layer, 50 mm or more thick, on top of the steel. The slag acts as an effective thermal blanket. Additionally or alternatively, a ladle cover may be used.

The solidification of continuously cast slabs, or of ingots, is also an obvious example of a heat transfer process. In this, the rate of the metal's solidification depends on the amount of heat extracted through the walls of the mould. For instance, by suitable adjustments of cooling rates in the upper section of static moulds (e.g., by use of hot top compounds and sidewall insulators), ingot yields can be vastly improved by avoiding (or limiting) the depth of pipe commonly formed as a result of metal contraction during freezing (Fig 1.6(b)).

Mass transfer phenomena in an ingot pouring operation are just as important, although less obvious. Thus, depending on the type and quality of steel desired (e.g., a rimming steel), the oxygen content of the steel being teemed is adjusted so that, during subsequent freezing in the mould, the decrease in oxygen solubility between solid and liquid iron can precipitate FeO droplets, so raising the local concentration product of dissolved carbon and oxygen. As a result, the partial pressure of carbon monoxide exceeds the hydrostatic (or ferrostatic) pressure in the liquid just ahead of the freezing interface, thereby causing CO bubbles to nucleate. The intention is to match the rate of evolution of carbon monoxide bubbles with the rate of contraction of the metal during freezing, so as to eliminate gross shrinkage cavities in an ingot and replace them with small dispersed voids. These voids are easily welded during subsequent rolling operations.

Many flow studies of the jet of steel entering the mould in continuous casting operations have been carried out by individual steel companies, to determine the optimum way in which recirculating flow patterns can be set up so as to minimize slag entrainment and to promote agglomeration and float-out of micrometre-sized impurities (e.g., iron manganese silicate reoxidation products).

For tackling such problems, an understanding of fluid flow phenomena and particle–liquid interactions is also of importance to the process metallurgist.

(g)   Soaking pits and slab reheat furnaces

Apart from the obvious heat transfer problem involved in heating eight or so 10-tonne ingots in a soaking pit (see Fig. 1.7) or a continuously cast slab in a slab-reheating furnace, to rolling temperature with maximum fuel efficiency, a less obvious mass transfer operation is involved. During the relatively slow solidification of thick ingots and slabs, microsegregation of solutes occurs

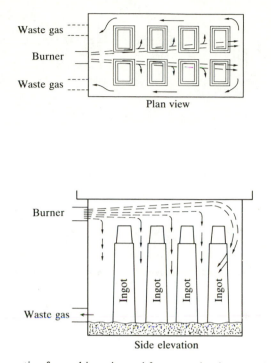

FIG. 1.7.   A schematic of a soaking pit used for preparing ingots prior to hot rolling operations.

within the steel matrix as a result of dendritic-type freezing. To eliminate this microsegregation, and thereby avoid its detrimental effects on the the steel's physical properties, time (e.g., 12–18 hours for thick ingots) must be allowed for the solutes to redistribute themselves evenly throughout the steel. Since this solute redistribution can only involve atomic diffusion mechanisms, long retention times at high temperatures are necessary, diffusion in solids being slow even under the best circumstances.

(h)   Hot rolling

As a final illustration of transport phenomena, the hot rolling of ingots received from the soaking pits, or the slab reheating furnace in continuous casting operations, is an excellent example of conductive heat transfer. As seen from Fig. 1.8, a roughing mill, followed by a series of finishing stands, roll the ingots down to strip (i.e., sheet). Enormous quantitites of stored heat are present and this is removed by conduction of heat from the sheet to the rolls. Thus, the roll surfaces have to be continuously cooled by intense water sprays

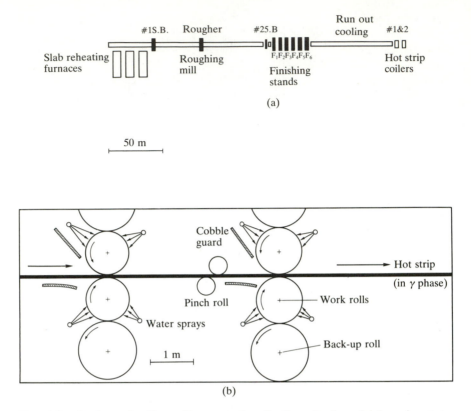

Fig. 1.8.   A schematic of hot rolling operations for the reduction of slabs or ingots to hot strip in the hot mill. (a) Typical layout (plan view); (b) Segment of finishing stands.

to prevent possible distortion, warping and hot checking of their surfaces. It is instructive to note that about 70 per cent of the total water requirement of an integrated steel plant is necessary for this operation and that of the subsequent heat treatment run-out cooling line.

(i)   Subsequent operations

The subsequent operations involved in sheet production include pickling, cold rolling, zinc-coating, electrolytic tin-plating, batch annealing, and continuous annealing. Various heat and mass transfer phenomena are involved but will not be treated here, since the reader will have appreciated by now that practically every metallurgical process involves the interphase transfer of heat or mass, and material transport.

(j)   Conclusions

Heat, mass, and momentum transport processes form an integral part of all metallurgical operations. A good understanding of the principles involved can lead to many useful predictions about and improvements in plant processing operations and practices, as well as being essential for any process design work.

1.2.   MECHANISMS OF MOMENTUM TRANSPORT

Let us now consider the case of the falling free jet of molten steel issuing from the teeming ladle, illustrated in Fig. 1.6(b), so as to illustrate the two basic equations governing the transport of momentum in Newtonian fluids, these being Newton's second law of motion, and Newton's law of viscosity.

(a)   Newton's second law of motion

Figure 1.9(a) depicts an element of liquid steel, of mass M, moving from position 1 to 2, as it accelerates towards the surface of the filling ladle. Newton's second law states that the sum of the normal forces acting on this body in the vertical z direction is equal to the mass of the body multiplied by its z acceleration, or time rate of z momentum change. i.e.

$$\Sigma F = Ma \quad \text{or} \quad \Sigma F = \frac{d}{dt}(MV). \tag{1.1a}$$

Referring to Fig. 1.9(a), the atmospheric pressure forces acting normally to the two horizontal surfaces of the volume element illustrated will be equal and opposite, over the small height in question, so that $\Sigma F = F_g$:

$$F_g = Ma. \tag{1.1b}$$

Replacing F_g with the weight of the volume element (i.e. Mg), one sees that velocity of the element will continuously increase at a rate of 9.8 metres per second every second, or g, the Earth's gravitational constant.

We may usefully delve a little further, and change our volume element from a Lagrangian to an Eulerian frame of reference. Consider therefore the Eulerian volume element (control volume) shown to the left side of Fig. 1.9(a), fixed in space, into which steel flows at z, and exits at $z + dz$. To convert from time to space coordinates in this way, we have the identities:

$$a = \frac{dV}{dt} = \frac{dV}{dz} \cdot \frac{dz}{dt} = V \cdot \frac{dV}{dz}.$$

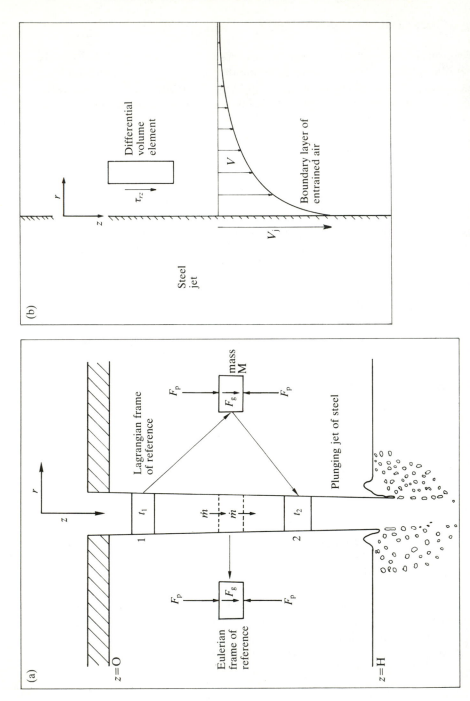

FIG. 1.9.   Momentum transport processes in fluids illustrated by a plunging jet of liquid steel. (a) Distinction between differential volume elements fixed in space (Eulerian), or moving (Lagrangian) with the flow. (b) Illustration of transmitted shear at the liquid–gas interface, causing a boundary layer of entrained air to develop.

Equation (1.1a) then becomes

$$F_g = MV \frac{dV}{dz},$$ (1.1c)

which can be integrated from $z=0$ to H and $V=V_0$ to V_f to give

$$F_g H = \tfrac{1}{2} M (V_f^2 - V_0^2).$$

Substituting $\rho A dz$ for M, where ρ is the fluid density,

$$\rho g H = \tfrac{1}{2} \rho V_f^2 - \tfrac{1}{2} \rho V_0^2.$$ (1.1d)

It is now instructive to compare eqns (1.1b) and (1.1c). One can readily show that the former represents a momentum balance over the control volume element shown in Fig. 1.9(a), and states that the nett force acting on the control volume, F_g is equal to the rate of momentum change.

$$F_g = MV \frac{dV}{dz} \equiv \rho A dz \, V \frac{dV}{dz} = \dot{m} dV = \dot{m} (V_f - V_0),$$

where $\dot{m}$ is the mass flowrate through the Eulerian volume element. On the other hand, the integrated form of this momentum balance shows that the decrease in potential energy of the steel in moving from $z=0$ to $z=H$ is exactly compensated by an increase in its kinetic energy. Equation (1.1c) therefore represents a particular form of the steady-flow energy equation, and eqn (1.1b) a particular form of the momentum equation, both deriving from Newton's second law of motion.

Taking a height of 1 metre, steel exiting a ladle at 5 m s^{-1} would (from eqn (1.1c) impact at 5.9 m s^{-1} with a specific volume kinetic energy of $1.2 \times 10^5 \text{ kg m}^{-1} \text{s}^{-2}$.

(b) Newton's law of viscosity

The second way in which momentum transport can occur is less obvious, but follows directly from the other fundamental law of fluid mechanics, again first enunciated by Newton. This is Newton's law of viscosity, which in its simplest terms states that a fluid element, in steady parallel flow, will resist the application of a shear stress on one of its surfaces, and that this resistance will increase in proportion to the *time rate* of deformation, i.e.:

$$F_s = A_s \mu \frac{dV}{dr} \quad \text{or} \quad \tau_{rz} = -\mu \frac{dV}{dr}.$$ (1.2)

The constant of proportionality μ is defined by this equation; μ, is known as the coefficient of molecular viscosity, the absolute viscosity, or the dynamic viscosity. The negative sign in eqn (1.2) results from the sign convention normally adopted, which specifies that the direction of increasing z is to be the direction of positive shear stress, τ_{rz}.

Consider Fig. 1.9b, where we see an element of gas, adjacent to the steel jet, moving tangentially downwards at constant velocity, V. The higher velocity gas, closer to the jet, will transmit a shearing force to this gaseous element, so as to cause a corresponding downward motion. Partly as a result of this shearing action, a boundary layer of air is formed, which can lead to entrainment of air in the liquid steel.

Specifically, expecting that the gas velocity gradient at the gas–liquid steel interface will be of the order of $10 \text{ m s}^{-1} \text{cm}^{-1}$ or 10^3 s^{-1}, and given that the molecular viscosity of oxygen at 2400 K is $10^{-6} \text{ kg m}^{-1} \text{s}^{-1}$, application of eqn (1.2) shows the surface shear stress transmitted will be $\simeq 0.001$ pascals (N m^{-2}). For simplicity of treatment, planar boundary layer flow is assumed, curvature effects in this case being minimal.

To interpret Newton's law of molecular viscosity as a momentum interchange phenomenon, one may draw an analogy from the kinetic theory of gases. Thus, in the very neighbourhood of the steel surface, the gas molecules will move at the surface velocity of the steel, adhering to surface atoms of iron (zero-slip condition). Farther out, the gas will be moving much more slowly and with much reduced tangential or z-momentum. We see then that z-momentum will be lost by faster gas molecules when they move away from the surface of the steel jet, and that this flux of z-momentum in the r direction will result in the generation of a shearing force F_s transmitted to the steel surface.

Consequently, τ_{rz} is customarily regarded as a shear stress acting in the z direction on a plane perpendicular to the r axis, or, alternatively, as the flux of z momentum in the r direction. Since shear stresses are tensors, nine components can exist, as opposed to three for vectors. Table 2.3 lists these components which can apply for complex, three-dimensional flows.

1.3. MECHANISMS OF HEAT TRANSFER

We shall now look more closely at the case of an ingot (or slab) cooling in air and identify the three main mechanisms by which heat is transferred from the ingot to its surroundings. These are conduction, convection and radiation, and are the three distinct modes of heat transmission generally recognized in heat transfer literature.

Consider the situation depicted in Fig. 1.10(a), in which the initial ingot temperature far exceeds room temperature, θ_A. Since thermal disequilibrium exists (i.e. $\theta_I \gg \theta_A$), heat is transferred from the ingot until the final ingot temperature θ_I^F becomes equal to θ_A^F, the final ambient temperature (second law of thermodynamics). Temperature profiles in the solid ingot at various times during the system's approach to equilibrium are shown in Fig. 1.10(b). It

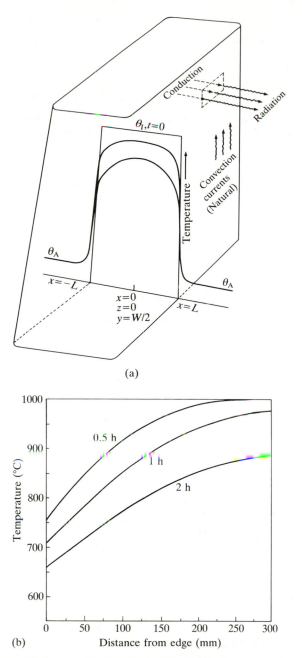

(a)

(b)

FIG. 1.10. Heat transport mechanisms in solids and fluids: conduction, convection and radiation. (a) Three mechanisms for heat transfer from an ingot to its surroundings. (b) Computed temperature profiles.

is important to notice that these temperature profiles are continuous at the 2-phase interface, as well as in the bulk of the two phases themselves.

(a) Conduction

Heat will be transported away from the interior portions of the ingot towards its surface solely as a result of heat *conduction*. The French scientist, J. B. J. Fourier, was the first to propose (in 1822) that the rate at which heat is *conducted* through a solid (or stationary medium) is proportional to the cross sectional area perpendicular to the heat flow, the temperature gradient in the solid at the location in question, and finally the nature of the material itself, (i.e. its thermal conductivity). Thus, the elementary equation for one-dimensional heat conduction is

$$\dot{q}_{x,t} = -kA\left(\frac{d\theta}{dx}\right)_{x,t} \tag{1.3}$$

where

$\dot{q}_{x,t}$ is the heat conducted per unit time in the x direction [watts or joules per second] at time t, across a plane of area A, located at x;

k is the thermal conductivity of material [constant of proportionality; $W\,m^{-1}\,K^{-1}$ or $cal\,m^{-1}\,k^{-1}\,s^{-1}$];

A is the cross-sectional area perpendicular to the x direction; [m^2];

$\left(\dfrac{d\theta}{dx}\right)_{x,t}$ is the temperature gradient at location x and time t [$k\,m^{-1}$].

The negative sign in eqn (1.3) again results from the sign convention normally adopted, which specifies that the direction of increasing x is to be the direction of positive heat flow.

Thus, referring to Fig. 1.10(b), the temperature gradient 100 mm from the edge of the cooling ingot is $\simeq 1220\,K\,m^{-1}$ after one hour of cooling in air. Taking the thermal conductivity of stainless steel to be $32\,W\,m^{-1}\,K^{-1}$, we see that the amount of heat being conducted at this location and instant in time will be 39 kW for every square metre of cross-sectional area. This numerical value is equal to the heat flux, $\dot{q}''_{x,t}$, since the heat flux is defined as the rate of heat transfer per unit cross-sectional area (i.e. $\dot{q}'' = 30\,kW\,m^{-2}$).

(b) Natural thermal convection

Part of the heat transfer from the cooling billet to the air occurs by way of thermal convection. This is the most common way by which thermal energy is transferred from a solid surface to a liquid or gas and takes place in several steps. First, heat will flow by conduction from the surface to adjacent particles of fluid on account of the temperature gradient.

These gas particles will heat up neighbouring particles close to the interface. The heated gas, being buoyant, will then tend to rise and make way for incoming colder (more dense) gas (air) from the surrounding atmosphere. In the process, the associated thermal energy is carried up and away by the motion of the gas. This aspect of the process gives rise to the term 'convection' which derives from the Latin *con veho* ('I carry with'). The particular type of motion just described is called natural (or free) convection since the mixing motion takes place solely as a result of the density differences caused by temperature gradients. When the mixing motion is induced by some external agency, such as a fan, pump or blower, the process is called forced convection.

The rate at which heat is transferred from the hot surfaces of the slab to the bulk of the surrounding atmosphere can be represented by the equation:

$$\dot{q}_c = hA(\theta_S - \theta_A) \qquad (1.4)$$

which was first proposed by Sir Isaac Newton in 1701.

$\dot{q}_c$ is the rate of heat transfer by convection [$W\,s^{-1}$];
h is the convective heat transfer coefficient [$W\,m^{-2}\,K^{-1}$];
A is the surface area [m^2];
θ_S is the surface temperature of the slab [K];
θ_A is the bulk or ambient temperature of the fluid [K].

The value of h, the surface heat transfer coefficient, is invariably semi-empirical, in that it is usually an experimentally determined constant whose value depends on the shape of the body (i.e. the ingot), the nature of fluid flow around it (e.g. the intensity and character of up-currents) and finally the thermal properties of the fluid. Unless otherwise stated, h represents the average heat transfer coefficient for the whole surface, rather than a local heat transfer coefficient which would be specific to one location on the solid's surface. A reasonable value for h in the present circumstances would be $10\,W\,m^{-2}\,K^{-1}$. Thus, the convective heat flux, $\dot{q}_c''$ due to natural convection would be $9.6\,kW\,m^{-2}$ ($\theta_S = 980\,°C$, $\theta_A = 20\,°C$).

(c) Radiation

As well as energy transmission by convective heat transport, a considerable amount of *radiant energy* will leave the ingot's surfaces at a rate proportional to the fourth power of the *absolute* temperature of the surface.

Quantitatively, a perfect radiator (or 'black body') emits radiant energy from its surfaces at a rate given by:

$$\dot{q}_r = \sigma A \theta_S^4 \qquad (1.5)$$

This relationship forms the basis of all radiant energy heat transfer calculations. The constant of proportionality, $\sigma (= 5.67 \times 10^{-8}\,W\,m^{-2}\,K^{-4})$ is known as the Stefan–Boltzmann constant in honour of two Austrian scientists who obtained the equation (J. Stefan experimentally and L. Boltzmann theoretically). However, since the ingot surface does not behave as a perfect radiator, its rate of energy emission to the surrounding enclosure (i.e., the walls of the melt shop) can be considered to be proportionally less at all wavelengths by a (dimensionless) factor ε known as the emissivity. Since we can consider the enclosure to act as a black body, the net rate of heat transfer then becomes:

$$\dot{q}_r = \varepsilon\sigma A(\theta_S^4 - \theta_A^4) \tag{1.6}$$

Thus, taking $\varepsilon = 0.8$ (for an iron oxide surface at 1255 K) and an enclosure temperature of 355 K, the radiant heat flux *emission* will be $0.8 \times 5.67 \times 10^{-8} \times (1255)^4$ or $113 \, \mathrm{kW \, m^{-2}}$ from the billet surface. In return, the billet will receive a negligibly small amount of radiant energy from the melt shop walls so that the net radiant heat flux will be only slightly less. It is important to note that the intervening atmosphere (air) absorbs little or no radiation. However, if the atmosphere comprised large amounts of CO_2 and/or H_2O, a significant part of the emitted energy would be absorbed. This is important in heat transmission from flames in metallurgical furnaces, where the exhaust gas compositions normally contain significant proportions of steam and carbon dioxide.

In summary, we see that the radiated heat lost from hot ingots far exceeds that convected away. This fact is readily appreciated by anyone caring to stand 6 metres away from a hot ingot (i.e., far out of range of thermal convection currents).

1.4. MECHANISMS OF MASS TRANSFER

Turning now to mass transfer, we may cite an example of mass transfer which is analogous to the case of the cooling ingot. Consider, for example, a piece of mild steel whose surface is to be carburized by the flow of carbon monoxide over it (Fig. 1.11(a)). Once again, we may illustrate the two basic modes of mass transfer referred to in the literature, these being *convective* and *diffusive* mass transfer, respectively. Thus, carbon monoxide will be transported to the steel interface as a result of convective mass transport, where it will dissociate according to the (Boudouard) reaction

$$2CO \leftrightarrows C + CO_2,$$

for which

$$K^{eq} = \frac{P_{CO_2, s}[wt\%C]}{P_{CO, s}^2}, \tag{1.7}$$

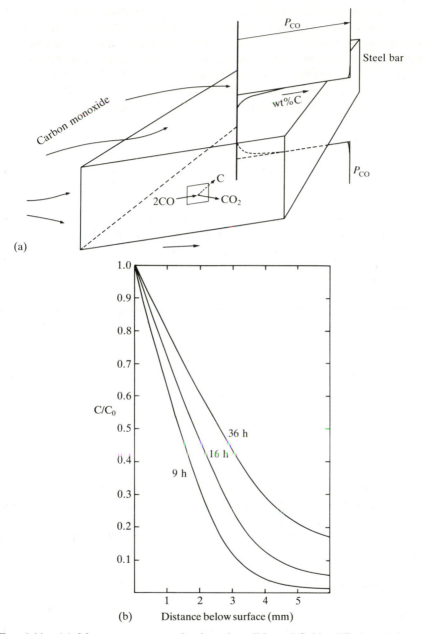

FIG. 1.11. (a) Mass transport mechanisms in solids and fluids: diffusion and convection mechanisms of mass transfer from a plate of mild steel being carburized by a flow of carbon monoxide over its surfaces. (b) Computed concentration profiles for dissolved carbon.

The carbon enters the steel as a result of *diffusive* mass transfer, and the gaseous product CO_2 leaves the interface and returns to the gaseous atmosphere as a result of convective mass transport. Carbon concentration profiles in the section during the carburization process are shown in Fig. 1.11(b). It is important here to remember that carbon will continue to pass into the sheet until its *activity* (or chemical potential) in the steel is equal to its chemical potential in the gaseous atmosphere surrounding it.

(a)   Diffusive mass transfer

Carbon will be transported inwards toward the centre of the plate solely as a result of molecular diffusion mechanisms. By analogy with the equation for heat conduction, Fick reasoned that the rate at which a solute diffuses through a stationary solvent (i.e. the steel section in this case) should be proportional to the cross sectional area perpendicular to the mass flow, the concentration gradient in the solid at the location in question, and finally some constant of proportionality (the diffusion coefficient) specific to the diffusing system involved. Thus, the elementary equation for one-dimensional diffusive mass transfer is:

$$\dot{N}_{A,x,t} = -D_{AB}A\left(\frac{dC}{dx}\right)_{x,t} \tag{1.8}$$

where

$\dot{N}_{A,x,t}$ is the number of moles of solute A diffusing per unit time in the x direction at time t across a plane located at x [mole s^{-1}];

D_{AB} is the binary diffusion coefficient of solute A in B [m^2 s^{-1}];

A is the cross-sectional area perpendicular to the x direction [m^2];

$\left(\dfrac{dC}{dx}\right)_{x,t}$ is the concentration gradient at location x and time t; [mol m^{-3} m^{-1}, i.e. mol m^{-4}].

Once again, the negative sign in eqn (1.8) results from the sign convention normally adopted which specifies that the direction of x is to be the direction of positive molar (or mass) flow.

Thus, referring to Fig. 1.11(b), the carbon concentration gradient 1 mm from the surface after 9 h is 3.2×10^3 kg mol m^{-4}. Taking the diffusion coefficient of carbon in iron (austenite) at 1000 K as 7.10^{-11} m^2 s^{-1}, the flux of carbon at that specific time and location will be 2.25×10^{-7} kg mol m^{-2} s^{-1}. This calculation assumes a carbon surface concentration of 1.6 wt.% C.

(b)   Convective mass transfer

Turning now to the mechanism of carbon dioxide transport from, and of carbon monoxide transport to, the plate surface (interface), we have an example of forced convective mass transfer.

As a result of the normal 'no-slip' condition between the gas and solid phase molecules at the interface, molecular diffusion processes will predominate in the immediate vicinity of the interface. However, a little way out (say, 0.1 mm) this mechanism of mass transfer will give way to one of convective mass transfer, since the gas mixture begins to move in a direction and with a speed approaching those of the bulk flow. The amount of CO_2 removed will depend upon the difference in CO_2 concentration between the interface and the bulk. By analogy with convective heat transfer, we can quantify the amount of CO_2 transported by the equation:

$$\dot{N}_A = kA(C_A^* - C_A^\infty), \tag{1.9}$$

where

$\dot{N}_A$ is the rate of mass transfer [mole s^{-1}];

k is the convective mass transfer coefficient [$m\,s^{-1}$];

A is the surface area [m^2];

C_A^* is the interfacial concentration of species A (CO_2) in the gas phase [mole m^{-3}];

C_A^∞ is the bulk concentration of species A (CO_2) in the gas phase [mole m^{-3}].

As in the case of heat transfer coefficients, the value of the mass transfer coefficient k is invariably semi-empirical in that it is an experimentally determined constant whose value depends on the shape of the body, the nature of fluid flow around it and various physical parameters which include the diffusion coefficient. Added complications are that k can be a function of the relative concentrations of solute and solvent, while in the case of multicomponent systems it may also be affected by solute–solute interactions. Fortunately, in many metallurgical extractive operations, impurities (i.e. solutes) are often present in sufficiently small amounts in the solvent that these complications can be ignored.

Unless otherwise stated, k represents the average mass transfer coefficient for the whole surface. A reasonable value for k would be $1 \times 10^{-5}\,m\,s^{-1}$ for CO flowing past the surface at a velocity of $0.3\,m\,s^{-1}$. Some typical values of heat and mass transfer coefficients in metallurgical processing operations are given in Table 1.1.

1.5. PROCESS KINETICS: RATE-LIMITING STEPS

Now that we have looked at the fundamental ways in which momentum, heat and mass are transported through solids and fluids, it is appropriate to

TABLE 1.1. *Some typical values of heat and mass transfer coefficients in metallurgical processing operations*

(a) Heat transfer coefficients; h (W m^{-2} K^{-1}, **where** $\dot{q}'' = h\Delta\theta$)

	Mode	Examples	h(Wm^{-2} K^{-1})
(i)	Natural convection in gases	Air currents around lower temperature equipment (<1000 K)	3–8
		Air currents around high-temperature ingots furnaces, etc. (>1000 K)	8–14
(ii)	Natural convection in liquid metals	Molten metals in contact with ladle or furnace sidewalls with modest temperature difference ($\Delta\theta < 1000$ K)	1000–3000
		Molten iron in contact with water-cooled plates (e.g. tuyères)	7000
(iii)	Forced convection in gases	Air flow in pipes (1–50 m s^{-1})	10–200
		Gas flow in packed beds (e.g. pre-heat furnaces, 50 mm solids, 1 m s^{-1} gas flow, $\Delta\theta = 1000$ K)	135
(iv)	Forced convection in water	Turbulent flow in pipes ($U \sim 1$ m s^{-1})	1000–8000
		Flow through pipes with incipient boiling	5000–100,000
		Quenching by immersion of hot metal into water or oil	50–500
		Metal cooling with water sprays/curtains (e.g. on to hot strip at 1000°C)	3000
(v)	Radiation from surfaces	Boiling water	10
		Clean liquid metals	30–150
		High-temperature refractories slags, oxides and luminous flames	200–500

(b) Mass transfer coefficients, $k(\text{m s}^{-1}$, where $N'' = k\Delta C)$

	Mode	Example	k (m s^{-1})
(i)	Gas phase mass transfer to liquid or solid surfaces	Laminar flow around falling metal drops	0.1–0.5
		Turbulent gas flows (with jets, tuyères, or bubbles)	1–100
(ii)	Liquid metal mass transfer to gas or solid interfaces	To a gas–liquid interface, with well-mixed liquid (e.g., rising bubbles, metal jets, bubble stirred interfaces, intensive natural convection, etc.)	10^{-4}–10^{-3}
		Around dissolving solids (e.g. alloy additions such as Fe–V, Fe–W, Fe–Mo, steel scrap in pig iron melts, etc.)	10^{-4}–10^{-3}
		To furnace/ladle walls with slow bulk stirring	$(1-5)\times10^{-5}$
		Vacuum refining of melts	k_e (m s^{-1})
(iii)	Surface evaporation coefficient, k_e $$k_e = \frac{\alpha P_i^0 \gamma_i M_s}{\rho_s \sqrt{2\pi RTM_i}}$$	Zinc (700 K)	8×10^{-6}
		Aluminium (1100 K)	8×10^{-6}
		Copper (1400 K)	1.3×10^{-8}
		Magnesium (1700 K)	1.5×10^{-1}
		Iron (1900 K)	1.3×10^{-6}

complete this section of Chapter 1 with a few words on the *rates* of processes involving the interchange of heat and/or mass between two phases.

It is self-evident from the examples considered so far that rates of processes will depend upon transport kinetics in both phases. For example, the ingot will cool down at a rate which depends on how fast heat can be extracted from the ingot, and also how fast it can be transmitted through the air. If either step is difficult, then this step will tend to predominate and become rate-limiting, or rate-controlling. Similarly, in the last example, if carbon diffusion through the steel is relatively difficult while carbon dioxide diffusion back into the gas phase is easy, then the former will be rate-controlling. This is actually the case, because diffusion rates through solids are so low.

However, there are a number of other possible controlling steps that must always be considered in any mass transfer operation. To appreciate this, let us consider the steps involved in a typical gas–molten metal interaction involving dissolved carbon and oxygen coming out of solution to form carbon monoxide, according to $(C)_{Fe} + (O)_{Fe} \rightarrow CO$. The steps might be:

(1) transport of the dissolved reacting species from the bulk of the molten metal phase to the liquid–gas interface by mass transfer mechanisms;

(2) adsorption of the species at the interface;

(3) chemical reaction at the interface to form the gaseous product;

(4) desorption of the gaseous product from the interface into the gas phase;

(5) transport of the gaseous product from the interface into the main bulk phase.

If any one of these steps takes much longer to complete than any of the others, that step will control the rate at which the reaction proceeds; hence the term, 'rate-controlling step'.

In many metallurgial processes operating at extremely high temperatures (e.g. 1873 K), interfacial steps are generally very rapid in comparison with the rates of transport (i.e., mass transfer) of the reacting species. Two important consequences follow:

1. The reaction may be considered to be transport-controlled.

2. Chemical equilibrium of the species dissolved in both phases may be assumed at the interface.

In such circumstances, reaction kinetics can be predicted quantitatively, in a reasonably straightforward way. However, it should be emphasized that transport control does not apply under all circumstances. For instance, if an insoluble new phase is formed at the interface, or if strongly adsorbed surface active molecules or atoms are present, (e.g. O and S in Fe), interpretation of rate data can become quite speculative and predictions can be hazardous. Such factors will be considered in greater detail later.

1.6. TRANSPORT PROPERTIES: EMPIRICAL

Now that the laws of diffusion, conduction and molecular transport of momentum (viscosity) have been presented, it is appropriate to present numerical values for their associated transport properties and to show, where possible, how they may be predicted, or estimated, from first principles.

Fluids of interest to process metallurgists include liquid metals, slags, molten salts, aqueous solutions and gases, at temperatures that often range between $0\,°C$ and $2000\,°C$. It is not surprising, therefore, to find large numerical differences in associated transport coefficients. For example, the thermal conductivities of metals are some 100 to 1000 times greater than those of refractory materials, metallurgical slags, or gases. Conductivities of metals vary from $400\,W\,m^{-1}\,K^{-1}$ for silver, to about $15\,W\,m^{-1}\,K^{-1}$ for stainless steel. Values for slags and refractory materials range from $1\,W\,m^{-1}\,K^{-1}$ for fireclay brick down to $0.01\,W\,m^{-1}\,K^{-1}$ for insulating sideboards and the like. Gases exhibit similarly low conductivity values ranging from $0.22\,W\,m^{-1}\,K^{-1}$ for hydrogen at $2000\,K$ down to $0.02\,W\,m^{-1}\,K^{-1}$ for nitrogen, oxygen, air, carbon monoxide, and carbon dioxide at room temperatures.

Something of the reverse holds true for diffusion coefficients and viscosities in the liquid phase. Rather than meeting extreme differences, one finds that diffusion coefficients in aqueous, ionic and metallic solutions all usually lie within the range $(1–10) \times 10^{-9}\,m^2\,s^{-1}$. Similarly, viscosities of liquid metals, ionic salts and aqueous solutions bunch quite closely around $1\,mPa\,s$ ($1\,cP$), with maximum variations of an order of ten around that value. Much higher viscosities are associated with metallurgical slags and glasses, with limiting values of some $10^7\,mPa\,s$ for plastic deformation processes in hot rolling operations.

The wide range in numerical values of the transport coefficients reflects major differences in the way energy, mass and momentum can be transferred through different substances or fluids. In gases, at normal pressures, heat and momentum are exchanged when randomly moving atoms or molecules collide with one another. In ionic liquids, energy is largely transferred through collisions of 'caged' molecules with nearest neighbours. In electronically conducting liquids (i.e. liquid metals), the major component of energy transport is by the movements of free electrons. In solids, molecules are truly caged in an ordered crystalline structure and their atomic oscillations lead to lattice vibrations which are propagated as waves (so-called phonons). These phonons are responsible for the transport of energy, together with any electronic component for metallic-type substances and any radiative components of conductance (photons) associated with oxides and glasses at the higher temperatures.

Figures 1.12, 1.13 and 1.14 summarize the typical range of transport coefficients met with in gases, liquids and solids of metallurgical interest. In Fig. 1.12 the most obvious point to note is that the viscosities of gases rise with temperature while those of liquids fall steeply. In Fig. 1.13, it is interesting to see that gases again show an increase, this time in thermal conductivity, with

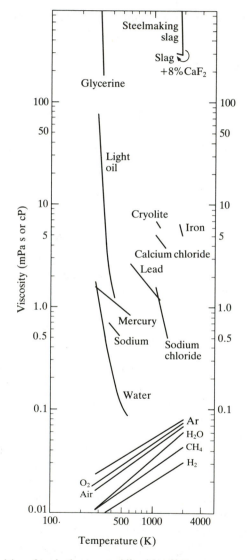

FIG. 1.12.   Viscosities of typical gases and liquids of interest to metallurgists, showing their variation with temperature.

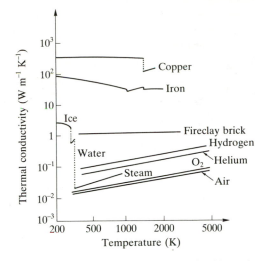

F<small>IG</small>. 1.13. Thermal conductivities of typical solids, liquids and gases of interest to metallurgists, showing their variation with temperature.

rising temperature. Many other materials exhibit drops after initial rises close to absolute zero (° K). Iron, which is commercially important shows a continuous drop in thermal conductivity beyond 20 K, as indeed does sapphire, which is not usually regarded as a good conductor of heat. Most metals, on liquefying, exhibit a further drop in thermal conductivity. Since the diffusion of solute atoms through iron is of such industrial significance, this system was chosen in Fig. 1.14 to show how diffusion coefficients vary with temperature and solid/liquid phases. For the gas phase, the estimated and experimental values for the binary diffusion coefficient of argon in iron vapour (or vice versa) at 1600 °C by Grieveson and Turkdogan (1964) are 3.7×10^{-4} and 3.3×10^{-4} m^2 s^{-1} respectively. By contrast, solute diffusion coefficients in liquid iron generally fall in the range $(1–15) \times 10^{-9}$ m^2 s^{-1} (i.e. 10^5 times lower than those in the gas phase), while diffusivity values in solid iron tend to be very much lower, far more temperature-dependent, and also sensitive to structure. Similarly, they are also affected by impurities, and by whether the solute is diffusing along grain boundaries or through the bulk of the crystal matrix in an interstitial or substitutional manner.

It should be remarked that experimental values for diffusion coefficients in liquid metals have sometimes proved to be notoriously unreliable. This is evidenced by typical variations for $D_{H/Fe}$ and $D_{O/Cu}$ shown in Figs 1.19(a) and 1.19(b). The present author's early work with Solar for the diffusion of hydrogen in liquid iron (1972) is seen to have stood the test of time, while diffusion data by Arkharov *et al.* must be rejected. However, for $D_{O/Fe}$ and

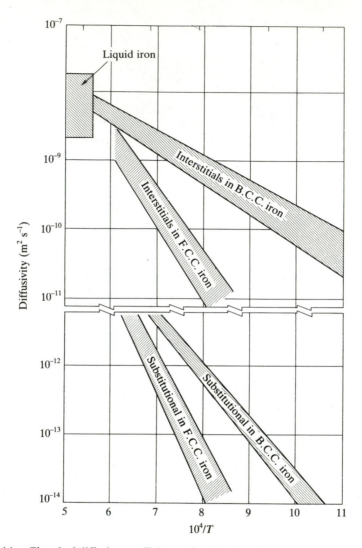

Fɪɢ. 1.14.   Chemical diffusion coefficients of carbon in solid and liquid iron, showing their variation with temperature, iron steel crystal structure (body-centred cubic/face-centred cubic) and relative size of diffusing element (interstitial or substitutional). (Adapted from Richardson 1974.)

$D_{C/Fe}$, estimates of 5×10^{-9} and 40×10^{-9} m^2 s^{-1} were proposed by Solar and Guthrie (1972), but these have had to be revised to 14×10^{-9} and 10×10^{-9} m^2 s^{-1} following the results obtained by Gourtsoyannis (1978, 'Kinetics of compound gas absorption by liquid iron and nickel'. Ph.D. thesis, McGill University Montreal, Québec, Canada) using an improved experimental version of the 'constant gas pressure/capillary reservoir technique'.

The purpose in mentioning these discrepancies is to show the need for good theoretical models and the care needed in choosing from the plethora of experimental values for D, k and μ that are to be found in the literature.

The unreliable experimental diffusivities in liquid metals or slags are more the result of notorious experimental difficulties than of a lack of care. For instance, it is difficult to control high temperatures and thermal gradients precisely. Similarly, unexpected convection currents, surface-tension-driven flows, side-reactions or reactions with containment vessels, can all contribute to the problems facing the experimentalist. It is largely for such reasons that the thermal conductivities of liquid iron and other high-melting-point metals

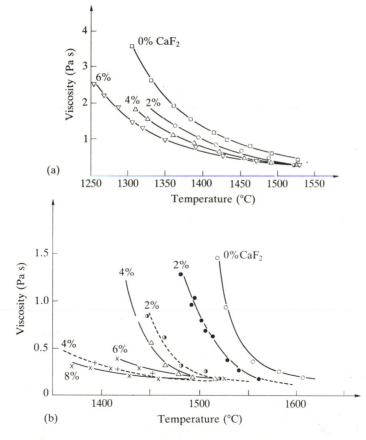

FIG. 1.15. The variation in slag viscosities with temperature, and the effects of fluoride additions. (a) CaF_2 in 44 wt%, SiO_2, 12 wt% Al_2O_3, 41 wt% CaO and 3 wt% MgO. (b) CaF_2 (solid lines) and MgF_2 (broken curves) in 32 wt% SiO_2, 13 wt% Al_2O_3, 52 wt% CaO, and 3 wt% MgO. (Adapted from Richardson 1974.)

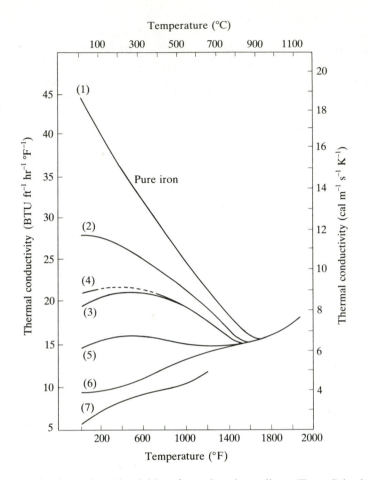

FIG. 1.16.   The thermal conductivities of pure iron base alloys. (From Schack 1965.)

	\multicolumn{6}{c}{Composition, wt%}					
	C	Si	Mn	Cr	Ni	Condition
1	Pure	iron				Annealed
2	0.23	0.11	0.63	—	—	Annealed
3	0.32	0.18	0.55	0.71	3.4	Annealed
4	3.50	2.20	0.30	—	1.3	As cast
5	0.13	0.17	0.25	13.00	0.1	Annealed
6	0.08	0.68	0.37	19.10	8.1	Quenched
7	0.46	1.30	1.20	15.20	26.9	—

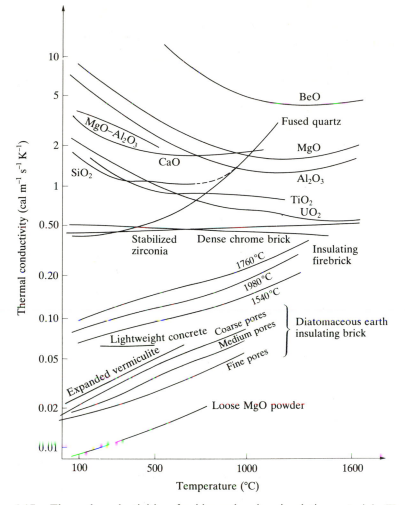

FIG. 1.17. Thermal conductivities of oxides and various insulating materials. (From Schack 1965.)

and alloys have yet to be measured experimentally. Fortunately our understanding of liquid metal systems is quite well advanced, and theoretical predictions are probably more reliable than experimental results.

Returning to experimentally determined transport coefficients, Fig. 1.15 provides typical information that is available on the way the viscosities of two metallurgical slags, containing silica, lime, alumina, and a little magnesia, vary when treated with fluoride additions. Evidently for good slag–metal separation, enhanced transport properties and more convenient physical processing, less viscous slags tend to be a desirable part of operational practices.

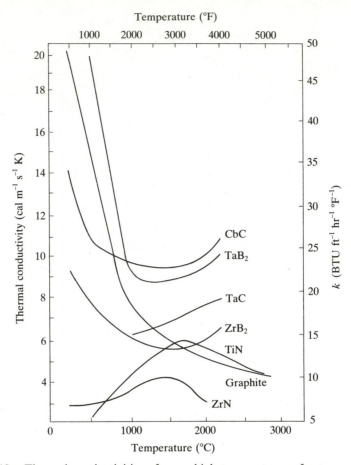

FIG. 1.18.   Thermal conductivities of some high-temperature, refractory materials.
(From Geiger and Poirier 1980.)

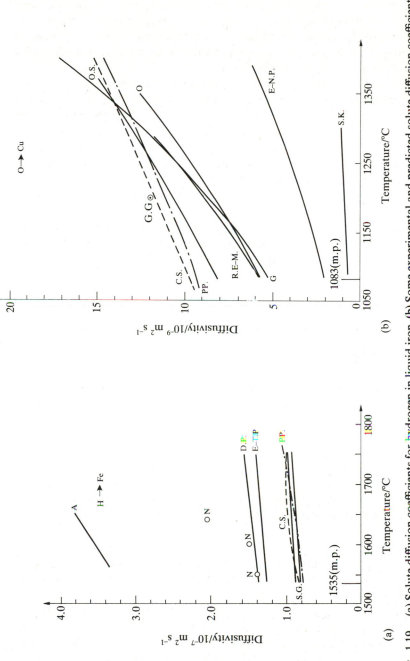
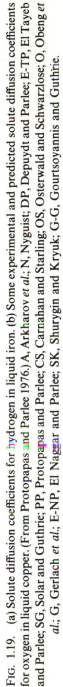

FIG. 1.19. (a) Solute diffusion coefficients for hydrogen in liquid iron. (b) Some experimental and predicted solute diffusion coefficients for oxygen in liquid copper. (From Protopapas and Parlee 1976.) A, Arkharov *et al.*; N, Nyguist; DP, Depuydt and Parlee; E-TP, El Tayeb and Parlee; SG, Solar and Guthrie; PP, Protopapas and Parlee; CS, Carnahan and Starling; OS, Osterwald and Schwarzlose; O, Obeng *et al.*; G, Gerlach *et al.*; E-NP, El Naggar and Parlee; SK, Shurygin and Kryuk; G-G, Gourtsoyannis and Guthrie.

The significant lowering of slag viscosity achieved by the fluoride additions typically used in commercial operations can be explained by a progressive breakdown of the chain size of the silica tetrahedral networks present in molten slag. Such ruptures are associated with strong bases such as the CaF_2 or MgF_2 compounds reported in Fig. 1.15.

Figures 1.16, 1.17 and 1.18 are self explanatory. They present experimental data on thermal conductivities of alloys of iron (Fig. 1.16) and of oxides and refractories of metallurgical interest (Fig. 1.17), including the more exotic high-temperature materials (Fig. 1.18). The latter are finding increasing applications in the metallurgical industry. Figure 1.19, as previously mentioned, concludes with experimental data for the diffusivities of (a) hydrogen in molten iron over the temperature range 1540–1760 °C and (b) oxygen in liquid copper over the temperature range 1083 (m.p.) to 1400 °C. For more complete lists of such data, the reader is referred to Appendix III or to reference books such as *Thermo-physical Properties of Matter* or the *Metals Handbook* of the American Society for Metals, or for liquid metals, to the *Physical Properties of Liquid Metals*, Iida and Guthrie (1988).

1.7.   THEORY OF TRANSPORT COEFFICIENTS IN GASES

(a)   Introduction

Now that empirical values of μ, k and D have been presented for a wide range of substances of interest to the metallurgist, it is appropriate to consider the underlying principles governing their numerical values.

Many theoretical models, some of considerable mathematical complexity and abstraction, are to be found in the literature concerning the properties of solids, liquids and gases. The approach in the present text has been to concentrate on theories deriving from the principles of Newtonian mechanics applied to atoms in gases or liquids. This is legitimate, as molecules, atoms, or ions obey Newtonian mechanics (or classical mechanics) at higher temperatures.

As gases at normal pressures have proved to be the most amenable to analysis and provide cluse for predicting transport coefficients in the liquid state, these are considered first. Transport coefficients for liquids and solids then follow.

Transport coefficients for gases can be estimated using classical kinetic theory. According to this, particles of matter in all forms of aggregation, are in a state of violent agitation, but in gases, the molecules are also in a state of random chaotic motion. Being small, and highly dispersed, they are generally free from the influence of any neighbouring atoms, spending most of their time in flight, except when collisions with other molecules or with a container wall take place.

The first step in the kinetic theory of gases is to derive an expression for the pressure exerted by a gas at a given temperature. This leads on to information about how fast molecules travel and collide, and then on to expressions for transport coefficients for momentum (μ), heat (k), and mass transfer (D_{AB}).

Suppose therefore that a mole of N gas molecules are contained within a box of volume V and sidewall areas equal to A as shown in Fig. 1.20(a). Every time a molecule of mass m strikes a wall normally at a velocity c, and rebounds elastically, its change in momentum will be $2mc$.

Given a concentration of n molecules per unit volume of space, the total number of molecules striking any one of the six sidewalls over a short time, δt, will be roughly $[(n/6) A c \delta t]$. For this simple analysis we suppose that on an average, one-sixth of all molecules impacting walls will be hitting one sidewall at an average, or root mean square, velocity of c. Some in fact will hit obliquely or be moving faster, but this need not concern us. The net impulse I exerted by these gas molecules on the wall is therefore

$$I = F \, \delta t = \left(\frac{n}{6}cA \, \delta t\right)(2mc),\qquad(1.10)$$

where F, the force of impact on the wall, divided by the wall area, gives an expression for P, the pressure exerted by the gas on the box's sidewalls:

$$P = \frac{F}{A} = \tfrac{1}{3}nmc^2.\qquad(1.11)$$

Recalling that we conveniently assumed that 1 mole of gas was contained in the box, N represents Avagadro's number, and $n = N/V$. Consequently, we can rewrite eqn (1.11) as

$$PV = \tfrac{2}{3}N(\tfrac{1}{2}mc^2),\qquad(1.12)$$

where $N = 6.023 \times 10^{23}$ molecules per mole.

Similarly, our knowledge of the ideal gas law allows us to extend this relationship to the following set of identities, where R is the gas constant, and θ is the absolute temperature of the gaseous system:

$$PV = R\theta = NK\theta = \tfrac{2}{3}N(\tfrac{1}{2}mc^2).\qquad(1.13)$$

Inspection of eqn (1.13) shows us that the Boltzmann constant K is equal to $\tfrac{2}{3}$ of the kinetic energy of a molecule per degree kelvin.

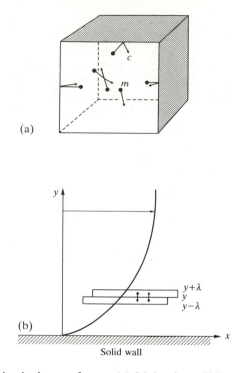

(a)

(b)

Solid wall

FIG. 1.20.   The kinetic theory of gases. (a) Molecules colliding with each other and off the walls of a square box of volume V, and sidewalls of area A. (b) Interchange of molecules between two layers of fluid, in which one layer is moving at a slightly higher imposed velocity.

We are now in a position to describe the r.m.s. velocity of a molecule in terms of known physical constants:

$$c=\left(\frac{3K\theta}{m}\right)^{\frac{1}{2}}. \tag{1.14}$$

Taking the atomic weight of hydrogen to be 1.008 g mole^{-1} for example, $m_{\mathrm{H}}=(1.008\times 2)/(6.023\times 10^{23})$ or 3.3×10^{-27} kg so that $c=1.27\times 10^2$ m s^{-1} for $\theta=300$ K. A more rigorous analysis shows that the factor of $\sqrt{3}$ should be $(8/\pi)^{1/2}$, suggesting that c is overestimated, by about 15%.

The next step in the present analysis is to estimate λ the mean free path of a molecule between collisions. Supposing that the diameters of molecules within the box are d, then one molecule will strike another if their centres are less than d apart. Put another way, a molecule speeding through space has an effective collision diameter d_{c} that is double its own diameter ($d_{\mathrm{c}}=2d$). The time taken for a molecule to sweep unit volume of the chamber (during which

n collisions are expected is $[t = 1/(\pi d_c^2/4)c]$. Consequently the time t_c between individual collisions is t/n. It follows that the mean free path λ is equal to (ct_c), so that

$$\lambda = C \frac{1}{n} \frac{1}{(\pi d_c^2/4)c},\qquad(1.15)$$

from which,

$$\lambda \simeq \frac{1}{n\pi d^2},\qquad(1.16)$$

and

$$t_c = \frac{1}{v} = \frac{\lambda}{c} = \frac{1}{n\pi d^2 c},\qquad(1.17)$$

where v is the collision frequency.

For hydrogen, $d_{H_2} = 2.915$ Å, so that at 300 K, $n = 2.446 \times 10^{25}$ molecules m^{-3}, $\lambda = 1530$ Å, and $v = 8.3 \times 10^{29}$ s^{-1} or about 10^{30} Hz!

(b) Viscosity

We may now address the question of transport coefficients. Imagine the box of molecules shown in Fig. 1.20 to be of unit cross-sectional area and thickness equal to 2λ with a centrally located plane at y. The total number of molecules contained within the upper half of the box at any instant in time will be $[n(\lambda.1.1)]$, as will the number in the lower portion. Using similar arguments to those just presented for calculating pressure P, we can suppose that one-sixth of all random molecular motions taking place will transfer molecules in the upper portion of the box to the lower, and vice versa. Consequently, the number of molecules moving down per unit area and time will be equal to $\frac{1}{6}n\lambda v$, and this will be matched by an equivalent number moving upwards.

If we now suppose that atoms at plane y have an imposed macroscopic (or time-averaged) velocity of U ($U \ll c$), while those contained between y and $y+\lambda$ have a slightly higher value of $U + [(\partial U/\partial \lambda)\lambda/2]$ and those below an equivalently lower velocity, $U - [(\partial U/\partial \lambda)\lambda/2]$ then a molecule in the upper portion of the box will, on average suffer a momentum loss of $m(\partial U/\partial \lambda)\lambda$ in transferring down from the region of $y+\lambda$, to $y-\lambda$ (and vice versa).

According to the principle of conservation of momentum, this loss will induce a corresponding momentum gain of $m(\partial U/\partial \lambda)\lambda$ to the slower moving portion of fluid that will be manifested as a negative shear stress, $-\tau_{yx}$ exerted by fluid of lesser y on the fluid at plane y. The net rate of momentum transfer per unit area and time is therefore given by

$$\tau_{yx} = \left(+\frac{n\lambda}{6}vm\frac{\partial U}{\partial y}(-\lambda) + \frac{n\lambda}{6}vm\left(-\frac{\partial U}{\partial y}\right)\lambda \right)\qquad(1.18)$$

$$= -\tfrac{1}{3}n\lambda vm\frac{\partial U}{\partial y}\lambda \tag{1.19}$$

$$= -(\tfrac{1}{3}n\lambda mc)\frac{\partial U}{\partial y}. \tag{1.20}$$

Consequently, we have the result, first obtained by Maxwell in 1860, that

$$\mu_{\text{gas}} = \tfrac{1}{3}nmc\lambda \tag{1.21}$$

or, substituting for c and λ,

$$\mu_{\text{gas}} \simeq \frac{1}{n\sqrt{3}}\left(\frac{mK\theta}{d^4}\right)^{\frac{1}{2}} \tag{1.22}$$

A more rigorous (and lengthier) analysis of fluxes and mean free paths leads to a coefficient of $2/(3\pi)^{3/2}$ or 0.12, compared with 0.18 above.

Despite the free use of averages, eqn (1.22) correctly shows that the viscosity of a gas is independent of the gas pressure—a rather surprising result— and that μ should increase in proportion to the square root of the absolute, or ideal gas, temperature θ. Similarly, it predicts that gases composed of smaller molecules should exhibit lower viscosities, which again is found to be the case.

The derivation gives a good description of gas viscosity that is qualitatively correct. It will be seen from Fig. 1.12 that viscosities rise more rapidly than $\theta^{\frac{1}{2}}$, the temperature dependence for oxygen, for instance, rising as $\theta^{0.7}$.

For precise calculations, at moderate pressures, the more recent kinetic theory of gases by Chapman and Enskog is used. This describes transport properties in terms of the potential energy of interaction between a pair of molecules in a gas, and is discussed later.

(c) Thermal conductivity

Returning to our box of molecules shown in Fig. 1.20, let us suppose that each molecule at plane y has a kinetic energy e of $\tfrac{1}{2}mc^2$ or $\tfrac{3}{2}K\theta$. As before at $y + \lambda/2$, we suppose molecules have a slightly higher kinetic energy of $e + \{(\partial e/\partial\lambda)\lambda/2\}$ and at $y - \lambda/2$ a slightly lower value of $e - \{(\partial e/\partial\lambda)\lambda/2\}$ and that this is physically manifested by a negative temperature gradient.

As before, we can equivalently conclude that the net rate of energy interchange per unit area per unit time is

$$\dot{q}'' = -\tfrac{1}{3}n\lambda v\frac{\partial e}{\partial y}\lambda. \tag{1.23}$$

Making the substitution $e = \frac{3}{2}K\theta$,

$$\dot{q}'' = -\frac{1}{3}n\lambda\left(\frac{c}{\lambda}\right)\frac{3}{2}K\frac{\partial\theta}{\partial y}$$

$$= -\frac{1}{2}nc\lambda K\frac{\partial\theta}{\partial y}, \tag{1.24}$$

showing us that k, the thermal conductivity of a gas, should be given by

$$k = -\frac{1}{2}ncK\lambda = \frac{1}{3}\rho C_v c\lambda. \tag{1.25}$$

Alternatively, since

$$C_v = \frac{\mathrm{d}}{\mathrm{d}\theta}\left(\frac{1}{m}e\right) = \frac{1}{m}\frac{\mathrm{d}e}{\mathrm{d}\theta}, \tag{1.26}$$

$$\dot{q}'' = -\frac{1}{3}ncmC_v\lambda = \frac{1}{3}\rho C_v c\lambda. \tag{1.27}$$

Substituting for C and λ, we finally reach an explicit prediction for the thermal conductivity of a gas:

$$k = \frac{\sqrt{3}}{2\pi}\left(\frac{K^3\theta}{md^4}\right)^{\frac{1}{2}}. \tag{1.28}$$

As before, the description for gas thermal conductivity proves to be qualitatively correct, and very similar to that for μ, in that gas thermal conductivities do increase with temperature, as seen from Fig. 1.13 and are normally independent of pressure up to about 10 atmospheres. Inspection of eqns (1.22) and (1.28) shows that there is a simple relationship between μ and k, i.e.

$$\mu_{\text{gas}} = \left(\frac{2}{3}\frac{m}{K}\right)k_{\text{gas}},$$

This is alternatively expressed in terms of C_v, the heat capacity per gram at constant volume for which $\mu = k/C_v$. ($C_v = 3R$ or 3.0 cal mole^{-1}).

(d) Diffusivities

Again using our box of molecules, but now of dimensions, $1 \times 1 \times 2\lambda$, with a fictitious midplane at $\lambda = 0$, the number of molecules on each side is $n(1 \times 1 \times 2\lambda)$. Let us suppose that in this case two different species A and B of gas molecules exist, and that there are initially more A molecules in the upper section of the box, than in the lower section. We can expect that, as a result of molecular movements across the midplane, any such concentration differences will be eliminated unless there is an externally imposed gradient of A molecules in the negative y direction.

Mathematically, at $y + \lambda/2$ we will have $n\{X_A + (\partial X_A/\partial y)\lambda/2\}$ molecules of A, while in the lower box at $y - \lambda/2$ we will have only $n\{X_A - (\partial X_A/\partial y)\lambda/2\}$. Here X_A represents the fraction of A molecules in the total assemblage. As before, we can assume that one-sixth of all molecular movements will lead to atoms leaving the top portion of the box for the lower portion, and vice versa. Consequently, the total number of A molecules transferring per unit area per unit time, $\Sigma \dot{n}_A''$ is

$$\Sigma \dot{n}_A'' = \left\{ \frac{n\lambda v}{6} \frac{\partial X_A}{\partial y}(-\lambda) + \frac{n\lambda v}{6}\left(-\frac{\partial X_A}{\partial y}\right)\lambda \right\}. \tag{1.29}$$

Dividing by N, Avagadro's number, the molar flux of A molecules, $\dot{N}_A''$, is

$$\dot{N}_A'' = -\frac{1}{3}\frac{n}{N}\lambda^2 v \frac{\partial X_A}{\partial y}. \tag{1.30}$$

Here X_A, the mole fraction, is substituted for X_A, the fraction of A molecules, as they are numerically equivalent. Substituting $n/N = C_T$ moles/unit volume, and $v = c/\lambda$, we have

$$\dot{N}_A'' = -\frac{1}{3}\lambda c C_T \frac{\partial X_A}{\partial y}, \tag{1.31}$$

or

$$\dot{N}_A'' = -\frac{1}{3}\lambda c \frac{\partial C_A}{\partial y}. \tag{1.32}$$

We may therefore conclude that $D_{AB} = \frac{1}{3}\lambda c$, or, making further substitutions, that

$$D_{AB} \simeq \frac{1}{\pi\sqrt{3}}\left(\frac{K^3}{m_A}\right)^{\frac{1}{2}} \frac{\theta^{\frac{3}{2}}}{Pd_A^2}. \tag{1.33}$$

This analysis, which predicts that the binary diffusivity D_{AB} varies inversely with pressure, proves to be correct for pressures below about 10 atmospheres. The predicted dependence on temperature is, as before for μ and k, in the correct direction, but also too low. Finally, this simplified analysis ignores the fact that differently sized molecules will exhibit different mean free paths and it implicitly assumes equal diameters d, and values of λ, for A and B.

(e)   The Chapman–Enskog equation for μ, k, D_{AB}

For a more accurate description of transport coefficients in gases, it is necessary to consider the time intervals during which binary collisions take place, together with the nature of such collisions. Chapman and Enskog solved the Maxwell–Boltzmann equation describing the temporal, chaotic behaviour of gas molecules moving into, and out of, a control volume and their associated collisions, and showed the following relations (Chapman and Cowling, 1951):

Viscosity

$$\mu = 2.6693 \times 10^{-5} \frac{(M\theta)^{1/2}}{d_c^2 \Omega_\mu}, \tag{1.34}$$

where the units of μ are [g cm^{-1} s^{-1}], of θ are [K], and d_c, the collision diameter [Å] or [10^{-10} m].

Thermal conductivity

$$k = 1.981 \times 10^{-4} \frac{(M\theta)^{1/2}}{d_c^2 \Omega_k}, \tag{1.35}$$

where the units of k are [cal cm^{-1} s^{-1}], and of d_c are [Å].

Diffusivity

$$D_{AB} = 0.18583 \times 10^{-6} \left(\frac{1}{M_A} + \frac{1}{M_B} \right)^{\frac{1}{2}} \theta^{3/2} \Big/ P d_{c,AB}^2 \Omega_{D,AB}, \tag{1.36}$$

where the units of D_{AB} are [cm^2 s^{-1}], of θ are [K], of P are [atm.] and of $d_{c,AB}$ are [Å].

It is found that these equations fit the experimental data very closely, giving the correct temperature exponent of about 2.0 at low temperatures, and 1.65 at higher temperatures. Values of Ω were solved for numerically by Chapman and Enskog and are plotted in Fig. 1.21. There, Ω is seen to be a slowly varying function of the dimensionless parameter, $K\theta/\varepsilon$. The numerator, $K\theta$ is representative a molecule's kinetic energy, while ε, the denominator, denotes the potential energy of attraction of the two molecules colliding.

Obviously, ε will depend upon the type of molecules engaging, as will their collision diameters. While the theory strictly applies to monatomic gases, it works well for polyatomic gases as well. When data are not available for the parameters ε and d_c, they can be estimated from thermodynamic properties of the fluids in question, viz. molar volumes and phase transition temperatures (here in K).

Fluids at critical temperature θ_c.

$$d_c \simeq 0.841 \ V_c^{1/3}, \ \varepsilon \simeq 0.77 K \theta_c. \tag{1.37}$$

Liquids at normal boiling point temperature θ_b

$$d_c \simeq 1.1666 \ V_{b,1}, \ \varepsilon \simeq 1.15 K \theta_b \tag{1.38}$$

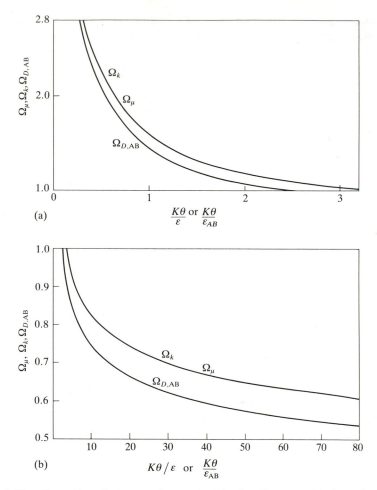

FIG. 1.21. Correction factors to be used with the Chapman–Enskog theory of gaseous transport coefficients. Ω_μ, Ω_k and $\Omega_{D,AB}$ plotted versus the dimensionless quantity $K\theta/\varepsilon$. (a) For $K\theta/\varepsilon$ ranging from 0 to 3. (b) For $K\theta/\varepsilon$ ranging from 0 to 80. (From data by Bird *et al.* (1960)).

Solids at melting point temperature θ_m:

$$d_c \simeq 1.1222\, V_{m,s} \quad \varepsilon \simeq 1.92\, K\theta_m. \tag{1.39}$$

Here V denotes molar volume, and subscripts c, b, m, l and s indicate values at the critical point, on boiling, on melting, of a liquid or solid respectively.

Occasionally it is possible to calculate collision diameters and potential energy wells, $-\varepsilon$, from first principles. As the next section discusses, the forces of interaction between atoms or molecules in liquids, it is worthwhile first becoming familiar with the forces involved in binary collisions of molecules, and which Chapman and Enskog accounted for in their kinetic theory of gases.

Thus, Fig. 1.22 shows the work of Slater and Kirkwood, (see Croxton 1974 in Further Reading), who calculated the potential energy of interaction between two helium atoms. To do this, they required a knowledge of features at the electronic level concerning the valence electrons' energy of repulsion and the repulsive exchange terms arising out of Fermi prohibition of overlapping closed core states at small separations. They obtained

$$\phi(r) = \left(770e^{-r/0.217} - \frac{1.49}{r^6}\right)10^{-12} \text{ ergs.} \tag{1.40}$$

The Lennard-Jones (LJ) model of interaction mimics this relationship with a simpler function commonly used by molecular physicists,

$$\phi(r) = 4\varepsilon\left\{\left(\frac{d_c}{r}\right)^m - \left(\frac{d_c}{r}\right)^n\right\}, \tag{1.41}$$

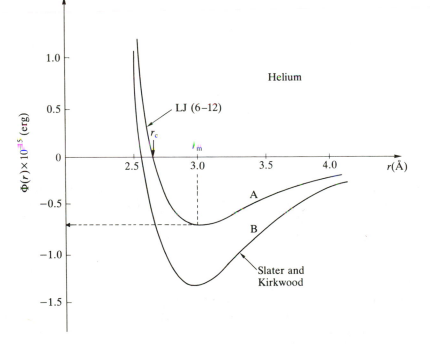

FIG. 1.22. Comparison of the theoretical helium atom–helium atom interaction using the Lennard–Jones (6–12) model function parametrically fitted. (After Croxton 1975.)

where ε and d_c represent the maximum energy of attraction and collision diameter and m and n are the exponents of repulsion and attraction respectively.

We see that curve A using an LJ (6–12) model ($m = 12$, $n = 6$) produces a similar potential energy curve to curve B. Calculated values of ε for He, Ne, Ar, Kr and Xe are (0.14, 0.49, 1.65, 2.29 and 3.03)10^{-14} ergs respectively.

Note that the collision diameter d_c is taken to be at $\phi = 0$, in view of the steeply rising character of the curve at smaller separation distances between atomic centres. Similarly note that $2r_m$, is the most likely centre-to-centre separation distance for two molecules if they reach a state of equilibrium with each other at the minimum of $-\varepsilon$, in the ϕ potential energy well. In the case of two helium atoms, therefore, their mean separation distance at equilibrium is about $6\,\text{Å}$ $(2r_m)$, while their collision diameter is about $5.15\,\text{Å}$ $(2r_c)$.

1.8. THEORY OF TRANSPORT COEFFICIENTS IN LIQUIDS

(a) Introduction

Unlike the case for solids or gases, a simple abstract model of the structure of a liquid is not available. Lying between that of a solid and a gas, its density is much closer to that of a solid and it occupies an identifiable volume as does a solid. However, as does a gas, it readily flows when submitted to shearing stresses. The intermediate status of a liquid is further illustrated in short-range ordering among neighbouring molecules, as in a solid, but in long-range disorder among more widely separated molecules, as in a gas.

Direct information regarding the structures of liquids can be obtained from their X-ray (or neutron) diffraction patterns. Such diffraction patterns allow one to deduce information regarding the local number density of atoms, n, surrounding the target or reference atom. The results can be presented in the form of curves of the pair distribution function $g(r)$. Such curves are shown for liquid iron, copper, lead and sodium in Figs. 1.23(a), (b), (c), and (d), respectively. The dimensionless factor $g(r)$, the pair distribution function, is plotted versus radial distance r from the centre of a reference atom located at $r = 0$.

If these liquid structures were gas-like, the atomic density at r, $n(r)$ ($= ng(r)$), would everywhere equal N/V (Avagadro's number/molar (atomic) volume of liquid), and $g(r)$ would be unity. However, we see for iron that there is a well-defined peak existing at $r_m \simeq 2.5 \times 10^{-10}$ m or (2.5 Å), and that this demonstrates a concentration, or lack of uniformity of iron atoms in a first coordination shell around the reference atom. This type of short-range order is less distinct than that in a solid, whose $g(r)$ curves appear as a series of vertical lines. Similarly, short-range order in liquids does not extend to any of the long-range ordering associated with solids. Evidence for this is the rapid decay in amplitude of the $g(r)$ curves for liquid iron, copper, lead and sodium in Fig. 1.23 with increasing distance r from the reference atom (from Iida and Guthrie 1988).

One should also note that the first maximum in the $g(r)$ curve corresponds to a minimum in the potential energy well of the atoms surrounding the reference atom, illustrated for a binary system of two helium atoms in Fig. 1.22.

Before describing appropriate transport theories for liquids, it will be helpful first to digress and recall the classical picture of the structure of a solid. In this, atoms are treated as being a geometric, three-dimensional assembly of harmonic oscillators, each atom vibrating around fixed lattice positions. These lattice positions correspond, again, to minima in the potential energy wells in which the atoms find themselves locked.

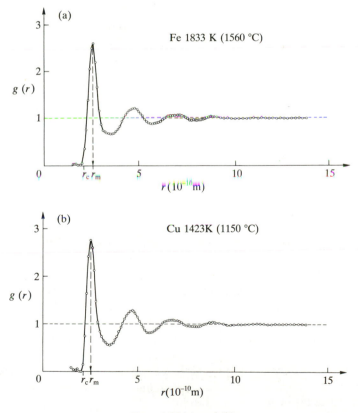

FIG. 1.23(a) and (b)

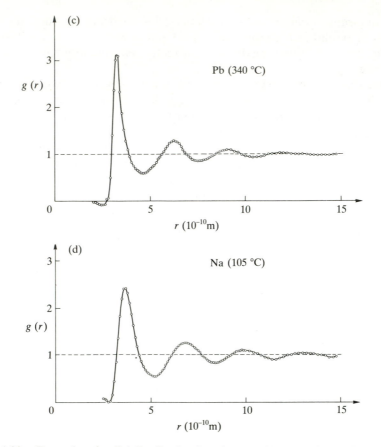

FIG. 1.23. Examples of radial distribution function, or $g(r)$, curves for (a) liquid iron; (b) liquid copper; (c) liquid lead; (d) liquid sodium; near their melting points. (After Iida and Guthrie 1988.)

The harmonic oscillator model was successfully used by Einstein in 1904 to explain why molar heat capacities of solids are asymtotically equal to $3R$ (or $3RN$) at higher temperatures.

In reality, measured frequency spectra show that solids rarely behave as perfect harmonic oscillators with a Gaussian frequency distribution. Figure 1.24, for instance, is instructive in showing the phonon (or vibration) spectra, $f(v)$, for solid and liquid sodium. One sees that at short wavelengths both solid and liquid sodium show peaks in their $f(v)$ curves, but that at long wavelengths ($v \to 0$), solid sodium exhibits a zero spectral density $f(v)$, whereas for the liquid $f(v)$ is non-zero. These low-frequency components of the phonon spectra reflect the development of diffusive modes of transport which are not present in the solid phase.

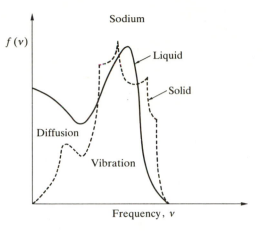

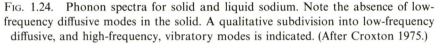

FIG. 1.24. Phonon spectra for solid and liquid sodium. Note the absence of low-frequency diffusive modes in the solid. A qualitative subdivision into low-frequency diffusive, and high-frequency, vibratory modes is indicated. (After Croxton 1975.)

A qualitative picture of the motion of an atom of argon in liquid argon has been obtained by modelling the simultaneous motions and interactions of an assembly of 1000 atoms—a significant computational task. The movement of a typical argon atom resulting from such computations is shown in Fig. 1.25. Note the effective representation of its three-dimensional motion through the use of open circles, which represent the central points of the argon atom which, of course, is proportionately very much larger ($d_c \simeq 6.84$ Å). Figure 1.25 confirms the modern view of liquid structures and shows that the argon atom will on average oscillate or vibrate (e.g. point A in the figure) about three times for every two diffusive jumps of the type shown at B, i.e. $v_v/v_j = 3/2$. For liquid metals, whose high-frequency spectra components are retained, vibratory motions will dominate.

We are now in a position to describe, albeit briefly, the development of theories relating to transport coefficients in liquids. Figure 1.26 shows the wandering particle caged by a field of nearest neighbours and the negative potential energy valley or well from which it is occasionally able to escape if its energy exceeds that required for it to 'boil' ($K\theta_B$).

(b) Viscosity

Since the major part of an atom's lifetime in a liquid metal is spent in vibrating and colliding with nearest neighbours rather than in jumping or diffusing from one location to another, (especially at, or near, the melting point) one can surmise that these collisions are the major way in which momentum is transferred among such atoms.

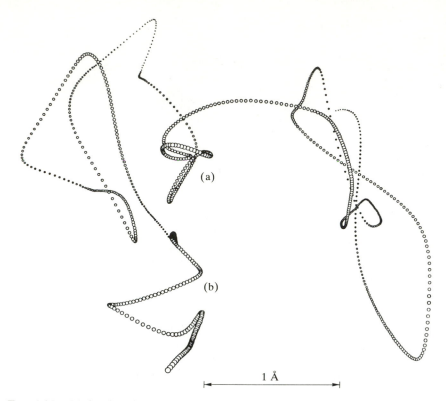

FIG. 1.25. Molecular dynamics trajectory of a liquid argon molecule, showing vibratory (a), and (b) diffusive, components. To account for the general features of the liquid argon spectral density, v_v/v_j is in the ratio 3:2 (v_v = vibration frequency, v_j = jump frequency). (After Croxton 1975.)

Early theories (by Andrade, Frenkel, Eyring), proposed that a simultaneous interchange of atoms and momentum took place and that this was made possible by the presence of voids, or holes, within a liquid, and into which the atoms jumped. On the basis of modern experimental evidence, these 'voids' are something of an abstraction, and represent an extrapolation of ideas about the structure of solids. There is no evidence for holes in real liquids, except perhaps close to their boiling points. Even there, homogeneous nucleation of bubbles is not commonplace.

Referring therefore to the pair distribution function, $g(r)$, curves for liquid iron, (or copper), say, one can estimate the number of atoms with which the reference atom is colliding (Iida and Guthrie 1988). The number of atoms in a spherical shell between r and $r + dr$ from an origin atom will be

$$dN = 4\pi r^2 \left(\frac{N}{V}\right) g(r) \, dr. \tag{1.42}$$

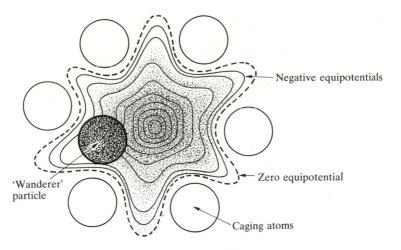

Negative equipotentials

Zero equipotential

'Wanderer' particle

Caging atoms

FIG. 1.26.   Caging of a 'wanderer' particle W in the field of nearest neighbours. (After Croxton 1975.)

Here, the function $4\pi r^2 ng(r)$ is frequently called the radial distribution function (RDF).

One finds for liquid metals that the number of atoms corresponding to the region under the first peak in the RDF curve, (i.e. the coordination number of nearest-neighbour atoms), is normally about 10–11. Incidentally, while $g(r)$ depends upon the magnitude $|r|$ of r, it is not a vector quantity, since real liquids are generally isotropic.

Following similar arguments to those leading to the equations of viscosity for gases in terms of momentum transport processes, Iida, Guthrie and Morita (1982) considered a reference atom A, colliding with atoms in neighbouring adjacent planes. The transfer of momentum per atom (the reference atom) per vibration is shown to be

$$\frac{4\pi}{3} ng(r) r^3 \, dr \cos(\phi) \frac{du}{dy}$$

where ϕ is the angle between the direction of motion and the y axis.

To calculate the transfer of momentum per atom per second, one needs the number of vibrations per second (i.e. frequency). This can be estimated assuming the atoms act as ideal harmonic oscillators and using a modified version of Lindemann's formula.

Finally, since a small portion of an atom's life is spent diffusing rather than vibrating (and this becomes increasingly the case as θ increases), one can estimate the probability $P(\theta)$ of an atom being in its vibrating mode during the time period under consideration. This is done using the Gaussian approximation. Using this approach, the authors showed that

$$\mu = 4.5g(r_m)[1 - r_0/r_m]\frac{v_0 P(\theta)_m}{a},$$ (1.43)

where μ is in units of mPa s (cP) and v_0 is the vibrational frequency of the atom in the absence of diffusion, r_0 is the closest distance between the reference atom and any of its surrounding nearest-neighbour atoms. By way of illustration, this has been depicted in the $g(r)$ curves for iron and copper, together with r_m, which is approximately the mean separation distance between iron atoms (or copper atoms) $r_m \simeq a$; a represents the average distance between atom centres.

To calculate the viscosity of liquid metals at their melting points, using this relationship, one can use a modified version of Lindemann's formula by Kasama *et al.* (1976):

$$v_0 P(\theta_m) = \beta v_L,$$ (1.44)

where v_L is the atomic frequency calculated from Lindemann's formula (eqn 1.47); substituting the left-hand term appearing in eqn 1.44 gives

$$\mu = 4.5\beta g(r_m)(1 - r_c/r_m)\frac{v_L m}{a}.$$ (1.45)

Average (experimental) values for β, $g(r_m)$ and r_0/r_m for liquid metals at their melting points are 0.55, 2.64 and 0.82 respectively, so that we can write roughly for liquid metals,

$$\mu \simeq 1.2\frac{v_L m}{a}.$$ (1.46)

This last relationship is very similar to that obtained by Andrade, the only difference being replacement of the factor 1.2 by a geometric factor of 4/3. However, eqn (1.43) for viscosity is more in keeping with the modern view of liquid metal structure, nor is it restricted to melting points. It explains the success of Andrade's equation with respect to predictions for liquid metal viscosities. A comparison of calculated and observed values for many liquid metal viscosities is provided in Table 1.2. Since experimental errors in viscosity data are typically $\simeq 10\%$, it is inappropriate to stress the *practical* merit of our factor of 1.2 in eqn (1.46) over Andrade's value of 4/3 for viscosities of liquid metals at their melting points. Both approaches make use of Lindemann's formula for an ideal harmonic oscillator:

$$v_L = C_1\left(\frac{\theta_m}{M V_m}\right)^{1/2},$$ (1.47)

where C_1 is a constant (2.8×10^{12} in c.g.s. units, 8.9×10^8 in SI units).

TABLE 1.2. *Comparison of the calculated and observed values for the viscosity of liquid metals at their melting points (in units of mPa s) and values of the parameters used for the viscosity calculation.*

Metal	T_m (K)	T_b (K)	v_0 (10^{12} s^{-1})	$P(T)$†	m (10^{-26} kg mol^{-1})	$\rho_N^†$ (10^{28} m^{-3})	$g(r_m)^a$	μ_{cal}‡	Viscosity (mPa s)	
									$\mu_{cal}^§$	μ_{obs}
Na	371	1151	1.99	0.966	3.82	2.43	2.46	0.68	0.62	0.70
Mg	923	1380	4.78	0.666	4.04	3.94	2.50	1.22	1.39	1.25
Al	933	2333	4.29	0.903	4.48	5.33	2.87	1.90	1.79	1.2–4.2
Kr	337	1035	1.10	0.966	6.49	1.28	2.36	0.50	0.50	0.54
Fe	1808	3003	5.56	0.716	9.27	7.58	2.64	6.37	4.55	6.92
Co	1765	3458	4.78	0.797	9.79	7.92	2.50	5.93	4.76	4.1–5.3
Ni	1728	3448	4.43	0.840	9.75	8.11	2.43	5.64	4.76	4.5–6.4
Cu	1356	2903	3.54	0.839	10.55	7.58	2.86	4.07	4.20	4.34
Zn	693	1203	3.16	0.738	10.86	6.06	2.50	2.65	2.63	3.50
Ga	303	2573	2.24	1.000	11.58	5.28	2.65	2.00	1.63	1.94
Ag	1234	2253	2.63	0.763	17.91	5.19	2.67	3.53	4.07	4.28
In	430	2373	1.48	1.000	19.07	3.69	2.67	1.97	1.97	1.80
Sn	505	2548	1.53	1.000	19.69	3.54	2.62	2.04	2.11	1.81
Sb	904	1913	1.58	0.833	20.21	3.20	2.35	2.23	2.68	1.43
Au	1336	2983	1.89	0.857	32.71	5.33	2.95	5.50	5.80	5.38
Hg	234	630	1.14	0.929	33.29	4.11	2.73	2.31	2.06	2.04
Tl	576	1730	1.06	0.959	33.92	3.35	2.82	2.55	2.85	2.64
Pb	601	2023	1.01	0.980	34.41	3.10	3.10	2.52	2.78	2.61
Bi	544	1833	0.91	0.980	34.70	2.90	2.57	2.13	2.54	1.63

*Values at the melting points; $g(r_m)$: extrapolated values.

†Values calculated from eqn 1.43

‡Values calculated from Andrade formula:

$$\mu_m = 1.8 \times 10^{-4} \frac{(MT_m)^{1/2}}{V_m^{2/3}} \quad \text{(in mPa s)}$$

where M: atomic weight, T_m: absolute temperature of melting point, V_m: atomic volume at V_m.

§This table is derived from Tables 6.3 and 6.4 of Iida and Guthrie (1988).

Rewriting eqn (1.46) in macroscopic quantities, by replacing m with (M/N), a with $(V/N)^{1/3}$ and v_L by the parameters in eqn (1.47), one obtains an equivalent form of relationship to that first obtained by Andrade for melting point viscosities of monatomic liquids:

$$\mu_m = C_2 \frac{(M\theta_m)^{1/2}}{V_m^{2/3}} \tag{1.48}$$

where μ_m is in mPa s, M is the atomic weight, θ_m is the melting point temperature (K), V_m is the atomic volume at melting point.

Figure 1.27 shows a plot of experimentally determined viscosities of liquid metals near their melting points plotted versus $(M\theta_m)^{1/2}/V^{2/3}$. The slope of 0.057 represents the empirical fitting factor C. As seen, the correlation is remarkably good. To account for decreases in viscosity with temperature, it is

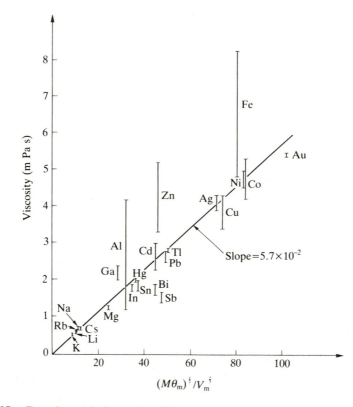

FIG. 1.27. Experimental viscosities of liquid metals at their melting points versus {atomic weight × melting point temperature)$^{1/2}$/(atomic volume)$^{2/3}$}. The bars denote the extremes of modern experimental values for μ. (After Iida and Guthrie 1988.)

found that an Arrhenius expression,

$$\mu = A \exp\left(\frac{H_\mu}{R\theta}\right) \tag{1.49}$$

can be used to approximate the data available, where

$$A = \frac{0.057 \, (\theta_m M)^{1/2}}{V_m^{2/3} \exp(H_\mu/R\theta_m)} \text{ and } H_\mu = 1.21 \, \theta_m^{1.2}.$$

(c) Thermal conductivity

(i) Non-metals

Heat can be regarded as being transported in non-metals by the vibrational motion of atoms or molecules in the liquid state. According to an early simple kinetic model of liquid non-metals by Osida (1939), the thermal conductivity of dielectric liquids at their melting points is given by

$$k = \frac{4K\nu}{a}, \tag{1.50}$$

where ν corresponds to the frequency of atomic vibrations at the non-metal's melting point. A modified Lindemann's formula is used for $\nu (= C(\sigma_m/M)^{1/2}$, $C = 6.8 \times 10^{11}$ in SI units)). Replacing a and ν with the appropriate expressions in bulk parameters, and substituting in eqn (1.50), one obtains

$$k = C_k \left(\frac{\sigma_m}{M}\right)^{1/2} \frac{1}{V_m^{1/3}}, \tag{1.51}$$

where σ_m is the surface tension at the melting point.

From the data plotted in Fig. 1.28 for a wide range of liquids, an empirical constant of $C_k = 0.012$ for SI units is seen to fit the data well and allow extrapolations to other liquids.

(ii) Metals

Heat transport in liquid metals, as well as solid metals, is dominated by electron transport processes. As a result, there is a close correspondence between the electrical energy and thermal energy that can be passed through a liquid metal by the 'nearly free electron' (NFE) cloud.

Lorenz derived the connection between the two transport coefficients:

$$\left(\frac{k}{\sigma_e \theta}\right) = \frac{\pi^2}{3}\left(\frac{K}{e}\right)^2 = 2.45 \times 10^{-8} \, (\text{W}\Omega\text{K}^{-2}) \tag{1.52}$$

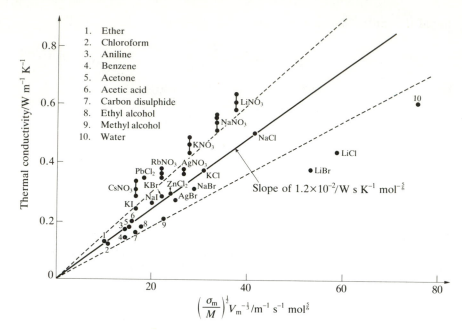

Fig. 1.28. Experimental thermal conductivities of liquid non-metals at their melting points, plotted versus the function $([\sigma/M]^{1/2} V_m^{-1/3}$. The broken lines denote $\pm 25\%$ error band. (After Iida and Guthrie 1988.)

where σ_e is the electrical conductivity of the liquid metal, e is the electronic charge, k is the thermal conductivity. The parameter $(\pi^2 K^2/3e^2)$ is known as the (theoretical) Lorenz number. Figure 1.29 shows the excellent correlation between the thermal conductivities of both solid and liquid metals and their corresponding electrical conductivities.

(d)   Diffusivities

Diffusion is the transport of mass from one region to another on an atomic scale. As for other transport coefficients, a wide variety of theories are available by authors such as Einstein, Sutherland, Walls and Upthegrove, Eyring, Reynik, Cohen and Turnbull, and Protopapas and Parlee (see Iida and Guthrie 1988).

For the purposes of the present text, which has dealt exclusively with theories of transport coefficients deriving from Newtonian kinetics on the motion of particles, only the Stokes–Einstein and Sutherland equations for diffusion coefficients will be presented. The justification for their sole inclusion is based on reasons of brevity, clarity of presentation and their proven track

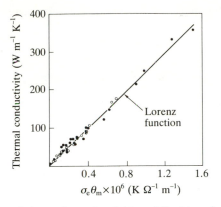

FIG. 1.29. Experimental thermal conductivities of liquid and solid metals at their melting points, plotted versus the product of their melting temperature and electrical conductivity (σ_e). (solid ●; liquid ○). (After Iida and Guthrie 1988.)

record versus other theories. A brief critique of these other theories in relation to solute diffusion in liquid metals is presented by Solar and Guthrie (1972).

The Stokes–Einstein equation, therefore, derives from earlier studies concerning the Brownian motion of a particle. The general treatment is based on the equation of Langevin, who proposed that the forces acting on a Brownian particle of mass m consists of two parts: a systematic frictional force proportional to the velocity and acting in the opposite direction, and a randomly fluctuating force arising out of collisions with a host of adjacent vibrating molecules.

The development involves solving a differential equation for the mean square displacement of a randomly moving Brownian particle from its original position at $r=0$ at time zero. Here ζ represents a friction frequency term and $R(v)$ the randomly fluctuating force on the particle:

$$mr\frac{dU}{dt} = -\zeta mrU + rR(v). \tag{1.53}$$

Assuming that the mean kinetic energy of the particle is related, as for molecules, to Boltzmann's constant and the absolute temperature θ, according to $mU^2 = 3K\theta$, solution yields

$$|r_t^2| = \left(\frac{6K\theta}{\zeta m}\right)t. \tag{1.54}$$

This shows that the displacement of the Brownian particle from its original position increases with time t, is greater at higher temperatures, and is inversely proportional to its mass and friction generated as a result of its motion in the liquid. Comparing this result with Einstein's relation for D in terms of the mean square displacements for a Brownian particle,

$$D = \mathrm{Lim}_{t \to \infty} \frac{1}{6t} \left(\overline{|r_t - r_0|^2} \right),$$

(1.55)

one has an expression for the diffusion coefficient

$$D = \frac{K\theta}{\xi m}.$$

(1.56)

We also have from Stokes' law that a rigid particle of diameter d travelling through a liquid at a steady velocity of U experiences a force

$$F = 3\pi\mu U d.$$

(1.57)

Comparison with eqn (1.53) $(\overline{rR(v)}) = 0)$ shows us that, since $F = m\,dU/dt = -\xi m U$, or $F/U = -\xi m$, we can write eqn (1.56) in the form

$$D = \frac{K\theta}{3\pi\mu d},$$

(1.58)

which is the famous Stokes–Einstein equation for diffusion.

A slightly modified version by Sutherland using a free slip boundary condition to extend this model to atoms (or ions) moving through liquids (for which $F = 2\pi\mu U d$) yields

$$D = \frac{K\theta}{2\pi\mu d} \quad \text{or} \quad \frac{K\theta}{4\pi\mu r},$$

(1.59)

where d is the diameter of the diffusing atom and r is its radius.

The extension of this hydrodynamic-kinetic model to molecular particles is made in the proposition that the motion of a diffusing atom, or molecule, can be treated in the same way as that of a far larger Brownian molecule. One envisages that the diffusing particle jostles its way through its nearest neighbours as it escapes from one 'cage' and moves on to another and that the phenomenon can be described in terms of Newtonian physics. As we have seen already, the low-frequency portion of the phonon spectrum that is associated with liquids corresponds to the physical, or diffusive, motion of atoms from one region of liquid to another.

Equation (1.59) can provide good estimates for both self- and solute diffusion coefficients, but values are very sensitive to r, the radius of the diffusing particle. While ionic radii are often used, there must be some doubt whether this is valid for liquid metals. For instance, the diffusion coefficient of hydrogen through molten iron is, as we have seen about $100 \times 10^{-9} \mathrm{m}^2\,\mathrm{s}^{-1}$. The radius of the proton core of hydrogen is so minute that if we assumed that hydrogen diffuses as an ion, its diffusion coefficient would be almost infinite! Nevertheless, for less extreme cases, and those in which the solute atom's diameter is similar to that of the solvent atom (ion), useful predictions are possible. For such calculations, the effective hard-sphere radius can be estimated on the basis of the hard-sphere model of liquids from which we have, for typical voidages in an assembly of atoms in the liquid state, that

$$r \simeq 0.47 \left(\frac{V}{N}\right)^{1/3}. \tag{1.60}$$

Substituting this into eqn (1.59), we have

$$D = \frac{K\theta}{5.9\mu(V/N)^{1/3}}. \tag{1.61}$$

This is well known as a modified Stokes–Einstein relation. Combining eqn (1.61) with eqn (1.49) for μ, one can construct an Arrhenius type relation to describe the increases in self diffusion coefficients with increasing temperature. The result (Iida and Guthrie 1988) is (in c.g.s.)

$$D_s = \frac{3.5 \times 10^{-6} V_m^{2/3} \exp(H_D/R\theta_m)}{(M\theta_m)^{1/2}} \frac{\theta}{V^{1/3}} \exp\left(\frac{-H_D}{R\theta}\right), \tag{1.62}$$

where D_s is the self-diffusivity of the liquid. The form of eqn (1.62) suggests that a plot of D_s versus $\{(\theta/M)^{1/2} V^{1/3}\}$ at $\theta = \theta_m$ should yields a linear relationship. Figure 1.30 demonstrates the adequacy of this relationship for self-diffusion coefficients in a variety of liquid metals. As seen the slope of 3.5×10^{-6} $(K^{-1/2} g^{1/2} cm s^{-1} mol^{-1/6})$ best fits the data available. H_D represents the activation energy for diffusion, given by $(H_D)_n = 2.5 \theta_m^{1.15}$ for normal metals (Iida & Guthrie (1988)).

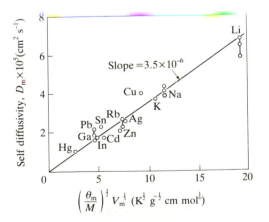

FIG. 1.30.   Self-diffusivities of liquid metals at their melting points, plotted as a function of $(\theta_m/M)^{1/2} V_m^{+1/3}$. (After Iida and Guthrie 1988.)

In the case of solutes diffusing through a liquid, their atomic (ionic) diameters will not usually be the same as those of the solvent atoms, so that 'friction forces' impeding their motion (i.e. the effect of μ) may be greater or smaller depending on their 'size. Iida and Morita (1978) proposed that the friction constant $4\pi\mu r$ be replaced by a modified friction coefficient given by $4\pi\mu_A{}^*r_A{}^* \simeq 4\pi\mu_A r_A^*$. Replacing the denominator in eqn (1.56) and substituting an appropriate viscosity relation for dilute mixtures, they showed that the solute diffusivity for solute A moving through solvent B atoms took the form:

$$D_{AB} = \frac{4.6 \times 10^{-14}\theta}{\{P(\theta)g(r_c)r\}_B k_D (M\theta_m)_A^{1/2}}, \tag{1.63}$$

where $P(\theta)$ represents the probability of a solvent atom being in vibration, and k_D is a complex function taking into account the size ratios of the solute and solvent atoms, their electronegativity and their different vibrational dependencies. In terms of the self-diffusivity, the solute diffusivity of a dilute binary alloy can be expressed in the simpler, implicit form:

$$D_{AB} = \frac{1}{k_D} \frac{(M\theta_m)_B^{1/2}}{(M\theta_m)_A^{1/2}} D_B \tag{1.64}$$

Figure 1.31 gives a comparison of predicted and observed diffusivities of dilute nickel solute (A) in liquid copper, as a function of temperature, using this approach.

(e)   Predictions

This section on transport coefficients in liquids would hardly be complete without a few predictions for transport coefficients in liquid iron, for which such experimental data are either lacking or very scanty.

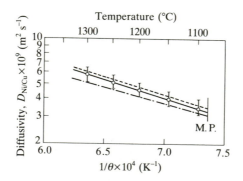

FIG. 1.31. Comparison of predicted and observed diffusivities of nickel in liquid copper versus reciprocal of temperature. (Iida and Morita 1975.)

In view of the frequent occurrence of molecular volumes V at liquid metal melting points in the predictive equations presented for μ, k and D, Fig. 1.32 has been provided for equivalent calculations on other systems of interest.

Figures 1.33, 1.34, 1.35 and 1.36 provide predictions for the viscosity of pure iron (versus temperature), liquid steel (versus carbon content), liquid steel (versus silicon content) and finally liquid steel with 1 at.% of various solutes. Table 1.3 provides a comparison of predicted and observed solute diffusion coefficients in liquid iron at 1600°C, while Fig. 1.36 predicts, among others, the diffusivities of niobium and tantalum in liquid iron, at low dilution, using eqn (1.64). No experimental data presently exist for these.

Predictions for the thermal conductivity of liquid iron are readily made on the basis of the Lorenz relation, suggesting a value of $38 \text{ W m}^{-1} \text{ K}^{-1}$, taking the electrical resistivity of iron, $\rho_e = 1.2 \times 10^{-6} \Omega \text{ m}^{-1}$ at 1600°C ($\rho_e = \sigma_e^{-1}$).

1.9. THEORY OF TRANSPORT COEFFICIENTS IN SOLIDS

The mechanisms by which heat and mass are transported through solids are reasonably well understood. However, the complex and specific nature of

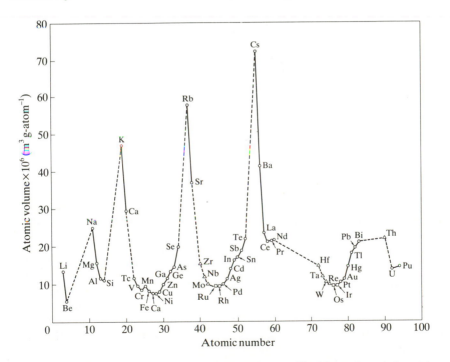

FIG. 1.32.   The periodicity of V_m, the atomic volumes of liquid metals at their melting points (From Iida and Guthrie 1988.)

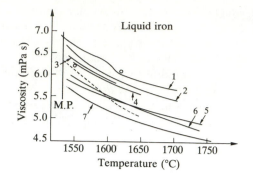

FIG. 1.33. Comparison of calculated and observed values (1–7) for the viscosity of pure liquid iron. Calculated (-----), from Iida and Guthrie (1988); experimental (——), from Iida and Morita (1975).

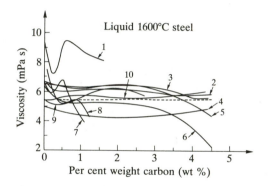

FIG. 1.34. Comparison of calculated and observed values (1–10) for the viscosity of Fe–C alloys at 1873 K (1600 °C). Calculated (-----), from Iida and Guthrie (1988); experimental (——), from Iida and Morita (1975).

associated interatomic forces at work among the atoms of a given substance generally precludes quantitative predictions for conductivity or diffusivity at present.

As previously mentioned, atoms can ideally be regarded as harmonic oscillators transferring kinetic energy (heat) from one to another through the lattice. The respective capabilities of solid compounds or substances as energy transmitters is reflected in their relative thermal conductivities. Solid metals, as liquid metals, have additional heat transport capabilities to those of ionically or covalently bonded compounds, owing to the great mobility, and energy carrying capacity, of their nearly free electron clouds.

The thermal conductivity of all pure metals (but not alloys) decreases with temperature (above their Debye temperatures) owing to an increase in the

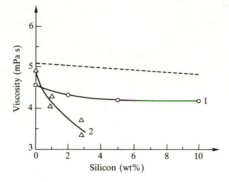

FIG. 1.35. Comparison of calculated and observed values for the viscosity of Fe–Si alloys at 1873 K (1600 °C). Calculated (------); experimental (———). (From Iida and Morita 1975.)

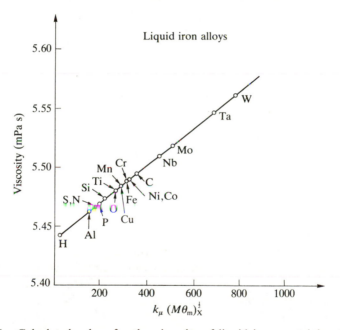

FIG. 1.36. Calculated values for the viscosity of liquid iron containing 1 Atomic Percent of Solute X at 1873 K (1600 °). (From Iida and Morita 1975).

amplitude of vibrations of the lattice hindering the passage of electrons. This lowering of conductivity is reflected in a simultaneous increase in electrical resistivity. One should note that the formation of any intermetallics tends to lower electrical and thermal conductivities of alloy systems quite significantly (e.g. ferro-alloys such as Fe–Si, Fe–Cr, etc.).

TABLE 1.3. *Comparison of calculated and experimental values for the solute diffusivity in liquid iron at 1873 K (1600 °C)*

Solute	Calculated[*] $(10^{-9}\,m^2\,s^{-1})$	Observed[†] $(\times 10^{-9}\,m^2\,s^{-1})$
C	4.8	2.0–20
Si	6.0	2.5–12
Mn	4.6	3.5–20
P	7.4	2.5
S	8.7	4.5–20
O	8.5	2.5–20
H	150	80–200
N	12	6–20
Ni	4.2	4.5–5.6
Co	4.2	3.5–5.0
Cr	4.3	3.0–5.0
Al	8.6	10
V	4.4	4.0–5.0
Mo	2.6	3.8–4.1

[*] Iida and Morita (1975).
[†] Quoted from: Ono (1977) and Gourtsoyannis (1978).

Various forms of diffusion through solids can be identified, and various diffusion coefficients defined as a result. The diffusion coefficient of interest to technologists is generally the chemical diffusion coefficient D_{AB}, occurring as a result of a concentration (or more properly) an activity gradient of the chemical species A in a solvent B. Further aspects of D_{AB} are considered in Sections 4 and 6 of Chapter 4. Other diffusion coefficients are those relating to the diffusion of tracers, intrinsic diffusion and self-diffusion.

The mechanisms by which diffusion takes place in solids can be quite varied and are discussed in most texts concerned with materials science or physical metallurgy. A predominant mechanism in solid metals is the 'vacancy mechanism; in which an atom is able to jump from one 'hole' or 'vacancy' in the lattice to another. The number of vacancies or holes in solids are known to increase dramatically with temperature, and this is reflected in an equally dramatic increase in values of experimentally derived solute, self-, or tracer diffusivities. Furthermore, variations in diffusivities can be adequately represented by an Arrhenius equation of the form

$$D = D_0 e^{-\Delta E_D / R\theta} \tag{1.65}$$

in which ΔF_D represents an activation energy presumably associated with the threshold energy required for atoms to escape their potential energy wells, while D_0 is thought of as a frequency of jumping factor. Both prove to be constant over wide ranges of temperature. It is for this reason that solute diffusivities in iron have been plotted versus the customary $1/T$ shown in Fig. 1.14.

In the case of solute atoms whose diameters are small in comparison with the dimensions of a lattice, diffusion can take place via an 'interstitial' as opposed to 'substitutional' mechanism. Such considerations apply to the gaseous elements dissolved in metals. For instance, the diffusivity of hydrogen dissolved in steel is relatively high because of its small atomic diameter relative to iron. As a result, dissolved hydrogen, even at room temperature, can diffuse to point defects within a solid metal matrix. There hydrogen atoms condense to form hydrogen molecules. Trapped, on account of their size, hydrogen over-pressures of many thousand atmospheres can build up and this may be sufficient to shatter the article in question. This has led to tremendous technological difficulties with respect to cracking in high-strength low-alloy steels for 'sour gas' pipelines, problems with railways steels, and many others. Not surprisingly, the control of gas content, and its elimination, can represent an important part of a steelmaker's responsibilities.

Diffusion in solid non-metals is also of considerable technological import-ance, particularly with respect to the silicon chip of the microelectronics industry. While similar considerations apply, diffusion being possible owing to the presence of vacancies in the atomic lattice work, the requirement of electroneutrality adds a local restriction to motion that is not imposed in metallic systems with their clouds of NFEs. Mention should therefore be made of Shottky defects or holes whose number must, for electroneutrality, be such that the number of cationic vacancies in a crystalline lattice compensate for an equal number of anionic vacancies. Similarly, for Frenkel defects, the number of lattice defects within a species sublattice must correspond to the number of interstitial sites occupied by the same species.

Such defects are made use of in probes for measuring dissolved elements such as oxygen in liquid steel and copper. The technique involves preparation of a pellet of alumina, impregnated with zirconia of valence two or additions of lime (CaO) which create Shottky defects in the alumina lattice. One side of the pellet contacts a liquid metal, the other is contacted with a $Cr–Cr_2O_3$ pellet to provide a source of oxygen at a known reference partial pressure. The electromotive force for movement of oxygen anions through the lattice via Shottky defects is measured. From this information, the concentration of oxygen dissolved in the metal can be deduced using the well-known relation-ship

$$E = E_0 + \frac{R\theta}{zeF} \ln \left| \frac{a_{0,\text{metal}}}{a_{0,\text{reference}}} \right| \tag{1.66}$$

As the electrical current is carried exclusively by the ionic components of the pellet (rather than by electronic conduction), one appreciates how closely solid-state diffusion in non-metals must be linked to the electrical structure of a substance.

1.10. RADIATION PROPERTIES

All electromagnetic radiation is transmitted at 3.0×10^8 m s^{-1}, the speed of light (*in vacuo*) and can be regarded as having wave-like properties. Figure 1.37 shows the electromagnetic spectrum, in which it is seen that *thermal radiation*, in which we are primarily interested in this text, extends between wavelengths of 10^{-7} to 10^{-4} metres (or 0.1 to 100 μm). The frequency v of this radiation is related to the speed of light (c) and wavelength (λ) according to

$$c = \lambda v. \tag{1.67}$$

Radiative thermal energy is transmitted through space in the form of discrete quanta of energy, each quantum packet having an energy level of

$$E = hv. \tag{1.68}$$

Here h is Planck's constant (6.625×10^{-34} J s). These discrete bursts of energy are known as photons. Following Einstein's famous equation, $E = mc^2$, photons can be regarded as having a mass m equal to $h/c\lambda$, and a momentum, mc equal to h/λ.

By considering radiation as a 'photon gas', the principles of quantum-statistical thermodynamics can be applied to derive an expression for the energy density, u_λ, of radiation per unit volume per unit wave length:

$$u_\lambda = \frac{8\pi hc\lambda}{\mathrm{e}(hc/\lambda K\theta)} - 1. \tag{1.69}$$

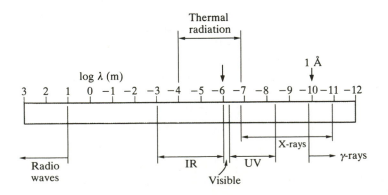

FIG. 1.37.   The electromagnetic spectrum.

When u_λ is integrated over all wavelengths, the total energy emitted is shown to be proportional to the absolute temperature to the fourth power, i.e.:

$$E_b \simeq \sigma\theta^4, \tag{1.70}$$

or

$$\dot{q}''_b = \sigma\theta^4,$$

as already presented in Section 1.3. The Stefan–Boltzmann constant σ, has the magnitude and units of $5.669 \times 10^{-8}\,\mathrm{W\,m^{-2}\,K^{-4}}$ or $0.1714 \times 10^{-8}\,\mathrm{BTU\,ft^{-2}\,{}^{\circ}R^{-4}\,h^{-1}}$, while E_b is the energy radiated by any ideal radiating surface per unit time and area (with units $\mathrm{W\,m^{-2}}$ or $\mathrm{BTU\,ft^{-2}\,h^{-1}}$).

Note that the subscript b used in the Stefan–Boltzmann equation refers to ideal or 'black body' radiation. Materials which obey this law often appear black to a human observer since they reflect no radiation. However, since the human eye registers radiation over a very narrow band ($\simeq 0.3$–$0.7\,\mu m$), and this is a small part of the total thermal spectrum, materials such as snow or white paints, which appear white to the eye, can act in fact as black bodies with respect to longer wavelength infra-red thermal radiation.

For non-black bodies the heat flux, or total energy emitted (see eqn 1.7) is less, and given by

$$\dot{q}'' = \varepsilon\sigma\theta^4, \tag{1.71}$$

where ε = total normal emissivity. This is needed in order to estimate the energy flux emitted from real surfaces. Emissivities of highly polished metallic surfaces are very low (~ 0.02), whereas oxidized surfaces typically range from 0.4 to 0.9. Similarly, slags and refractories tend to have high emissivities $\simeq 0.80$–0.95. Appendix 3 provides a list of values in terms of total emissivities, normal to the flat surfaces of a wide range of materials.

Before we conclude this section, the reader's attention is drawn to the plot of $E_{b,\lambda}(=cu_\lambda/4)$ shown Fig. 1.38, in which a black body's emissive power is shown as a function of wavelength for temperatures $1650\,^{\circ}C$ and $1090\,^{\circ}C$. One sees that the total energy flux E_b that is emitted rises sharply since

$$\dot{q}''_b \equiv E_b = \int_0^\infty E_{b,\lambda}\,d\lambda = \sigma\theta^4. \tag{1.72}$$

The areas under the two curves are representative of this fact. The other point to note is that the peak value of $E_{b,\lambda}$, as well as rising rapidly, shifts to the left (i.e. to smaller wavelengths). The displacement of the maximum is related to λ by Wien's law of displacement:

$$\lambda_{max}\theta = 2898\,(\mu m\,K). \tag{1.73}$$

This explains how practical steelmakers are able to gauge temperatures so accurately by visual estimates of colour. At about $500\,^{\circ}C$, a steel ingot will

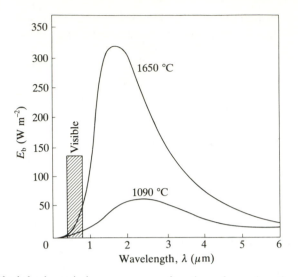

FIG. 1.38. Black-body emissive power as a function of wavelength and temperature.

appear as dull red; with progressive changes in temperature, it will appear red, red-orange and then yellow as λ_{max} shifts to shorter wavelengths. It also appears much brighter with rising temperature, since a greater fraction of the photons emitted are in the visible range. However, as a caveat, molten slag will look hotter than liquid steel at the same temperature because of its higher emissivity. Similarly, the reflection of incident radiation by a cold polished surface can lead to the erroneous conclusion that the surface is hot.

Since radiative heat transfer phenomena encountered in industry invariably involve heat exchanges between different surfaces, Section 4.13 introduces the subject of radiation view factors and demonstrates a few sample calculations involving radiation. For more detailed treatments, the reader is referred to textbooks devoted to heat transfer.

1.11. CONCLUSIONS

This chapter has served as an introduction to the subject of process engineering metallurgy. At the one extreme, it has described typical daily operations taking place around a steel-making plant and pointed out where, and how, the heat, mass and momentum transport phenomena are involved. While steel-making plants are an important example, it is emphasized that they are but one of a myriad of metallurgical operations. A few texts recommended for other metal production operations are listed below.

This chapter has also served, at the other extreme, to introduce fundamental transport equations and their transport coefficients. Some emphasis has been placed on molecular physics, so that experimental values of transport coefficients, with their wide range of values, can be understood and, where data are not available, reasonable guesses made.

FURTHER READING

1. **General extractive metallurgy**
   Pehlke, R. D. (1975). *Unit processes of extractive metallurgy.* American Elsevier Publishing Co., Amsterdam.
   Rosenqvist, T. (1974). *Principles of extractive metallurgy.* McGraw-Hill, New York.
2. **Iron and steel**
   The making, shaping and treating of steel. US steel handbook (9th edn.)
   Pehlke R. D. *et al.* (1977). *B.O.F. steelmaking Pt I–V.* Iron and Steel Society of AIME.
   Ward, R. G. (1962). *The physical chemistry of steelmaking.*
3. **Copper**
   Biswas, A. K. and Davenport, W. G. (1980). *The extractive metallurgy of copper.* Pergamon Press, New York.
4. **Nickel**
   Boldt, J. R. and Queneau, P. (1967). *The winning of nickel.* Longmans, Toronto.
5. **Aluminium**
   Hatch, J. E. (1984). *Aluminium properties and physical metallurgy.* American Society for Metals, Metals Park, Ohio.
6. **Magnesium**
   Emley, E. F. (1966). *Principles of magnesium technology* (1st edn). Pergamon Press, Oxford.
7. **Tin**
   Wright, P. A. (1966). *The extractive metallurgy of tin.* Elsevier, Amsterdam.
8. **Transport phenomena or coefficients in systems of metallurgical interest**
   Richardson, F. D. (1974). *Physical chemistry of melts in metallurgy,* Vols. 1, 2. Academic Press, New York.
   Turkdogan, E. T. (1980). *Physical chemistry of high temperature technology.* Academic Press, New York.
   Bird, R. B., Stewart, W. E., and Lightfoot, E. N. (1960). *Transport phenomena.* Wiley, New York.
   Iida, T. and Guthrie, R. I. L. (1988). *The properties of liquid metals.* Oxford University Press.
   Geiger, G. H. and Poirier, D. R. (1973). *Transport phenomena in metallurgy.* Addison-Wesley, Philippines.
   Croxton, C. A. (1974). *Introduction to liquid state physics.* Wiley, London.
   Cole, G. H. A. (1967). *Introduction to the statistical theory of classical simple dense fluids.* Pergamon Press, New York.

9. Reference books on physical and thermal properties

Weast, R. C. and Asle, M. J. (1981–82). *C.R.C. handbook of chemistry and physics*, (62nd edn). C.R.C. Press, Boca Raton, FL.

Bolz, R. E. and Tuve, G. L. (1970). *C.R.C. handbook of tables for applied engineering science*. C.R.C. Press, Boca Raton. FL.

Perry, R. H. and Chilton, C. H. (1973). *Chemical engineers' handbook* (5th edn). McGraw-Hill, New York.

Touloukian, Y. S. and others. (1967). *Thermophysical properties of high temperature solid materials*; Vol. 1, Elements; Vol. 2, Pt 1, Non-ferrous binary alloys; Vol. 2, Pt 2, Non-ferrous multiple alloys; Vol. 3, Ferrous alloys; Vol. 4, Pt 2, Oxides and glasses.

Smithells, C. J. (1983). *Smithells metal reference book* (5th edn). Butterworths, London.

Iida, T. and Guthrie, R. I. L. (1988). *The properties of liquid metals*. Oxford University Press.

2

FLUID STATICS AND FLUID DYNAMICS

2.0. INTRODUCTION

As noted in Chapter 1, many metallurgical operations require fluid to be transferred from one location to another. Often the simultaneous transport of heat or mass as a result of such fluid convection processes can be of prime significance.

The main purpose of this chapter is to describe basic concepts and equations governing the science of hydrostatics and hydrodynamics. The subject is presented from both the microscopic and the macroscopic viewpoints. However, since the latter approach is generally more valuable for general process engineering problems, and is certainly more amenable to treatment, its use will be emphasized through the presentation of a wide variety of metallurgical examples.

2.1. FLUID STATICS

The science of fluid statics, or rather hydrostatics, was fully developed in the Ancient world. It dates back at least to Egyptian and Grecian times and provides us with some of the concepts or building blocks needed for the more general subject of fluid dynamics or hydrodynamics.

(a) Pressure at a point

In hydrostatics one is typically concerned with variations in the pressure (or force/unit area) within a static fluid and with a study of pressure forces on finite surfaces. To demonstrate the fundamental concept that the pressure at a point within such a fluid is equal in all directions, consider a small segment, or wedge, of fluid of unit width, as shown in Fig. 2.1. For a fluid at rest no shear stresses can exist. Consequently, the only forces that can act must be normal (rather than tangential) surface forces, together with gravity. Writing force balances in the x and y directions, we have

$$P_x \Delta y = P_s \Delta s \sin(\theta), \tag{2.1}$$

$$P_y \Delta x - P_s \Delta s \cos(\theta) - \rho g \frac{\Delta x \Delta y}{2} = 0. \tag{2.2}$$

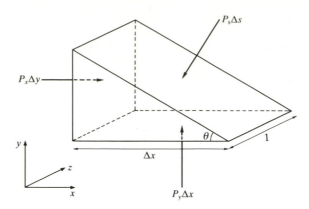

FIG. 2.1.   Free body diagram of wedge-shaped body in fluid.

Noting that $\Delta y = \Delta s \sin \theta$, and $\Delta x = \Delta s \cos \theta$, we see that

$$P_x = P_s \tag{2.3}$$

$$(P_y - P_s)\Delta x = \rho g \frac{\Delta x}{2} \Delta y. \tag{2.4}$$

In the limit as $\Delta x = \Delta y \to 0$, the second-order term $\Delta x \Delta y$ can be neglected and

$$P_y = P_x = P_s. \tag{2.5}$$

However, once the fluid is in motion and shear stresses are generated, the normal stresses are generally no longer the same in all directions at a point and the pressure is defined as the average of any three mutually perpendicular normal compressive stresses at a point,

$$P = \frac{P_x + P_y + P_z}{3} \tag{2.6}$$

Note that in 'frictionless fluids', pressures at a point are equal in all directions. This is useful to remember, since in many instances low-viscosity fluids such as gases, aqueous media, and liquid metals exhibit some of the characteristics of 'frictionless fluids'.

(b)   Pressure variations in static fluids

In a fluid at rest, it is self-evident that any points within it that lie in the same horizontal plane, must be at equal pressure. If not, fluid motion would occur as a result of horizontal pressure gradients. However, vertical pressure gradients can be sustained within a contained fluid.

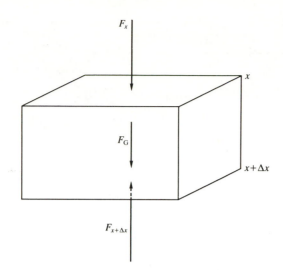

FIG. 2.2.   Volume element of liquid showing the normal vertical forces acting on it.

Figure 2.2 shows a body of liquid, at a distance x below the surface. Since no shear forces exist, the three normal forces shown in Fig. 2.2 must be in equilibrium, so that

$$P_x A + \rho g A \Delta x = \left(P_x + \frac{dP_x}{dx} \Delta x \right) A \qquad (2.7)$$

$$\frac{dP}{dx} = \rho g \qquad (2.8)$$

or

$$\Delta P = P_x - P_0 = \rho g x \qquad (2.9)$$

for an incompressible liquid or fluid. Note that x is taken to be positive downwards, so that the familiar hydrostatic law of pressure variation is brought out.

(c)   Centre of pressure

The centre of pressure is defined as that point on a surface where the line of action of the resultant forces acts. Unlike horizontal surfaces, the centre of pressure does not act at the centroid (i.e. centre of mass) for inclined surfaces. Consider the case shown in Fig. 2.3, depicting the resultant force acting on an inclined rectangular plate of width w and inclination θ, at the centre of

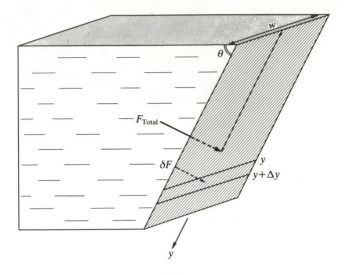

FIG. 2.3.   Resultant force acting on an inclined rectangular plate.

pressure. By moments, we have

$$y_P = \frac{\int_0^L y\,\delta F}{F} = \frac{\int_0^L y\,\delta F}{\int_0^L dF} = \frac{\int_0^L y\,\rho g y \sin(\theta)\,w\,dy}{\int_0^L \rho g y \sin(\theta)\,w\,dy}$$

$$= \frac{\int_0^L y^2\,dy}{\int_0^v y\,dy} = \frac{2}{3}\frac{L^3}{L^2} = \tfrac{2}{3}L. \tag{2.10}$$

This may be generalized to yield

$$y_P - \bar{y} = \frac{I_G}{\bar{y}A}. \tag{2.11}$$

Since the moment of inertia about the centroid at $\bar{y}$ is always positive, we see that the centre of pressure will always lie below the centroid of the surface as demonstrated in the simple example chosen. I_G represents the moment of inertia of the surface, of arbitrary shape and surface area A, about its centroid.

(d)   Buoyancy forces

Consider the submerged body shown in Fig. 2.4; the fluid, of density ρ, in which it is immersed, will generate a net resultant force directed vertically upwards through the centroid.

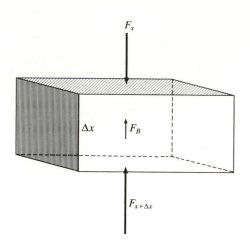

FIG. 2.4. Hydrostatic forces acting on the top and bottom surfaces of a rectangular parellepiped, the imbalance registering as a buoyancy force (Archmides buoyancy force).

Resolving forces, we have

$$F_B = F_{x+\Delta x} - F_x = A \frac{\partial P}{\partial x} \Delta x = \rho A \, \Delta x \, g. \tag{2.12}$$

In other words, a vertical buoyancy force equal to the weight of displaced fluid acts on submerged or a partially submerged body. This fundamental law of hydrostatics was first enunciated by Archimedes of Syracuse in about 260 BC.

2.2. EXAMPLES IN HYDROSTATICS

(a) Hydrostatic pressure developed at the bottom of a ladle

A cylindrical ladle of steel, of 3 m internal diameter and 3.5 m high, is filled to capacity, leaving a freeboard of 0.15 m. Calculate the pressure developed on its bottom surface, the resultant force, and the maximum rate at which the crane operator can accelerate or decelerate the vessel without spilling liquid steel onto the shop floor (see Fig. 2.5).

Solution

Part (1). Taking the top surface as the reference point, we have

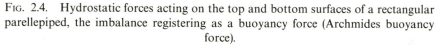

$$\int_{P_0}^{P} dP = \int_{0}^{H} \rho g \, dz$$

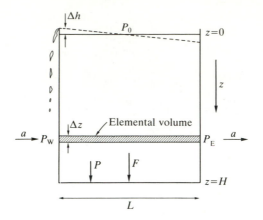

FIG. 2.5. Illustration of the movement of the top surface of liquid steel during continuous transverse acceleration of a ladle.

giving

$$P_H = P_0 + \rho_s g H$$

P_0 will be atmospheric pressure $= 1.01 \times 10^5$ Pa, so that

$$P_H = 1.01 \times 10^5 + 7000 [\text{kg m}^{-3}] \, 9.81 [\text{m s}^{-2}] 3.35 [\text{m}]$$

or

$$P_H = 3.31 \times 10^5 \text{ Pa or 3.3 atm.}$$

The resultant force will be $(P_H A - P_0 A)$ since the atmospheric pressure acts on the bottom surface of the ladle, i.e. $F = 16.3 \times 10^5$ N.

Part (ii). Acceleration will cause the steel to move up towards the trailing top edge of the ladle (and vice versa for deceleration). Thus, once the surface of liquid steel has moved to its new position and the liquid is everywhere stagnant, no shear forces exist and Newton's second law can be applied to the horizontal elemental volume of liquid shown in this figure:

$$F = (P_W - P_E) A = ma.$$

Making appropriate substitutions:

$$\rho g \, 2\Delta h (1 \times \Delta z) = \rho (L \times 1 \times \Delta z) a$$

or

$$a = \frac{g 2 \Delta h}{L} = \frac{9.81 \times 0.30}{3} \text{ m s}^{-2}$$

Maximum acceleration (or deceleration) $= 0.98$ m s^{-2} (a force of 0.1 g).

(b)   The growth and detachment of bubbles nucleating in melts

It is postulated that gas bubbles can nucleate at gas-filled crevices within the surface of a refractory or container vessel, and that this is the mechanism by which CO bubbles nucleate at the bottom of an open hearth steel-making furnace, or in a basic oxygen furnace during the late stages of refining when bath carbon is low ($\leq 0.1\%$C). Assuming an effective crevice diameter of 1 mm, calculate:

(1) the dissolved gas overpressure (with respect to ambient pressure) needed for continued nucleation and growth of bubbles on the bottom of a 1 m deep body of liquid;

(2) the size of bubble that detaches.

Compare results for water, liquid aluminium and liquid steel, given the following data and referring to Fig. 2.6.

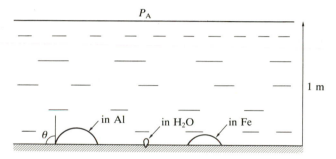

Fɪɢ. 2.6.   Bubbles nucleating heterogeneously at fissures in the bottom surface of a furnace or holding vessel.

Surface tension,

	$\sigma(\mathrm{N\,m^{-1}})$	Contact angle, θ	Density $(\mathrm{kg\,m^{-3}})$
H_2O	0.07	0	1000
Al	0.90	90	2300
Fe	1.76	90	7000

Solution

Part (i). Carrying out a static force balance between surface tension forces around the equator of an assumed growing hemisphere, and the surface tension overpressure, ΔP_σ within the bubble.

$$2\pi r\sigma = \Delta P_\sigma \pi r^2 \quad \text{or} \quad \Delta P_\sigma = \frac{2\sigma}{r}$$

$$\Delta P_{\text{gas}} = \Delta P_{\text{static}} + \Delta P_\sigma \quad \text{or} \quad \Delta P_{\text{gas}} = \rho g h + \frac{2\sigma}{r}.$$

Inserting appropriate data, one obtains gas overpressures, ΔP_{gas} of 0.1×10^5, 0.26×10^5, and 0.76×10^5 Pa respectively for H_2O, Al and Fe.

Part (ii). Since the aqueous system is wetting, the rim of the forming bubble remains attached to the periphery of the crevice, and detachment will occur once buoyancy forces exceed adhesion forces. For release,

$$(\rho_L - \rho_g)gV_b \geq 2\pi r_0 \sigma, \quad V_b \geq \frac{2\pi r_0 \sigma}{\rho_L g} \geq 0.0224 \text{ cm}^3$$

where r_0 is the radius of the crevice.

For the liquid metal systems, which are non-wetting, gas will spread across the refractory surface since the bubble rim will not be anchored to the crevice's periphery. We therefore have a maximum radius r_m for a growing hemispherical bubble given by

$$(\rho_L - \rho_g)g\{\tfrac{4}{3}\pi r_m^3 \times \tfrac{1}{2}\} = 2\pi r_m \sigma,$$

whence

$$r_m = \left(\frac{3\sigma}{\rho_L g}\right)^{1/2},$$

so that

$$V_m = \tfrac{4}{3}\pi \left(\frac{3\sigma}{\rho_L g}\right)^{3/2}$$

Consequently, $V_m = 0.02$, 5.8 and 2.8 cm^3 respectively for water, liquid aluminium, and liquid steel.

2.3.   EXERCISES IN FLUIDS STATICS

(1)   Metal casting forces in moulds

A hollow mould for producing a die-cast sphere is filled with molten zinc alloy through a small hole in the top. Find the resultant vertical forces exerted by the metal on the top and bottom half of the inside surface of the mould. Express the result in terms of the total weight of metal.

(2) Buoyancy of ladle alloy additions

In ladle alloy addition operations, lumps of various alloys are added to the liquid steel for alloying purposes. Determine the equilibrium positions of 0.1 m cubes of aluminium and ferromanganese in a stagnant bath of liquid steel covered with 0.15 m of liquid slag. (Ignore dissolution/melting phenomena). Data: $\rho_{Fe} = 7200$, $\rho_{slag} = 3000$, $\rho_{Al} = 2700$, $\rho_{Fe-Mn} = 6500$ (kg m^{-3}).

(3) Forces on dam, walls

The thrust on a dam constructed for a waste tailings disposal lake is to be taken by three *equally* loaded horizontal steel beams. If water acts on one side only of the vertical steel plate to a depth of 9 m, determine the loading per metre run on each external horizontal steel beam. Determine the respective positions of the three beams below the water line.

(4) Height of Earth's atmosphere

Calculate the height of the Earth's atmosphere, taking the surface temperatures as 0°C and pressure as 1.01×10^5 Pa using three models of gas behaviour: (a) constant density of gas; (b) constant temperature of gas; (c) $PV^n = $ constant, $n = 1.2$.

(5) Centrifuging of molten metals

In a molten metal mixing process, a paddle wheel contained within a vertical cylindrical column rotates liquid aluminium in a forced vortex fashion (i.e., radial and vertical fluid velocity components are zero). Determine the angular speed (revs s^{-1}) required for the metal level at the centre of the vortex to drop to half its original level. Ignore possible wall effects as well as viscosity. Vessel diameter $= 1$ m; density of aluminium $= 2300$ kg m^{-3}, liquid depth $= 1.0$ m.

2.4. INTRODUCTION TO FLUID DYNAMICS

Practically all operations in extractive metallurgy involve fluid in motion (e.g., all combustion systems, all reactor vessels, and all fluid transport systems incorporating blowers, pumps, and fans). Many opportunities exist for improving the performance of metallurgical operations by paying attention to these flows. It is therefore essential for the practising metallurgist to know something about fluids and the equations of continuity and momentum governing their behaviour.

Fluid flow problems can be divided into many categories, but the best for the present limited treatment, is to make the all-important distinction between *laminar flows* and *turbulent flows*. Steady laminar flow refers to those conditions in which the liquid or gas flows in layers and in which streamlines and path lines within the fluid coincide. (Path lines trace the motion of individual particles of fluid through the flow system; streamlines represent lines drawn tangential to the instantaneous velocity vectors in the flow). Turbulent flows are typically characterized by random, high-frequency, low-amplitude motions of fluid superimposed on a bulk motion that is steady on a time-averaged basis. For turbulent flows, therefore, the path lines traced by individual fluid elements lying along a streamline would not be coincident, even though on average the flows were steady.

Owing to the physical size and scale of most industrial processing vessels and operations, most of the flows that one encounters in industry are turbulent in nature. However, one does occasionally meet situations in which laminar flow takes place. These cases occur when velocities are very low, or viscosities are very high, or sizes are very small. A succinct way of putting this, is to require that the Reynolds number, Re (representing the product of density, velocity and length-scale divided by viscosity, $\rho U L/\mu$), be low.

In Chapter 3 it is shown that Re represents the ratio of inertial forces to viscous forces. When Re is low, viscosity in the denominator is relatively important and laminar flow *may* apply. Examples of low Reynolds number flows include settling of small particles in clarifiers, electrostatic collection of fine dusts, flow of viscous slags through troughs, lubrication of bearings by thin films of oil, and various other boundary layer flows over solid surfaces (or vice versa).

Since the treatment of laminar flows, and their mathematical description, lead on to ways of analysing (in the engineering sense) turbulent flows, we now direct our attention to the former, and simultaneously demonstrate the use of shell momentum balances.

(a)   Laminar flow and the shell momentum balance

Example of laminar flow: Slag as a lubricant in continuous casting operations

One will see from Fig. 2.7 (or Fig. 1.6(a)) that it is customary in continuous casting operations to use mould additions (powdered slag) for covering the surface of liquid steel in the mould. The purpose of the powder addition is:

(1) to cut down heat losses from the steel surface;

(2) to prevent surface oxidation of the steel;

(3) to melt and form a slag of high fluidity so as to help with the absorption of non-metallic inclusions floating out of the steel;

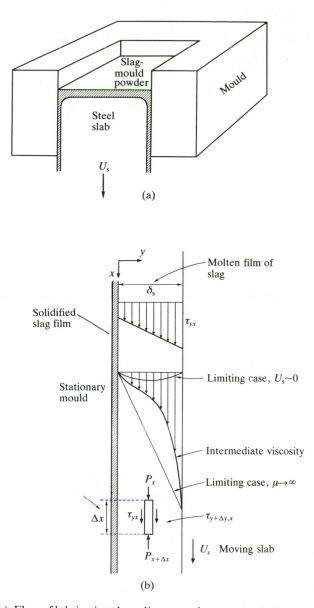

FIG. 2.7. (a) Flow of lubricating slag adjacent to the water-cooled copper mould of a continuous casting machine for slabs. (b) Velocity and momentum fields generated by flow of slag lubrication layer.

(4) to act as a lubricant preventing steel from sticking to the inside surfaces of the water-cooled copper mould.

We will now demonstrate how the feed rate of slag needed to form a surface coating of thickness δ_s can be calculated from first principles.

Before embarking on this, or any other fluid analysis however, we must first decide whether the phenomenon involves laminar or turbulent flow. Supposing that we are interested in δ_s values up to 0.1 mm, slab exit velocities up to 3 m min^{-1}, and slag viscosities as low as 10 mPa s (10 cP), then the maximum Reynolds number, Re, we should expect is about $[3000 \times (3/60) \times (1 \times 10^{-4})]/[10 \times 10^{-3}] = 1.5$.

We will later see that the onset of turbulence in flows through pipes and ducts occurs at about Re $\sim 10^{+3}$, indicating that we should be quite safe in supposing that this is a laminar flow problem.

Question

Develop an expression to show how much slag powder should be added to a continuous caster so as to maintain a coherent lubricating film of thickness δ_s between the mould and the slab.

Solution

Referring to Fig. 2.7(a), the physical situation shows a section of the water-cooled copper mould, together with the slab and an overlaying fluid slag lubricant. Depending on the thickness of the film, we could expect it to drain out by itself under the influence of gravity, (provided Taylor instability associated with surface tension is not important). Superimposed on this natural efflux will be the 'drag-out' effect of the downward moving slab, dragging slag down towards the bottom of the cooled copper mould. This happens because of the non-slip condition between a solid surface and the adjacent molecules of fluid. Similarly, since the solidified steel slab is still at a high temperature, and the slag composition is chosen for its fluidity and low freezing range, steel-makers contrive to ensure that most of the film remains liquid during its descent through the water-cooled copper mould.

In Fig. 2.7(b), a suitable coordinate system has been chosen for solving the problem. An x–y coordinate system was chosen with x as the vertical axis, taken as positive in the downward direction.

The general approach required for solving fluid flow phenomena is to write down the governing differential equations for continuity of mass, and for momentum, for a volume element of fluid. For simple problems such as this, we can demonstrate the use of the shell momentum balance. We will apply this to a typical volume element of slag so as to deduce the differential equations of

motion governing its flow. Suitable boundary conditions then allow velocity profiles for the specific problem to be solved, and thence the flow rates. For a steady flow such as this, we can state that:
for mass continuity,

(Rate of mass in)—(rate of mass out)=(rate of mass accumulation)=0,

$$(2.13)$$

for momentum,

(rate of momentum in)—(rate of momentum out)

+(Sum of forces acting on system)=0. $$(2.14)$$

We shall not dwell on applying the mass continuity equation, since the solution is trivial. It is sufficient to note that since the flow is one-dimensional, U_y and $U_z=0$. Similarly, since slag is incompressible $\partial\rho/\partial t=0$, and $\partial(\rho U_x)/\partial x=0$.

Turning our attention to the momentum equation and choosing a volume element of thickness Δy, of length Δx and unit perimeter length, we can carry out the following accounting procedure for our shell momentum balance.

Rate of x momentum *in* across top surface at x	$(L\Delta y)\rho U_x \cdot U_x$
Rate of x momentum *out* from bottom surface at $x+\Delta x$	$(L\Delta y)\rho U_{x+\Delta x} \cdot U_{x+\Delta x}$
Rate of x momentum *in* across side surface of element between x and $x+\Delta x$	$(L\Delta x)\tau_{y,x}$
Rate of x momentum *out* across side surface of element between x and $x+\Delta x$	$(L\Delta x)\tau_{y+\Delta y,x}$
Gravity force acting on fluid element (i.e. component of weight)	$(L\Delta x\Delta y)\rho g_x$
Pressure force acting on top surface of element	$(L\Delta y)P_x$
Pressure force acting on bottom surface of element	$(L\Delta y)P_{x+\Delta x}$

Assembling all these terms in the momentum balance, and making appropriate expansions of the type $\left(U_{x+\Delta x}=U_x+\dfrac{\partial U_x}{\partial x}\cdot\Delta x\right)$

we obtain

$$-\left\{L\Delta y\rho\frac{\partial}{\partial x}(U_x\cdot U_x)\Delta x+L\Delta x\frac{\partial\tau_{y,x}}{\partial y}\Delta y\right\}+L\Delta x\Delta y\rho g_x+L\Delta y\frac{\partial P}{\partial x}\cdot\Delta x=0$$

$$(2.15)$$

Dividing by $L\Delta y\Delta x$, remembering that $\partial U_x/\partial x=0$ for one-dimensional, steady-state, incompressible flow, and realizing that $\partial P/\partial x=0$, (since $P=P_{\text{At}}$

both at the top of the mould and at the bottom where the molten slag is exiting)

$$\frac{d\tau_{yx}}{dy} = \rho g_x.$$ (2.16)

Integrating,

$$\tau_{yx} = \rho g_x y + A.$$ (2.17)

Substituting Newton's law of viscosity for laminar, one-dimensional flow,

$$-\mu \frac{dU_x}{dy} = +\rho g_x y + A.$$ (2.18)

Integrating once more,

$$-\mu U_x = \rho g_x \frac{y^2}{2} + Ay + B$$ (2.19)

Applying the customary non-slip boundary conditions that $U_x = 0$ at $y = 0$ and that $U_x = U_s$ at $Y = \delta_s$, we have for the constant of integration, B, that

$$-\mu(0) = 0 + 0 + B.$$ (2.20)

B must therefore be zero since viscosity is finite. Evaluating A:

$$-\mu U_s = \rho g \frac{\delta_s^2}{2} + A\delta_s$$ (2.21)

or

$$A = \frac{-\mu U_s}{\delta_s} - \frac{\rho g \delta_s}{2}$$ (2.22)

Substituting in eqn (2.19), we finally reach an expression for the way in which the velocity U_x varies with distance, across the slag film. The velocity profile, drawn in Fig. 2.7(b), reflects the relative importance of gravity forces and viscous drag forces caused by the downward motion of the billet surface:

$$U_x = U_s \frac{y}{\delta_s} + \frac{\rho g}{2\mu}(\delta_s y - y^2)$$ (2.23)

This expression shows that, for very high viscosities, the second term vanishes and a linear velocity profile across the slag film will be generated by the drag effect of the moving billet. At the other extreme, for low viscosity, larger slag thickness and low casting speeds, the right-hand side term dominates, and

$$U_x \simeq \frac{\rho g y^2}{2\mu}\left(\frac{\delta_s}{y} - 1\right).$$ (2.24)

Differentiating and setting $\partial U_x / \partial y = 0$, we see that a maximum in the self-draining velocity will occur at $Y = \delta_s/2$, equal to $\rho g y^2 / 2\mu$. Consequently, as seen in Fig. 2.7, a set of velocity profiles is possible, depending upon the relative magnitudes of the parameters appearing in eqn (2.24).

We may now move on to show how much slag needs to be fed per unit length of slab periphery. Clearly the slag flowrate, $\dot{Q}'$ ($m^3\ s^{-1}\ m^{-1}$) is given by

$$\dot{Q}' = \int_0^{\delta_s} U_x\,dy \tag{2.25}$$

$$= \left(\frac{U_s\delta_s}{2} + \frac{\rho g\delta_s^3}{12\mu}\right). \tag{2.26}$$

Our expression for the mass flow of slag per linear metre finally becomes

$$\dot{m}' = \rho\left(\frac{U_s\delta_s}{2} + \frac{\rho g\delta_s^3}{12\mu}\right). \tag{2.27}$$

As we would expect, the mass flow of slag is proportional to slag density and slab withdrawal velocity, and inversely proportional to slag viscosity. It has a relatively complex dependence on slag film thickness δ_s. Taking a numerical example; $\rho = 3000\ kg\,m^{-3}$, $U_s = 0.05\ m\,s^{-1}$, $\mu = 3$ Poise or $0.3\ kg\,m^{-1}\,s^{-1}$, and $\delta_s = 0.1$ mm.

$$\dot{m}' = 3000\left(\frac{1/20 \times 10^{-4}}{2} + \frac{3000 \times 9.81 \times 10^{-12}}{12 \times 0.3}\right)$$

$$= 3000(2.5 \times 10^{-6} + 0.008 \times 10^{-6}),$$

which gives

$$\dot{m}' = 0.1\ kg\,s^{-1}\,m^{-1}.$$

Note that, by convention, quantities are always taken as being 'in' and 'out' in the direction of the positive x and y axes. It is also important to take care to ensure that correct signs are taken for all forces and momentum fluxes with respect to the axes one chooses in solving such problems. For example, because we took the x direction as being positive downwards, the gravity force acting on the fluid element was positive.

It is left as an exercise to show that the momentum flux distribution, τ_{yx}, is correctly drawn. It should be evaluated from differentiating the velocity profile with respect to y and applying Newton's law of viscosity.

(b) The continuity and Navier–Stokes equations

An alternative to shell momentum balances that is often safer, easier and quicker to apply, and ensures that the correct set of differential equations are to be solved, is to start with the general equations of continuity and motion listed in Tables 2.1, 2.2 and 2.3. These are presented in a number of textbooks dealing with fluid mechanics (*Transport Phenomena*, Bird, Stewart and

Lightfoot, is a excellent descriptive source, see Chapter 1). The velocity vectors and momentum fluxes, and notation relevant to Tables 2.1, 2.2, and 2.3, are illustrated in Fig. 2.8. Note that these momentum fluxes may be considered stresses. Thus τ_{xx} is the normal stress on the x-face, and τ_{yx} is the x-directed shear stress on the y-face resulting from viscous forces.

As will be seen from Tables 2.1, 2.2 and 2.3, the important equations of continuity, momentum and stress relationships are tabulated for rectangular, cylindrical and spherical coordinates. While the derivation of the differential continuity equations are readily obtained by considering mass input, output,

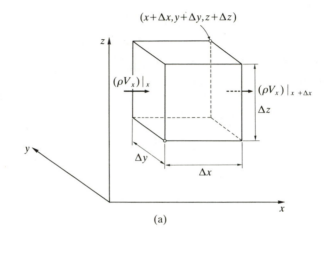

(a)

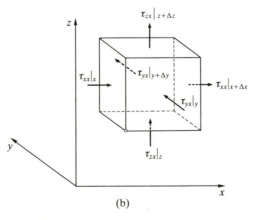

(b)

Fig. 2.8. Volume element of fluid $\Delta x \, \Delta y \, \Delta z$, fixed in space and time. The arrows indicate the direction in which the x component of mass (upper element) and x momentum (lower element) are transported through the surfaces.

TABLE 2.1. *The equation of continuity in several coordinate systems*

Rectangular coordinates (x, y, z):

$$\frac{\partial \rho}{\partial t} + \frac{\partial}{\partial x}(\rho V_x) + \frac{\partial}{\partial y}(\rho V_y) + \frac{\partial}{\partial z}(\partial V_z) = 0 \tag{A}$$

Cylindrical coordinates (r, θ, z):

$$\frac{\partial \rho}{\partial t} + \frac{1}{r}\frac{\partial}{\partial r}(\rho r V_r) + \frac{1}{r}\frac{\partial}{\partial \theta}(\rho V_\theta) + \frac{\partial}{\partial z}(\rho V_z) = 0 \tag{B}$$

Spherical coordinates (r, θ, ϕ):

$$\frac{\partial \rho}{\partial t} + \frac{1}{r^2}\frac{\partial}{\partial r}(\rho r^2 V_r) + \frac{1}{r \sin \theta}\frac{\partial}{\partial \theta}(\rho V_\theta \sin \theta) + \frac{1}{r \sin \theta}\frac{\partial}{\partial \phi}(\rho V_\phi) = 0 \tag{C}$$

and accumulation in a volume element of appropriate geometry (slab, cylindrical or spherical shell), the derivation of the equation of motion (or momentum) is not straightforward, the tensor terms associated with the viscous shear and normal stresses being particularly tricky, as will be apparent from inspection of Table 2.3.

To illustrate the method, the last problem will be reconsidered and the appropriate differential equation to be solved will be deduced. Since the problem was set up in rectangular coordinates, the two general equations of relevance to us are:

Continuity: (eqn A, Table 2.1)

$$\frac{\partial \rho}{\partial t} + \frac{\partial}{\partial x}(\rho V_x) + \frac{\partial}{\partial y}(\rho V_y) + \frac{\partial}{\partial z}(\rho V_z) = 0 \tag{2.28}$$

$$\quad\ (1)\qquad (2)\qquad\ \ (3)\qquad\ \ (4)$$

Motion (x component, Newtonian fluid, ρ, μ, constant): (eqn D, Table 2.2)

$$\rho\left(\frac{\partial V_x}{\partial t} + V_x \frac{\partial V_x}{\partial x} + V_y \frac{\partial V_x}{\partial y} + V_z \frac{\partial V_x}{\partial z}\right) =$$

$$\quad\ (5)\qquad\ (6)\qquad\ \ (7)\qquad\ \ (8)$$

$$-\frac{\partial P}{\partial x} + \mu\left(\frac{\partial^2 V_x}{\partial x^2} + \frac{\partial^2 V_x}{\partial y^2} + \frac{\partial^2 V_x}{\partial z^2}\right) + \rho g_x$$

$$\quad\ (9)\qquad (10)\quad\ (11)\quad\ (12)\qquad (13)$$

Each term has been labelled beneath with a number. Let us proceed with an evaluation of which terms are relevant to our last problem.

TABLE 2.2(a)   *The equation of motion in rectangular coordinates* (x, y, z)

In terms of τ

x-component:

$$\rho\left(\frac{\partial V_x}{\partial t} + V_x\frac{\partial V_x}{\partial x} + V_y\frac{\partial V_x}{\partial y} + V_z\frac{\partial V_x}{\partial z}\right) = -\frac{\partial P}{\partial x} - \left(\frac{\partial \tau_{xx}}{\partial x} + \frac{\partial \tau_{yx}}{\partial y} + \frac{\partial \tau_{zx}}{\partial z}\right) + \rho g_x \quad (A)$$

y-component:

$$\rho\left(\frac{\partial V_y}{\partial t} + V_x\frac{\partial V_y}{\partial x} + V_y\frac{\partial V_y}{\partial y} + V_z\frac{\partial V_y}{\partial z}\right) = -\frac{\partial P}{\partial y} - \left(\frac{\partial \tau_{xy}}{\partial x} + \frac{\partial \tau_{yy}}{\partial y} + \frac{\partial \tau_{zy}}{\partial z}\right) + \rho g_y \quad (B)$$

z-component:

$$\rho\left(\frac{\partial V_z}{\partial t} + V_x\frac{\partial V_z}{\partial x} + V_y\frac{\partial V_z}{\partial y} + V_z\frac{\partial V_z}{\partial z}\right) = -\frac{\partial P}{\partial z} - \left(\frac{\partial \tau_{xz}}{\partial x} + \frac{\partial \tau_{yz}}{\partial y} + \frac{\partial \tau_{zz}}{\partial z}\right) + \rho g_z \quad (C)$$

In terms of velocity gradients for a Newtonian fluid with constant ρ *and* μ

x-component:

$$\rho\left(\frac{\partial V_x}{\partial t} + V_x\frac{\partial V_x}{\partial x} + V_y\frac{\partial V_x}{\partial y} + V_z\frac{\partial V_x}{\partial z}\right) = -\frac{\partial P}{\partial x} + \mu\left(\frac{\partial^2 V_x}{\partial x^2} + \frac{\partial^2 V_x}{\partial y^2} + \frac{\partial^2 V_x}{\partial z^2}\right) + \rho g_x \quad (D)$$

y-component:

$$\rho\left(\frac{\partial V_y}{\partial t} + V_x\frac{\partial V_y}{\partial x} + V_y\frac{\partial V_y}{\partial y} + V_z\frac{\partial V_y}{\partial z}\right) = -\frac{\partial P}{\partial y} + \mu\left(\frac{\partial^2 V_y}{\partial x^2} + \frac{\partial^2 V_y}{\partial y^2} + \frac{\partial^2 V_y}{\partial z^2}\right) + \rho g_y \quad (E)$$

z-component:

$$\rho\left(\frac{\partial V_z}{\partial t} + V_x\frac{\partial V_z}{\partial x} + V_y\frac{\partial V_z}{\partial y} + V_z\frac{\partial V_z}{\partial z}\right) = -\frac{\partial P}{\partial z} + \mu\left(\frac{\partial^2 V_z}{\partial x^2} + \frac{\partial^2 V_z}{\partial y^2} + \frac{\partial^2 V_z}{\partial z^2}\right) + \rho g_z \quad (F)$$

These equations can be summarized in vector form:

$$\frac{\partial}{\partial t}(\rho V) = \qquad -[\nabla \cdot \rho VV] \qquad -\nabla P \qquad -[\nabla \cdot \tau] \qquad +\rho\mathbf{g}$$

rate of increase of momentum per unit volume	rate of gain of momentum by convection per unit volume	pressure force on element per unit volume	rate of gain of momentum by viscous transfer per unit volume	gravitational force on element per unit volume

or, in even condensed (vector) form:

$$\rho\frac{DV}{Dt} = \qquad -\nabla P \qquad -[\nabla \cdot \tau] \qquad +\rho\mathbf{g}$$

mass per unit volume times acceleration	pressure force on element per unit volume	viscous force on element per unit volume	gravitational force on element per unit volume

TABLE 2.2(b).   *The equation of motion in cylindrical coordinates* (r, θ, z)

In terms of τ

r-component:[a]

$$\rho\left(\frac{\partial V_r}{\partial t} + V_r\frac{\partial V_r}{\partial r} + \frac{V_\theta}{r}\frac{\partial V_r}{\partial \theta} - \frac{V_\theta^2}{r} + V_z\frac{\partial V_r}{\partial z}\right) = -\frac{\partial P}{\partial r}$$

$$-\left(\frac{\partial}{\partial r}(r\tau_{rr}) + \frac{1}{r}\frac{\partial \tau_{r\theta}}{\partial \theta} - \frac{\tau_{\theta\theta}}{r} + \frac{\partial \tau_{rz}}{\partial z}\right) + \rho g_r \qquad (A)$$

θ-component:[b]

$$\rho\left(\frac{\partial V_\theta}{\partial t} + V_r\frac{\partial V_\theta}{\partial r} + \frac{V_\theta}{r}\frac{\partial V_\theta}{\partial \theta} + \frac{V_r V_\theta}{r} + V_z\frac{\partial V_\theta}{\partial z}\right) = -\frac{1}{r}\frac{\partial P}{\partial \theta}$$

$$-\left(\frac{1}{r^2}\frac{\partial}{\partial r}(r^2\tau_{r\theta}) + \frac{1}{r}\frac{\partial \tau_{\theta\theta}}{\partial \theta} + \frac{\partial \tau_{\theta z}}{\partial z}\right) + \rho g_\theta \qquad (B)$$

z-component:

$$\rho\left(\frac{\partial V_z}{\partial t} + V_r\frac{\partial V_z}{\partial r} + \frac{V_\theta}{r}\frac{\partial V_z}{\partial \theta} + V_z\frac{\partial V_z}{\partial z}\right) = -\frac{\partial P}{\partial z}$$

$$-\left(\frac{1}{r}\frac{\partial}{\partial r}(r\tau_{rz}) + \frac{1}{r}\frac{\partial \tau_{\theta z}}{\partial \theta} + \frac{\partial \tau_{zz}}{\partial z}\right) + \rho g_z \qquad (C)$$

In terms of velocity gradients for a Newtonian fluid with constant ρ and μ

r-component:[a]

$$\rho\left(\frac{\partial V_r}{\partial t} + V_r\frac{\partial V_r}{\partial r} + \frac{V_\theta}{r}\frac{\partial V_r}{\partial \theta} - \frac{V_\theta^2}{r} + V_z\frac{\partial V_r}{\partial z}\right) = -\frac{\partial P}{\partial r}$$

$$+\mu\left\{\frac{\partial}{\partial r}\left(\frac{1}{r}\frac{\partial}{\partial r}(rV_r)\right) + \frac{1}{r^2}\frac{\partial^2 V_r}{\partial \theta^2} - \frac{2}{r^2}\frac{\partial V_\theta}{\partial \theta} + \frac{\partial^2 V_r}{\partial z^2}\right\} + \rho g_r \qquad (D)$$

θ-component:[b]

$$\rho\left(\frac{\partial V_\theta}{\partial t} + V_r\frac{\partial V_\theta}{\partial r} + \frac{V_\theta}{r}\frac{\partial V_\theta}{\partial \theta} + \frac{V_r V_\theta}{r} + V_z\frac{\partial V_\theta}{\partial z}\right) = -\frac{1}{r}\frac{\partial P}{\partial \theta}$$

$$+\mu\left\{\frac{\partial}{\partial r}\left(\frac{1}{r}\frac{\partial}{\partial r}(rV_\theta)\right) + \frac{1}{r^2}\frac{\partial^2 V_\theta}{\partial \theta^2} + \frac{2}{r^2}\frac{\partial V_r}{\partial \theta} + \frac{\partial^2 V_\theta}{\partial z^2}\right\} + \rho g_\theta \qquad (E)$$

z-component:

$$\rho\left(\frac{\partial V_z}{\partial t} + V_r\frac{\partial V_z}{\partial r} + \frac{V_\theta}{r}\frac{\partial V_z}{\partial \theta} + V_z\frac{\partial V_z}{\partial z}\right) = -\frac{\partial P}{\partial z}$$

$$+\mu\left\{\frac{1}{r}\frac{\partial}{\partial r}\left(r\frac{\partial V_z}{\partial r}\right) + \frac{1}{r^2}\frac{\partial^2 V_z}{\partial \theta^2} + \frac{\partial^2 V_z}{\partial z^2}\right\} + \rho g_z \qquad (F)$$

---

[a] The term $\rho V\theta^2/r$ is the centrifugal force. It gives the effective force in the r direction resulting from fluid motion in the θ direction. This term arises automatically on transformation from rectangular to cylindrical coordinates; it does not have to be added on physical grounds.

[b] The term $\rho V_r V_\theta/r$ is the Coriolis force. It is an effective force in the θ direction when there is flow in both the r and θ directions. This term also arises automatically in the coordinate transformation. The Coriolis force arises in the problem of flow near a rotating disc. (See, for example, Schlichting.)

TABLE 2.2(c). *The equation of motion in spherical coordinates (r, θ, ϕ)*

In terms of τ

r-component:

$$\rho\left(\frac{\partial V_r}{\partial t}+V_r\frac{\partial V_r}{\partial r}+\frac{V_\theta}{r}\frac{\partial V_r}{\partial \theta}+\frac{V_\phi}{r\sin\theta}\frac{\partial V_r}{\partial \phi}-\frac{V_\theta^2+V_\phi^2}{r}\right)$$

$$=-\frac{\partial P}{\partial r}-\left(\frac{1}{r^2}\frac{\partial}{\partial r}(r^2\tau_{rr})+\frac{1}{r\sin\theta}\frac{\partial}{\partial \theta}(\tau_{r\theta}\sin\theta)+\frac{1}{r\sin\theta}\frac{\partial\tau_{r\phi}}{\partial \phi}-\frac{\tau_{\theta\theta}+\tau_{\phi\phi}}{r}\right)+\rho g_r \quad (A)$$

θ-component:

$$\rho\left(\frac{\partial V_\theta}{\partial t}+V_r\frac{\partial V_\theta}{\partial r}+\frac{V_\theta}{r}\frac{\partial V_\theta}{\partial \theta}+\frac{V_\phi}{r\sin\theta}\frac{\partial V_\theta}{\partial \phi}+\frac{V_rV_\theta}{r}-\frac{V_\phi^2\cot\theta}{r}\right)$$

$$=-\frac{1}{r}\frac{\partial P}{\partial \theta}-\left(\frac{1}{r^2}\frac{\partial}{\partial r}(r^2\tau_{r\theta})\right)+\frac{1}{r\sin\theta}\frac{\partial}{\partial \theta}(\tau_{\theta\theta}\sin\theta)+\frac{1}{r\sin\theta}\frac{\partial\tau_{\theta\phi}}{\partial \phi}+\frac{\tau_{r\theta}}{r}-\frac{\cot\theta}{r}\tau_{\phi\phi}\right)+\rho g_\theta$$

$$(B)$$

ϕ-component:

$$\rho\left(\frac{\partial V_\phi}{\partial t}+V_r\frac{\partial V_\phi}{\partial r}+\frac{V_\theta}{r}\frac{\partial V_\phi}{\partial \theta}+\frac{V_\phi}{r\sin\theta}\frac{\partial V_\phi}{\partial \phi}+\frac{V_\phi V_r}{r}+\frac{V_\theta V_\phi}{r}\cot\theta\right)$$

$$=-\frac{1}{r\sin\theta}\frac{\partial P}{\partial \phi}-\left(\frac{1}{r^2}\frac{\partial}{\partial r}(r^2\tau_{r\phi})+\frac{1}{r}\frac{\partial\tau_{\theta\phi}}{\partial \theta}+\frac{1}{r\sin\theta}\frac{\partial\tau_{\phi\phi}}{\partial \phi}+\frac{\tau_{r\phi}}{r}+\frac{2\cot\theta}{r}\tau_{\theta\phi}\right)+\rho g_\phi \quad (C)$$

In terms of velocity gradients for a Newtonian fluid with constant ρ and μ:[a]

r-component:

$$\rho\left(\frac{\partial V_r}{\partial t}+V_r\frac{\partial V_r}{\partial r}+\frac{V_\theta}{r}\frac{\partial V_r}{\partial \theta}+\frac{V_\phi}{r\sin\theta}\frac{\partial V_r}{\partial \phi}-\frac{V_\theta^2+V_\phi^2}{r}\right)$$

$$=-\frac{\partial P}{\partial r}+\mu\left(\nabla^2 V_r-\frac{2}{r^2}V_r-\frac{2}{r^2}\frac{\partial V_\theta}{\partial \theta}-\frac{2}{r^2}V_\theta\cot\theta-\frac{2}{r^2\sin\theta}\frac{\partial V_\phi}{\partial \phi}\right)+\rho g_r \quad (D)$$

θ-component:

$$\rho\left(\frac{\partial V_\theta}{\partial t}+V_r\frac{\partial V_\theta}{\partial r}+\frac{V_\theta}{r}\frac{\partial V_\theta}{\partial \theta}+\frac{V_\phi}{r\sin\theta}\frac{\partial V_\theta}{\partial \phi}+\frac{V_rV_\theta}{r}-\frac{V_\phi^2\cot\theta}{r}\right)$$

$$=-\frac{1}{r}\frac{\partial P}{\partial \theta}+\mu\left(\nabla^2 V_\theta+\frac{2}{r^2}\frac{\partial V_r}{\partial \theta}-\frac{V_\theta}{r^2\sin^2\theta}-\frac{2\cos\theta}{r^2\sin^2\theta}\frac{\partial V_\phi}{\partial \phi}\right)+\rho g_\theta \quad (E)$$

ϕ-component:

$$\rho\left(\frac{\partial V_\phi}{\partial t}+V_r\frac{\partial V_\phi}{\partial r}+\frac{V_\theta}{r}\frac{\partial V_\phi}{\partial \theta}+\frac{V_\phi}{r\sin\theta}\frac{\partial V_\phi}{\partial \phi}+\frac{V_\phi V_r}{r}+\frac{V_\theta V_\phi}{r}\cot\theta\right)$$

$$=-\frac{1}{r\sin\theta}\frac{\partial P}{\partial \phi}+\mu\left(\nabla^2 V_\phi-\frac{V_\phi}{r^2\sin^2\theta}+\frac{2}{r^2\sin\theta}\frac{\partial V_r}{\partial \phi}+\frac{2\cos\theta}{r^2\sin^2\theta}\frac{\partial V_\theta}{\partial \phi}\right)+\rho g_\phi \quad (F)$$

[a] In these equations

$$\nabla^2=\frac{1}{r^2}\frac{\partial}{\partial r}\left(r^2\frac{\partial}{\partial r}\right)+\frac{1}{r^2\sin\theta}\frac{\partial}{\partial \theta}\left(\sin\theta\frac{\partial}{\partial \theta}\right)+\frac{1}{r^2\sin^2\theta}\left(\frac{\partial^2}{\partial \phi^2}\right)$$

TABLE 2.3.   *Components of the stress tensor for Newtonian fluids*

$$\tau_{xx} = -\mu\left(2\frac{\partial V_x}{\partial x} - \tfrac{2}{3}(\nabla \cdot V)\right) \tag{A}$$

$$\tau_{yy} = -\mu\left(2\frac{\partial V_y}{\partial y} - \tfrac{2}{3}(\nabla \cdot V)\right) \tag{B}$$

$$\tau_{zz} = -\mu\left(2\frac{\partial V_z}{\partial z} - \tfrac{2}{3}(\nabla \cdot V)\right) \tag{C}$$

$$\tau_{xy} = \tau_{yx} = -\mu\left(\frac{\partial V_x}{\partial y} + \frac{\partial V_y}{\partial x}\right) \tag{D}$$

$$\tau_{yz} = \tau_{zy} = -\mu\left(\frac{\partial V_y}{\partial z} + \frac{\partial V_z}{\partial y}\right) \tag{E}$$

$$\tau_{zx} = \tau_{xz} = -\mu\left(\frac{\partial V_z}{\partial x} + \frac{\partial V_x}{\partial z}\right) \tag{F}$$

$$(\nabla \cdot V) = \frac{\partial V_x}{\partial x} + \frac{\partial V_y}{\partial y} + \frac{\partial V_z}{\partial z} \tag{G}$$

In cylindrical coordinates $(r,\ \theta,\ z)$

$$\tau_{rr} = -\mu\left(2\frac{\partial V_r}{\partial r} - \tfrac{2}{3}(\nabla \cdot V)\right) \tag{A}$$

$$\tau_{\theta\theta} = -\mu\left(2\left(\frac{1}{r}\frac{\partial V_\theta}{\partial \theta} + \frac{V_r}{r}\right) - \tfrac{2}{3}(\nabla \cdot V)\right) \tag{B}$$

$$\tau_{zz} = -\mu\left(2\frac{\partial V_z}{\partial z} - \tfrac{2}{3}(\nabla \cdot V)\right) \tag{C}$$

$$\tau_{r\theta} = \tau_{\theta r} = -\mu\left\{r\frac{\partial}{\partial r}\left(\frac{V_\theta}{r}\right) + \frac{1}{r}\frac{\partial V_r}{\partial \theta}\right\} \tag{D}$$

$$\tau_{\theta z} = \tau_{z\theta} = -\mu\left\{\frac{\partial V_\theta}{\partial z} + \frac{1}{r}\frac{\partial V_z}{\partial \theta}\right\} \tag{E}$$

$$\tau_{zr} = \tau_{rz} = -\mu\left\{\frac{\partial V_z}{\partial r} + \frac{\partial V_r}{\partial z}\right\} \tag{F}$$

$$(\nabla \cdot V) = \frac{1}{r}\frac{\partial}{\partial r}(rV_r) + \frac{1}{r}\frac{\partial V_\theta}{\partial \theta} + \frac{\partial V_z}{\partial z} \tag{G}$$

(continued)

TABLE 2.3 (*continued*)

In spherical coordinates (r, θ, ϕ)

$$\tau_{rr} = -\mu\left(2\frac{\partial V_r}{\partial r} - \tfrac{2}{3}(\mathbf{\nabla}\cdot\mathbf{V})\right) \tag{A}$$

$$\tau_{\theta\theta} = -\mu\left\{2\left(\frac{1}{r}\frac{\partial V_\theta}{\partial \theta} + \frac{V_r}{r}\right) - \tfrac{2}{3}(\mathbf{\nabla}\cdot\mathbf{V})\right\} \tag{B}$$

$$\tau_{\phi\phi} = -\mu\left\{2\left(\frac{1}{r\sin\theta}\frac{\partial V_\phi}{\partial \phi} + \frac{V_r}{r} + \frac{V_\theta\cot\theta}{r}\right) - \tfrac{2}{3}(\mathbf{\nabla}\cdot\mathbf{V})\right\} \tag{C}$$

$$\tau_{r\theta} = \tau_{\theta r} = -\mu\left\{r\frac{\partial}{\partial r}\left(\frac{V_\theta}{r}\right) + \frac{1}{r}\frac{\partial V_r}{\partial \theta}\right\} \tag{D}$$

$$\tau_{\theta\phi} = \tau_{\phi\theta} = -\mu\left\{\frac{\sin\theta}{r}\frac{\partial}{\partial \theta}\left(\frac{V_\phi}{\sin\theta}\right) + \frac{1}{r\sin\theta}\frac{\partial V_\theta}{\partial \phi}\right\} \tag{E}$$

$$\tau_{\phi r} = \tau_{r\phi} = -\mu\left\{\frac{1}{r\sin\theta}\frac{\partial V_r}{\partial \phi} + r\frac{\partial}{\partial r}\left(\frac{V_\phi}{r}\right)\right\} \tag{F}$$

$$(\mathbf{\nabla}\cdot\mathbf{V}) = \frac{1}{r^2}\frac{\partial}{\partial r}(r^2 V_r) + \frac{1}{r\sin\theta}\frac{\partial}{\partial \theta}(V_\theta\sin\theta) + \frac{1}{r\sin\theta}\frac{\partial V_\phi}{\partial \phi} \tag{G}$$

Term 1 = zero because ρ = constant.
Term 2 = zero because ρV_x = constant, $V_x \neq f(x)$.
Term 3 = zero because V_y = 0.
Term 4 = zero because V_z = 0.
Term 5 = zero because $V_x \neq f(t)$.
Term 6 = zero because $\partial V_x/\partial x = 0$.
Term 7 = zero because V_y = 0.
Term 8 = zero because V_z = 0.
Term 9 = zero because $\partial P/\partial x = 0$.
Term 10 = zero because $\partial V_x/\partial x = 0$. $\partial^2 V_x/\partial x^2 = 0$
Term 11 = finite because $\partial^2 V_x/\partial y^2 = f(y)$, $\partial^2 V_x/\partial x^2$ not a non-zero term.
Term 12 = zero because $\partial^2 V_x/\partial z^2$ (zero since $\partial V_x/\partial z = 0$, $V_x \neq f(z)$.).
Term 13 = ρg_x, the body force term applies on Earth. We may therefore avoid setting up our continuity and shell momentum balances and write for:

Continuity:
$$\frac{\partial}{\partial x}(\rho V_x) = 0.$$

Motion:
$$0 = \frac{\mu\partial^2 V_x}{\partial y^2} + \rho g_x.$$

It will be seen that these are equivalent to eqn (2.16), so that we would be led to the correct differential equations with a minimum of delay, and a maximum of confidence.

(c) Laminar flow around a sphere

As previously mentioned, the behaviour of small particles in melts or fluids represents an important class of flow systems in which laminar flows can dominate.

While the classical problem of viscous flow around a sphere is not amenable to solution via a shell momentum balance, since the streamlines are curved, one can deduce the relevant differential equations requiring solution by reference to Tables 3.1, 3.2 and 3.3. As this and the solutions would fill several pages with complex mathematics, only the results will be quoted here. They refer to conditions corresponding to incompressible, steady flow in which all inertial force terms are treated as being very small in comparison with the viscous force terms.

These analytical solutions were obtained by Stokes and are covered by Lamb in *Hydrodynamics* (1945). Equivalent solutions for high Re cases dominated by inertial flows do not exist, although potential or inviscid flow solutions are available and these agree well with experiment for converging flow systems.

Figure 2.9 shows the coordinate system used for the solutions presented below. These solutions provide analytical expressions for the shear stress, pressure and radial and tangential velocities at any position within the liquid up to the spherical surface located at $r = R$.

Shear stress

$$\tau_{r\theta} = \frac{3}{2}\frac{\mu U_\infty}{R}\left(\frac{R}{r}\right)^4 \sin(\theta). \tag{2.30}$$

Pressure

$$P = P_0 - \rho g z - \frac{3}{2}\frac{\mu U_\infty}{R}\left(\frac{R}{r}\right)^2 \cos(\theta). \tag{2.31}$$

Radial velocity

$$U_r = U_\infty\left\{1 - \frac{3}{2}\left(\frac{R}{r}\right) + \frac{1}{2}\left(\frac{R}{r}\right)^3\right\}\cos(\theta). \tag{2.32}$$

Tangential velocity

$$U_\theta = -U_\infty\left\{1 - \frac{3}{4}\left(\frac{R}{r}\right) - \frac{1}{4}\left(\frac{R}{r}\right)^3\right\}\sin(\theta). \tag{2.33}$$

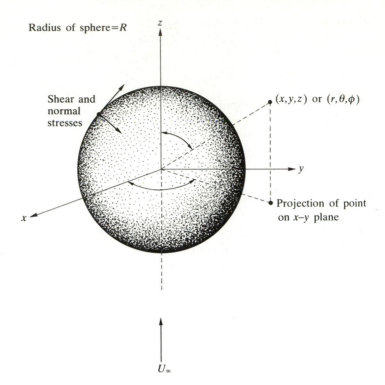

Radius of sphere=R

Shear and
normal
stresses

(x,y,z) or (r,θ,ϕ)

Projection of point
on x–y plane

U_∞

Free stream velocity

FIG. 2.9.   Coordinate system used to describe flow of a fluid about a rigid sphere.

In order to find the normal force acting on the sphere as liquid flows up around it, we need to resolve the pressure P_R at R in the vertical z direction and integrate it over the surface to obtain the resultant normal force, F_n,

$$F_n = \int_0^{2\pi} \int_0^{\pi} (-P|_{r=R}\, \cos(\theta))\, R^2 \sin(\theta)\, \mathrm{d}\theta\, \mathrm{d}\phi. \qquad (2.34)$$

Substituting for P from eqn (2.3), one obtains

$$F_n = \frac{4}{3}\pi R^3 \rho g + 2\pi\mu R U_\infty. \qquad (2.35)$$

The first term can be recognized as the buoyancy term equal to the weight of fluid displaced by the sphere, the second is the *form drag* arising from the shape and cross-sectional area normal to the moving fluid and about which it must separate.

As well as exerting a normal force, or form drag, the liquid will generate shear stresses as it flows over the surface of the sphere. This shear force can be similarly obtained, since

$$F_\tau = \int_0^{2\pi} \int_0^\pi (+\tau_{r\theta}|_{r=R} \sin(\theta)) R^2 \sin(\theta) d\theta d\phi, \tag{2.36}$$

and, by substituting for $\tau_{r\theta, r=R}$

$$F_\tau = 4\pi\mu R U_\infty. \tag{2.37}$$

Hence, the total *drag force* on the sphere caused by fluid having to separate and shear past it is given by

$$F_D = 6\pi\mu R U_\infty. \tag{2.38}$$

We have already used this famous result in discussing the Stokes–Einstein theory for diffusion coefficients in Chapter 1.

(d)   Non-metallic inclusions rising through steel

Calculate the rising velocity of a 1 μm slag particle, rising through stagnant steel at 1600°C, given its density is 3000 kg m^{-3}, the density of steel is 7000 kg m^{-3} and viscosity 7 mPa s (or 7 cP).

Solution:

Under steady-state conditions (i.e. no acceleration) the sum of forces acting on the body must be zero by Newton's second law:

$$\sum F = 0 = F_{\text{buoyancy}} - F_{\text{gravity}} - F_{\text{drag}},$$

$$0 = \left(\frac{\pi d^3}{6}\right)\rho g - \left(\frac{\pi d^3}{6}\right)\rho_{\text{sp}} g - 6\pi\mu\frac{D}{2} U_\infty;$$

therefore

$$U_\infty = \frac{(\rho - \rho_{\text{sl}})gd^2}{18\mu} = \frac{(7000 - 3000)9.81 \times 10^{-12}}{18 \times 0.7}$$

$$= 3.1 \times 10^{-9} \text{ m s}^{-1}.$$

The rising velocity of such impurities is only 10^{-2} mm/h!

2.5.   EXERCISES ON TYPICAL LAMINAR FLOW PHENOMENA

(1)   Shear stress—momentum flux distributions

Given the velocity profiles for one-dimensional laminar flows drawn in the diagrams below, sketch their associated momentum flux distributions, and explain where necessary.

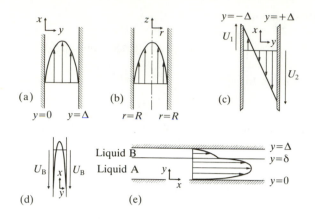

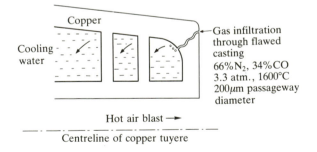

(2)   An incipient failure of a blast furnace tuyère

Blast furnace tuyères are constructed of copper of high thermal conductivity and, although water cooled, operate in an environment some 500 °C above their melting point. Owing to casting difficulties, gas entrapment during their fabrication can lead to microporosity. Suppose that a small gas hole of 200 μm has been exposed as a result of wear from the corrosive action of coke, slag and hot metal. Estimate the flow of blast furnace gas (litres min^{-1}) that will flow into the water system, given the dimensions and conditions shown in the figure, using appropriate values for the relevant properties.

N.B. The detection of gas bubbles in exhaust cooling water from tuyères is used as an early warning aid.

(3)   Removal of inclusions from liquid metals by settling/floating

During the preparation of liquid aluminium suitable for electrical grades, it is necessary to eliminate any dissolved titanium, so as to enhance the metal's electrical conductivity. This is accomplished by making boron additions to the melt contained in a gas-fired holding furnace, thereby nucleating particles of (Ti–V)B, which must subsequently be allowed to precipitate from the melt. Recent experiments with a particle detection device have shown that approximately 1.5 h is required for the inclusion count rate on 20 μm size particles to fall to negligible levels. Given a liquid metal depth of 1 m, and no convection currents, calculate the theoretical settling time (hours) needed for such particles to settle on the basis of Stokes' law. The density of liquid aluminium at 700°C can be taken as 2300 kg m^{-3} and the density of (Ti–V)B particles as 4000 kg m^{-3}.

(4)   Clarifier system for effluent water from Bayer alumina plant

The diagram shows a novel technique for clarifying liquids, in which a bank of circular tubes are set at an angle of 20° to the vertical, and the liquid to be clarified is slowly passed through each pipe under laminar conditions, at a flowrate of Q per pipe. This design allows particles to settle out, agglomerate and drop from the bottom of the pipes where they can be removed.

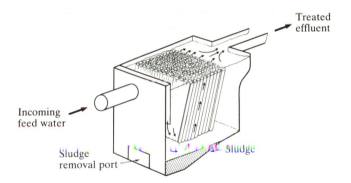

It is desired to calculate the length of 10 mm diameter pipe that will be needed to ensure that all Fe$_2$O$_3$ particles (red mud) greater than 2 μm in diameter are precipitated from the effluent water before discharge.

Given the requirements that 10 000 gallons of such effluent are to be treated per day, that the viscosity of water is 1 mPa s, and that the density of the spherical particles is 3000 kg m^{-3}, calculate the minimum number of tubes of specified length that will be required. Assume particle settling velocities obey Stokes' law: i.e.) $F_{\text{DRAG}} = 6\pi\mu r U$.

[*Hint*. Develop an expression for the velocity profile for laminar flow through a pipe (the Hagen–Poiseuille equation), assume the particles are carried along at the liquid's velocity, with a superimposed settling velocity as given by Stokes' law.]

(5)   Some tracer studies on oil injection through blast furnace tuyères

Viscous oil ($\mu = 1.0$ Pa s, density 980 kg m^{-3}) flows through a 25 mm I.D. pipe at a rate of 80 litres min^{-1}, as part of the delivery system to a blast furnace. In some tracer experiments, to track its course through the burden, the composition of the oil was changed to include an isotope upstream of its exhaust point into the blowpipe of the blast furnace. Assuming that the closest distance one can make the tracer addition is 10 m from the exhaust point, develop a general expression to show when the exhausting oil reaches a tracer concentration level equal to 95% of the input tracer concentration. What is this time for the conditions given, assuming laminar flow throughout the delivery system?

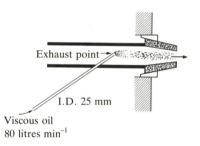

Exhaust point

I.D. 25 mm

Viscous oil
80 litres min^{-1}

(6)   The stability of gas envelopes trailed behind large bubbles in viscous liquids.

Large bubbles rising through highly viscous liquids adopt a characteristic spherical cap shape (i.e. mushroom) but also develop long, thin, trailing skirts of gas. A cross-section of such a bubble is shown in the diagrams; it is held by imposing a downward velocity of U_B equal to its rising velocity. Assuming that the gas within the 'skirt' is in laminar flow, that the gas pressure within the skirt rises in accordance with increases in external hydrostatic flow, that the skirt is at steady state (i.e. neither growing, nor shrinking) and that the weight of the gas can be ignored in relation to viscous forces, develop an expression* to show the variables on which the thickness of the skirt, Δ, should depend, given the conditions shown in the diagram. Assume a rectangular coordinate system, since $R_c \gg \Delta$.

---

* See Guthrie and Bradshaw (1969) for solutions. A similar phenomenon can occur when a laminar jet penetrates a quiescent tank of liquid, causing entrainment of air, similar to that illustrated in Fig. 1.9(a) for turbulent entrainment of air.

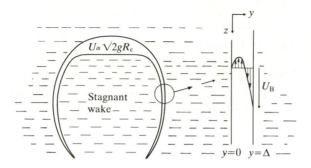

(7) Performance of an electric dust collector

A dust precipitator consists of a pair of oppositely charged plates between which dust-laden gases flow (see figure). It is desired to establish a criterion for the minimum length of the precipitator in terms of the charge of the particle, e, the electric field strength, H, the pressure difference $(P_0 - P_L)$ the particle mass m, the gas viscosity μ. That is, for what length L will the smallest particle present (mass m) reach the bottom plate just before it has a chance to be swept out of the channel? Assume that the flow between the two plates is laminar. Assume also that the particle velocity in the z direction is the same as the fluid velocity in the z direction; assume further that Stokes drag and gravity forces act on the particle as it is accelerated in the x direction can be neglected. Discuss the probable error in the neglect of the Stokes drag.

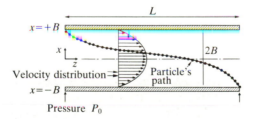

2.6. INTRODUCTION TO INERTIAL FLOWS

(a) Flow past spheres and cylinders at high Reynolds numbers

Now that we have considered flows in which viscous forces are all-important, it is appropriate to introduce the subject of inertially dominated flows. In these flows, the forces of acceleration and deceleration on fluid elements moving through a system outweigh viscous forces. A logical point to start such a discussion is to consider the flow patterns that would be generated if the velocity of fluid passing over a sphere, considered in the last section on laminar flows, is increased.

Figure 2.10 shows the striking series of events that were observed by Prandtl, when water surrounding a submerged cylinder was accelerated from rest. In Fig. 2.10(a), the flow is seen to be symmetrical about the cylinder's centre, fluid separating and then recombining symmetrically to the rear. In Fig. 2.10(b), recombination of flow behind the cylinder has been delayed and the flow has become asymmetrical. Similarly, the tracer particles used to reveal the flow are seen to congregate close to the rear of the cylinder, indicating low mixing rates between this region (a boundary layer) and the bulk. In Fig. 2.10(c), (corresponding to Re = 5), separation has occurred, and two standing vortices have been established behind the cylinder. These gradually grow larger as they absorb rotational energy, and by Fig. 2.10(f) have become so large that parts or whole vortices (as in the Von Karmen vortex street) are shed downstream. At much higher Reynolds number ($\sim 4 \times 10^5$), the separation point which one sees at about 90° with respect to the impacting fluid shifts backwards to about 130°.

If one plots the absolute drag force versus Reynolds number for the series of events depicted, one would observe a gradual change in slope with rising velocity, the exponent changing from 1 to 2. At low Re values, for instance, a doubling in velocity (or Re), would cause a doubling in the drag force on the sphere or cylinder, but at higher Re values (e.g., 1000), a doubling in velocity would lead to a quadrupling in the drag force.

This behaviour is shown in Fig. 2.11, where the steady drag force experienced by a 10 mm diameter sphere (e.g., a ferro-alloy lump) is computed for various relative velocities in liquid steel.

This, and all other such data, can be neatly reduced to dimensionless form through the definition of a drag coefficient C_D, given by

$$F_D = C_D A_\perp (\tfrac{1}{2}\rho U_\infty^2), \tag{2.39}$$

where $A_\perp$ represents the projected area of the particle or object perpendicular to the flow, and $\tfrac{1}{2}\rho U_\infty^2$ represents the dynamic pressure of the fluid.

Introducing Stokes' law for the drag force on spheres at low Re, and equating it to this last equation, we have:

$$3\pi\mu D U_\infty = C_D \left(\frac{\pi D^2}{4} \right) \tfrac{1}{2}\rho U_\infty^2,$$

or

$$C_D = \frac{24\mu}{\rho U_\infty D} \quad \text{or} \quad \frac{24}{\text{Re}}. \tag{2.40}$$

One finds that at higher velocities, C_D becomes constant at about 0.44. This constant corresponds to the region where drag forces increase as the square of the velocity, and which is known as the Newton's law regime.

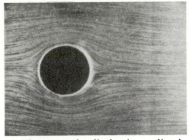

(a) Flow round cylinder immediately after starting (potential flow).

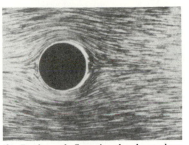

(b) Backward flow in the boundary layer behind the cylinder; accumulation of boundary layer material.

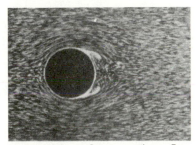

(c) Formation of two vortices; flow breaking loose from cylinder.

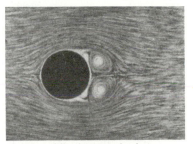

(d) The eddies increase in size.

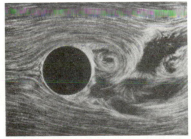

(e) The eddies grow still more; finally the picture becomes unsymmetrical and disintegrates.

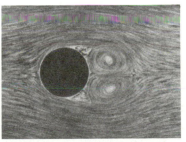

(f) Final picture obtained a long time after starting.

The direction of flow in all photographs is from left to right.

FIG. 2.10. Formation of vortices in flow past a circular cylinder after acceleration from rest (Prandtl and Tietjens 1957).

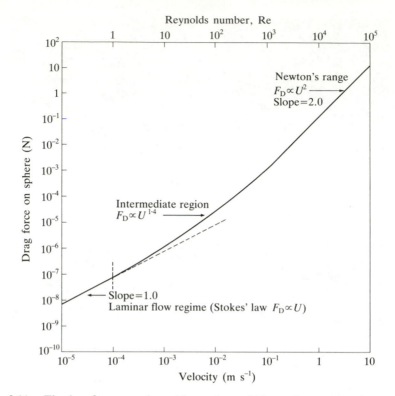

FIG. 2.11.   The drag force experienced by a sphere of 10 mm diameter as a function of
its relative velocity through a bath of liquid steel.

Figure 2.12 provides a useful summary of drag coefficients experienced by submerged bodies; the data allow one to calculate drag forces, energy requirements, steady translational velocities, and so on, of objects submerged in fluids. The limiting case of $C_D = 24/Re$ is seen to hold up to Re values of 0.1.

We have spent some time describing the change in flow as the velocity past the object increases, but the question remains why the separation of flow from the rear of the cylinder takes place. At its most basic level, the phenomenon can be explained in terms of fluid adjacent to the cylinder's surface being slowed by viscous effects. This viscous shear dissipates energy irreversibly, in the form of low grade heat. Consequently, the fluid close to the sphere carries an energy deficit which increases proportionately with increased velocity.

One will notice that the flow becomes separated at $\theta = 90°$ and that this point corresponds to the location at which fluid travelling around the periphery of the cylinder, having first accelerated away from the nose of the sphere, must start to decelerate back to zero at a stagnation point. This

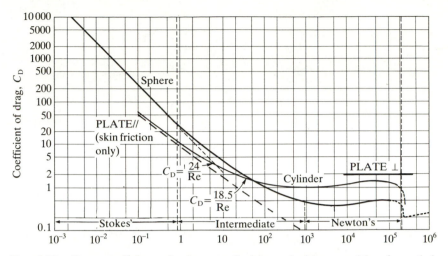

FIG. 2.12. Drag coefficient of spheres, cylinders and plates, resulting from their steady translatory motion through a liquid. Plot of C_D versus Re. Dashed line represents plate parallel to flow, solid line for $C_D = 2.0$ is for plate perpendicular to flow. Characteristic lengths for L in Re; sphere, D (diameter); cylinder, D (diameter); plate ∥ to flow, L (length); plate ⊥ to flow, H (plate height).

deceleration requires an attendant increase in pressure energy as the rear stagnation point is approached. The fluid, with its energy deficit, is unable to provide this and the situation of $U = 0$ and $\partial U/\partial r = 0$ at $\theta = 90°$ leads to the probability of reverse flow, and an underpressure immediately to the rear of the sphere.

The vortices formed to fill the incipient void behind the sphere swirl around in a turbulent manner, interacting and dissipating energy.

This account provides a qualitative explanation for the rapid increase in drag force and separation that are observed in practice.

(b) Turbulence

The flows produced behind spheres and cylinders are just two of many situations in which laminar flows are disrupted, and swirling motions, known as eddies, are generated. Figure 2.13 presents a collage of examples showing situations in which turbulence is encountered. Most will be familiar to the reader. They include the following:

(1) flow discontinuities, such as the submerged jet in which the kinetic energy contained within the entering jet $(1/2mU_0^2)$ is dissipated in a turbulent expanding jet containing a wide spectrum of sizes of swirling, interacting eddies;

(a) Velocity discontinuities

(b) Bluff objects

(c) Diverging flows

(d) Flow through pipes

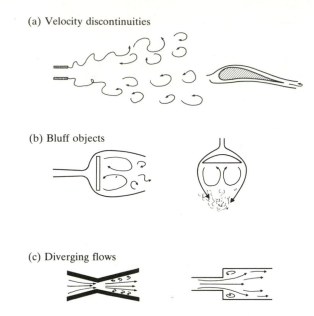

FIG. 2.13.   A collage of examples of shearing flows demonstrating situations in which turbulence is induced and sustained.

(2) bluff objects, where the requirement of rapid increases in pressure cannot be accommodated, and turbulence is again generated in the form of a downstream wake;

(3) diverging flows, where divergences of more than 11° again cannot normally be accommodated without flow separation;

(4) flow through pipes. (A Lagrangian frame of reference is used in Fig. 2.13 so as to illustrate the motion of eddies superimposed on the bulk downstream flow).

All these flows have one key feature in common. They all involve *shearing flows*, without which turbulence can neither be generated nor sustained. At the same time, these flows have defied both analytical and numerical solutions, so that much of the subject still rests on empirical correlations (e.g., the C_D versus Re curve), and the macroscopic energy and mass balances to be presented in the next section.

(c) Differential equations for turbulent flow

Meanwhile, turbulence—which has its origins in chaos—represents one of the last great challenges to physics as far as its quantitative description at the macroscale is concerned. The exact differential equations describing the flows just illustrated have been known for more than a century. They are, of course, the continuity and Navier–Stokes equations for incompressible flow. However, owing to the microscale (millimetre and submillimetre) level at which turbulent motions are generated and turbulent energy is dissipated, any attempts to model such eddying motions using the discretization procedures to be outlined in Chapter 6 for the numerical solution of the continuity and N–S equations, would need far too fine a gridwork for the storage capacity of today's computers.

The alternative has been to develop models of turbulence which mimic, at the macroscale, the effects of turbulence on bulk flow phenomena. The approach is to time-average the N–S and continuity equations, to extract the fluctuating components of velocity and to express these in terms of turbulence equation(s), or tubulence parameters.

Let us first imagine what we would observe using a sensitive velocity probe (preferably of a non-disturbing type such as a laser Doppler anemometer (LDA)) to monitor fluid motions within such regions of turbulence. Figure 2.14 shows the sort of information that can be obtained for the case of water flowing across a planar surface at an average velocity of 370 mm s^{-1}, provided our probe is sufficiently sensitive and able to respond to pressure fluctuations of about 10–100 Hz frequency. Thus, at any time instant, we would observe an instantaneous velocity somewhere between 300 and 420 mm s^{-1}.

Once we had obtained sufficient data, arithmetic averaging would provide us with a mean velocity of 370 mm s^{-1}, while the sum of the areas above the 370 mm s^{-1} line in Fig. 2.14 would equal the sum of the areas below the line.

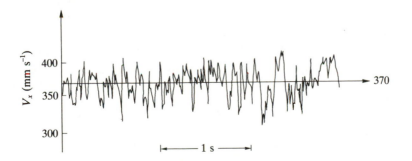

FIG. 2.14. Eddy fluctuations in water at Re = 6500, at the centre of a pipe of 22 mm I.D. measured by laser Doppler anemometry. (After Lewis *et al.* 1968.)

In time-averaging the continuity and N–S equations, it is more efficient to rephrase the equations appearing in Tables 2.1 and 2.2 in index form. Thus, in Cartesian tensor notation, in which i takes successive values of 1, 2 and 3 to represent vector components in the x_1, x_2, x_3 (or x, y, z) directions:

Continuity equation:

$$\frac{\partial U_i}{\partial x_i} = 0. \tag{2.41}$$

N–S equation:

$$\frac{\partial U_i}{\partial t} + U_j \frac{\partial U_i}{\partial x_j} = -\frac{1}{\rho}\frac{\partial P}{\partial x_j} + v\frac{\partial^2 U_i}{\partial x_j \partial x_j} + g_i. \tag{2.42}$$

Time averaging in the manner suggested, such that

$$\bar{U} = \frac{1}{t_2 - t_1}\int_{t_1}^{t_2} U\,dt, \quad \bar{P} = \frac{1}{t_2 - t_1}\int_{t_1}^{t_2} P\,dt, \tag{2.43}$$

where $(t_2 - t_1) \gg$ time period of turbulent flutuations, one arrives at the time-averaged forms of the continuity and N–S equations:

$$\frac{\partial \bar{U}_i}{\partial x_i} = 0, \tag{2.44}$$

and

$$\frac{\partial \bar{U}_i}{\partial t} + \bar{U}_j \frac{\partial \bar{U}_i}{\partial x_j} = -\frac{1}{\rho}\frac{\partial \bar{P}}{\partial x_i} + \frac{\partial}{\partial x_j}\left(v\frac{\partial \bar{U}_i}{\partial x_j} - \overline{u_i' u_j'}\right) + g_i. \tag{2.45}$$

In reading the second term of the so-called turbulent Navier–Stokes, or Reynolds equation, the notation:

$$U_j \frac{\partial U_i}{\partial x_j} = \sum_{j=1}^{j=3} U_j \frac{\partial U_i}{\partial x_j} = U_1 \frac{\partial U_i}{\partial x_1} + U_2 \frac{\partial U_i}{\partial x_2} + U_3 \frac{\partial U_i}{\partial x_3}. \tag{2.46}$$

Here, and for the rest of this book, overbars denoting time-averaged velocities are dropped for convenience. It is left as an exercise to verify the equivalence of the convective pressure and gravity terms between those appearing in eqn (2.45), and those in Table 2.2.

As seen, the major difference between the time-averaged and instantaneous forms of the N–S equation is the appearance of $\overline{u_i' u_j'}$ in eqn (2.45). These represent the Reynolds shear stress components, $\rho\,\overline{u_i' u_j'}$ discussed in later sections.

In reading the (dyadic) product of the velocity vectors, $u_i u_j$, one obtains a set of nine tensors:

$$u_i u_j = \begin{vmatrix} u_1 u_1 & u_1 u_2 & u_1 u_3 \\ u_2 u_1 & u_2 u_2 & u_2 u_3 \\ u_3 u_1 & u_3 u_2 & u_3 u_3 \end{vmatrix}. \tag{2.47}$$

However, since the Reynolds shear stress terms $u_i u_j = u_j u_i$ are symmetric, it will be seen that only six tensors are involved.

(d) Turbulent flow in pipes

For planar, shearing flows, the major stress components that arise are generated by turbulent fluctuations perpendicular to the main flow. As such, they are among the most simple to analyse and quantify. In this respect, mention must be made of the classical experiments of Osborne Reynolds, a British engineering professor, during his early development of the subject, in 1883.

Reynolds noticed that a dye tracer continuously injected into a streamline of fluid entering a pipe would maintain its identity for a considerable way downstream. However, as shown in Fig. 2.15, he found that a point was sometimes reached at which the dye would begin to be dispersed as a result of transverse velocity fluctuations occurring within the fluid in that region.

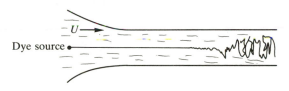

FIG. 2.15. Continuous injection of a dye into water entering the faired entrance of a horizontal, smooth-walled, tube, reveals a transition in flow behaviour downstream of the entrance as the dye becomes dispersed.

He discovered that a critical value of the dimensionless parameter $\rho U d/\mu$ had to be reached before such disturbances could be observed. The grouping $\rho U d/\mu$ is known as the Reynolds number in his honour. Its accepted value to mark the onset of turbulent flow within a pipe is $\mathrm{Re}_{d,\,crit} \simeq 2300$.

It is now well known that the random fluctuations observed by Reynolds are the result of eddy formation and propagation within the pipe further upstream, together with the attendant formation of a thickening turbulent boundary layer. This finally reaches into the centre of the pipe after about 50–100 pipe diameters, leading to the phenomena illustrated in Fig. 2.15.

Consider a typical boundary layer shearing flow of the type depicted in Fig. 2.16. The figure illustrates the effect of random perturbation can have on a fluid that is initially in laminar flow. Thus in Fig. 2.16(b), a random disturbance is shown, the density and velocity gradient effects of which are shown by the arrow. In Fig. 2.16(c) we see that more momentum is fed into the disturbance than can be damped out by viscous shear and that as a result an

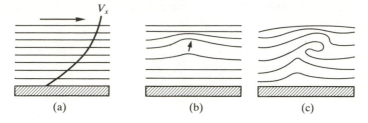

Fig. 2.16.   A typical mechanism whereby turbulence can be generated in pipe flow. A random disturbance (originating at a slight protuberance on the pipe wall, or generated by another disturbance within the body of flowing fluid) is not damped by viscous forces as it moves towards the central axis of the pipe.

eddy forms. Small-perturbation theory leads to the conclusion that below a critical Reynolds number, all disturbances will be damped, while above Re_{crit} certain frequencies may be amplified while others continue to be damped.

One can imagine the cascading effect as one initial disturbance, or 'spot', generates a multiplicity of disturbances downstream. Experimentally, a slight roughness in a pipe's surfaces can precipitate a breakdown of laminar flow into turbulent flow. Similarly, turbulence can be delayed to higher velocities if the pipe is very smooth.

(e)   The spectrum of eddy lengths

In fully developed turbulence (when Re is well above its critical value), there is a wide spectrum of eddy lengths. This is illustrated in Fig. 2.17, where the kinetic energy of turbulence is plotted in schematic form against the reciprocal of eddy length. The largest turbulence eddies are generally of a size comparable to the diameter of the pipe or vessel, their character depending on the nature of the flow. Similarly, their fluctuation velocities can be of the order of the maximum velocity at the centre of the tube. These large-scale eddies can contain as much as 20% (say) of the kinetic energy of the turbulent motion, deriving their energy from that of the main flow. They dissipate little energy by viscous effects and are relatively long-lived. The interactions of these large eddies with each other generate smaller eddies, of length l_e, containing most of the kinetic energy of turbulence, and for which isotropic turbulence applies (see later). With decreasing eddy size, viscous dissipation becomes increasingly important.

Thus, within the smaller eddies, the local Reynolds number (whose characteristic length is equal to the eddy length) becomes small (e.g., $Re \sim 6$–8; Davies 1972), and there is then considerable dissipation of kinetic energy into heat through viscous forces.

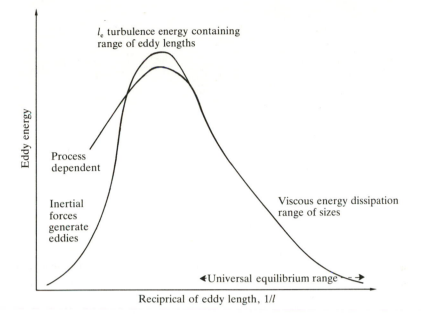

FIG. 2.17. Schematic of the turbulence kinetic energy spectrum as a function of reciprocal eddy size. The largest eddies (closest to the origin) are a function of the flow system, whereas the energy-containing eddies, l_e, and those to the right of the peak, tend to form a 'universal equilibrium' range of isotropic turbulence.

One postulates a universal equilibrium range of eddy sizes, over which energy is continually drawn from the larger eddies as these interact with each other, i.e. there is a continuous transfer of energy from the bulk motion of the fluid into large recirculatory flows (eddies), and from these to intermediate eddies which in turn supply kinetic energy to still smaller eddies, where the energy is finally dissipated as heat. The process is neatly summed up by a verse:

> *Big-size whirls have little whirls*
> *That feed on their velocity.*
> *Little whirls have lesser whirls*
> *And so on to viscosity.*

In the special case of eddies in a fluid flowing in a pipe, the eddies formed cannot be very large: indeed, at the wall ($y=0$), a laminar sublayer exists in which there can be no movement and no eddies, i.e. $l=0$. The simplest relation for a (Prandtl) eddy of length l, not too far from a solid wall, is (see Fig. 2.18).

$$l = 0.4y, \qquad (2.48)$$

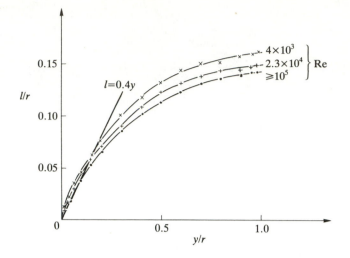

Fɪɢ. 2.18.   The eddy mixing length l as a function of vertical distance away from the wall of pipe; $l = 0.4y$ is limiting form, strictly valid between the surface of the pipe and 10 per cent of the radial distance towards the pipe centre. $\times$ Re = 4000, + Re = 23 000, $\cdot$ Re = 105 000. (After Nikuradse 1932.)

while at the centreline, the eddy mixing length l reaches a maximum of about 0.16 times the pipe's radius, or 0.08 times the pipe's diameter. This eddy mixing length refers only to those eddies absorbing energy and defined in accordance with Prandtl's mixing length theory of turbulence, presented below.

(f)   Prandtl's theory of turbulence for boundary layers

Thus Prandtl, a German engineering professor, proposed in the 1930s, that the transfer of turbulent energy within a fluid was analogous to the phenomenon of energy dissipation in shearing *laminar* flows. Rather than molecules jumping from one layer to another, exchanging momentum and generating attendant shear stresses, he postulated that discrete packages of fluid jumping discrete distances l' were responsible for turbulent momentum transport and turbulent shear stresses. This concept leads directly to a proposition analogous to Newton's law of molecular viscosity that:

$$\tau_{t,\,y,\,x} = \mu_t \frac{\mathrm{d}\bar{U}_x}{\mathrm{d}y}. \tag{2.49}$$

Here $\tau_{t,\,y,\,x}$ represents the x-directed turbulent shear stress acting on fluid at y by fluid at lesser y, as the result of an x velocity gradient within the shear layer.

The constant of proportionality is the turbulent, or eddy, viscosity, analogous to Newton's dynamic, or kinematic, viscosity, but typically 10^3–10^5 times greater!

Referring to Fig. 2.19, we see that a package of fluid moving across the flow into a faster layer of liquid, at a mass flow rate of $(\rho v')$/area × time, would generate locally an x velocity deficit of $-u'$, and thereby develop an instantaneous shear stress of $-\rho u'v'$ at y. Taking a time-average of such $-\rho u'v'$ stress variations, the resulting mean shear stress generated as a consequence, i.e. $\tau_{t,y,x}$ is commonly termed the *Reynolds stress*. This appears in eqn (2.45) as just one of six possible components.

Assuming these 'packages' jump eddy mixing lengths l', one can replace their corresponding velocity deficits u' by

$$u' = \frac{dU}{dy} l'.$$

In order to avoid upward and downward velocity fluctuations and l' displacements from cancelling ($\Sigma u' = \Sigma v' = 0$, $\Sigma l' = 0$), it is customary to sum the squares of the velocity fluctuations and consider time-averaged r.m.s. values (i.e. $\tilde{u} = \sqrt{\Sigma u'^2}$, $l = \sqrt{\Sigma l'^2}$); l is termed the Prandtl eddy length, or mixing length.

Summarizing, we have

$$\tau_{t,y,x} = -\mu_t \frac{dU_x}{dy} = -\overline{\rho u'v'} \qquad (2.50)$$

It is worth noting that random values of u' and v' would not produce a nett shear stress. However, there is a strong correlation between $+v'$ and $-u'$, and

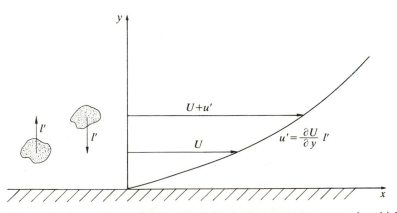

FIG. 2.19. Illustration of Prandtl's turbulence energy stress concepts, in which a dynamic exchange of fluid packages, radial to the tangential flow of fluid in a pipe, leads to the proposition of a turbulent viscosity, μ_t.

$-v'$ and $+u'$ fluctuations, thereby lending some support to Prandtl's concept of jumping packages. Thus, making the replacement

$$\tilde{u} = l\frac{dU_x}{dy},\qquad(2.51)$$

where

$$\tilde{u} = \frac{1}{T}\left(\int_0^T u'^2\,dt\right)^{1/2}\quad\text{and}\quad\overline{u'v'} = \left(\frac{1}{T}\int_0^T u'v'\,dt\right),$$

over a period, $T \gg t$, where T represents a total time, one can also write:

$$\tau_{t,y,x} = -\rho\tilde{v}l\frac{dU_x}{dy}\qquad(2.52)$$

whence, by definition,

$$\mu_t = \rho\tilde{v}l\qquad(2.53)$$

The mean fluctuating cross-flow velocity component, $\tilde{v}$, is difficult to measure. Prandtl went on to propose that analogous to $\tilde{u}$,

$$\tilde{v} = l\frac{dU_x}{dy},$$

leading to the proposition

$$\mu_t = \rho l^2\frac{dU_x}{dy},\qquad(2.54)$$

where l is defined, and determined, through experiment (i.e. empirically). This can be accomplished by measuring velocity gradients and turbulent shear stresses through measurements of u', v' and w'. The results are given in Fig. 2.18 for flow in pipes.

One difficulty with this formulation should be immediately apparent. It suggests that the eddy viscosity must drop to zero at the centreline of the pipe, where symmetry requires

$$\frac{dU_x}{dy} = 0.$$

While this does not pose a practical problem (i.e. $\mu_t \neq 0$, at $r=0$), since the shear stress will still reduce to zero because

$$\tau_{yx} = -\mu_t\frac{dU_x}{dy}\quad\text{and}\quad\frac{dU_x}{dy} = 0$$

by symmetry, it does lead one to a certain disquiet with Prandtl's proposition. In practical terms, the Prandtl formulation proves to be quite successful in modelling the effects of turbulence in shearing flows, allowing one to recast the Reynolds equation (2.45) in the form

$$\frac{dU_i}{\partial t} + U_j \frac{\partial U_i}{\partial x_j} = -\frac{1}{\rho}\frac{\partial P}{\partial x_i} + \frac{\partial}{\partial x_j}\left\{(v+v_t)\frac{\partial U_i}{\partial x_j}\right\} + g_i, \tag{2.55}$$

where v_t is the turbulent kinematic viscosity, μ_t/ρ, and $v_t \gg v$. For many industrial-scale turbulent flow problems, in which inertial forces far outweigh viscous forces and the pressure gradient terms also outweigh viscous forces (i.e., for the Froude dominated flows to be described in Chapter 3), one can take v_t to be constant, and produce quite realistic flow fields (e.g., Salcudean and Guthrie 1979).

One should note that the corresponding equations for the turbulent transport of mass (e.g. for C_i or X_i), or of energy (θ, temperature) are given, in index notation by

$$\frac{\partial \phi}{\partial t} + U_i \frac{\partial \phi}{\partial x_i} = -\frac{\partial}{\partial x_i}\left(\lambda \frac{\partial \phi}{\partial x_i} - \overline{u_i'\phi}\right) + S_\phi, \tag{2.56}$$

where ϕ represents temperature, or concentration of species i and λ denotes the molecular conduction or diffusive transport coefficients. By analogy with the Prandtl mixing length hypothesis, $\overline{u_i'\phi}$ can be replaced by $\Gamma(\partial\phi/\partial x_i)$ where $\Gamma \gg \lambda$. Γ is an eddy diffusivity or transport coefficient for the diffusion of ϕ within the flow. Since the processes of turbulence (i.e. movement of fluid packages) should generally carry heat and/or mass, over equivalent Prandtl mixing length distances it is customary to assume that the turbulent Prandtl–Schmidt numbers, defined by the general equation:

$$\frac{\Gamma_t}{v_t} = \sigma_t \simeq 1, \tag{2.57}$$

are equal to unity. Here Γ_t, the turbulent transport coefficient, is equivalent to D_t, ($\Gamma_t \equiv D_t$), the turbulent eddy diffusivity for mass transport, or to α_t, the turbulent thermal diffusivity for heat transport. σ_θ and σ_M represent the turbulent Prandtl (v_t/α_t) or Schmidt (v_t/D_t) numbers respectively. In practice, the assumption of $\sigma_t = 1$ proves to be quite accurate in many cases, provided molecular conduction processes are relatively unimportant in comparison to turbulent transport processes.

The transport of heat and mass will be covered in further detail in Chapters 4, 5 and 6; suffice it here, as a *caveat*, to recognize that eqn [2.56) lacks any pressure gradient term equivalent to that in the turbulent N–S equation.

In consequence, while a precise description of turbulence is often not needed in predicting flows within large-scale vessels, (and viscosity can even be ignored sometimes), the same cannot be held true of associated heat and mass transfer phenomena. There, a more precise model of turbulence is needed to be able to replace the all-important $\overline{u_i'\phi}$ diffusive term balancing the convective terms to the left of the equation.

None the less, Prandtl's mixing length hypothesis has met with great success in describing the effects of turbulence in a variety of parabolic shearing flows. Thus, in pipe flows;

$$l \simeq 0.4y, \qquad y \leqslant 0.2r, \tag{2.58}$$

$$l \simeq 0.16r \qquad 0.2r < y \leqslant r; \tag{2.59}$$

and in wall boundary layers of thickness δ:

$$l \simeq 0.44y, \qquad y \leqslant 0.2\delta, \tag{2.60}$$

$$l \simeq 0.09\delta, \qquad 0.2\delta < y \leqslant \delta; \tag{2.61}$$

while for free shear layers, the following table applies:

Flow type	Plane mixing layer	Plane jet	Round jet	Radial (or fan) jet	Plane wake
l/δ	0.7	0.9	0.075	0.125	0.16

where δ represents the distance from the axis of jet symmetry to the 1% velocity point at the outer edge of the jet's boundary layer.

These Prandtl mixing length relations allow one to close the set of transport equations, through replacement of the Reynolds stress terms $\overline{\rho u'v'}$ and energy/mass transport equivalents, $\overline{\rho u_i'\phi}$ with suitable empirical prescriptions for turbulence viscosity, μ_t, and turbulence diffusivity, α_t or D_t.

However, while these models have been successful for boundary layer/thin shear flows, they are not sufficiently sophisticated for handling recirculating flow situations, or more complex two-dimensional flows, in which six Reynolds stress terms rather than one may be significant, and in which the 'convection' of eddying motions from one place to another must be allowed for.

(g) Differential models of turbulence for bulk convecting flows

Prandtl (1945) and Kolmogorov (1968) independently suggested that bulk turbulent viscosity should be determined by way of differential rather than algebraic equations, and that the most meaningful way of characterizing velocity fluctuations in high-Re turbulent flows (i.e. far from walls or shear flow layers) would be through 'square-rooting' the kinetic energy of turbulence (per unit mass). The turbulence kinetic energy is defined by

$$k = \tfrac{1}{2}(\tilde{u}^2 + \tilde{v}^2 + \tilde{w}^2), \qquad (2.62)$$

which for isotropic conditions ($\tilde{u} = \tilde{v} = \tilde{w}$), typical of the eddies containing the major portion of the kinetic energy of turbulence, reduces to

$$k = \tfrac{3}{2}\tilde{u}^2. \qquad (2.63)$$

We see that the characteristic turbulent velocity component, $\tilde{u}_k$ becomes

$$\tilde{u}_k \equiv k^{1/2} \quad \text{or} \quad \left(\frac{3\tilde{u}^2}{2}\right)^{1/2}. \qquad (2.64)$$

A transport equation for k was therefore obtained by suitable manipulation of the exact, and time-averaged, forms of the Navier–Stokes equation, so as to bring k into prominence. This so-called 'one-equation' model represented an important step in the development of more general models of turbulence, in that the direct link between the velocity gradient and the fluctuating velocity component was removed. However, difficulties in specifying a length-scale l still remained, as with algebraic models.

Further developments in the field have lead to the development, among others, of a two-equation, k–ε, model of turbulence. This has been used for the computational flows described in Chapter 6. In this model, the rate of turbulent energy dissipation $\dot{\varepsilon}$ (which for local isotropy and high Re is equal to the molecular viscosity × fluctuating vorticity, $\mu(\partial u_i/\partial x_j)^2$) is expressed in the form of a further transport equation (eqn 6.47), in conjunction with a transport equation for the kinetic energy of turbulence (eqn 6.46). Written in index notation, the transport equations for k and $\dot{\varepsilon}$ take the form:

For turbulence energy transport:

$$\frac{\partial k}{\partial t} + U_i\frac{\partial k}{\partial x_i} = \frac{\partial}{\partial x_i}\left(\frac{v_t}{\sigma_k}\frac{\partial k}{\partial x_i}\right) + v_t\left(\frac{\partial U_i}{\partial x_j}+\frac{\partial U_j}{\partial x_i}\right)\frac{\partial U_i}{\partial x_j} - \dot{\varepsilon}; \qquad (2.65)$$

rate of generation	convection	diffusion	generation, P	destruction

and, for turbulence energy dissipation:

$$\frac{\partial \dot{\varepsilon}}{\partial t} + U_i\frac{\partial \dot{\varepsilon}}{\partial x_i} = \frac{\partial}{\partial x_i}\left(\frac{v_t}{\sigma_\varepsilon}\frac{\partial \dot{\varepsilon}}{\partial x_i}\right) + C_1\frac{\dot{\varepsilon}}{k}(P) - C_2\frac{\dot{\varepsilon}^2}{k}, \qquad (2.66)$$

rate of change	convection	diffusion	generation	destruction

where

$$v_t = \mu_t/\rho \quad \text{and} \quad \Gamma = \frac{v_t}{\sigma}$$

and

$$\mu_t = (\rho k^{1/2} l) = C_D\frac{\rho k^2}{\dot{\varepsilon}}. \qquad (2.67)$$

Again, the concept of a turbulent viscosity is central to the model and, following the Prandtl–Kolmogorov formulation, $\mu_t = \rho k^{1/2} l$, where $l \equiv C_D k^{3/2}/\dot\varepsilon$. The use of $\dot\varepsilon$ as an indirect expression for eddy mixing length has proved popular owing to its appearance in the turbulence energy equation, and the relative ease by which the exact equation for $\dot\varepsilon$ can be derived. Both contain convection, diffusion, generation and destruction (i.e., source) terms, and are of similar form to the momentum, mass conservation, and energy equations.

As seen, a number of empirical constants $C_D, C_1, C_2, \sigma_k, \sigma_\varepsilon$ are contained in this turbulence model. C_D can be deduced on the basis of near-wall turbulence, where convection and diffusion of turbulence can be ignored. If so, only the last two terms of the k equation (2.65) apply, and

$$\frac{\mu_t}{\rho}\left(\frac{\partial U}{\partial y}\right)^2 = \frac{C_D \rho k^2}{\mu_t}, \tag{2.68}$$

or

$$\tau = \rho k C_D^{1/2}. \tag{2.69}$$

Thus a plot of the kinematic shear stress versus the kinetic energy of turbulence k gives C_D as the constant of proportionality. The quotient lies between 0.25 and 0.3, suggesting a value for C_D of about 0.09.

In the decay of turbulence behind a grid for a fluid in planar flow, the diffusion and generation terms in the k equation are zero, so that this reduces to

$$U\frac{\partial k}{\partial x} = -\dot\varepsilon, \tag{2.70}$$

while the $\dot\varepsilon$ equation reduces to

$$U\frac{\partial \dot\varepsilon}{\partial x} = -C_2 \frac{\dot\varepsilon^2}{k}. \tag{2.71}$$

Experiments show that dk/dx drops off in proportion to $1/x^2$, and that a value of 1.8–2.0 for C_2 (the unknown quantity) is appropriate.

Finally, in near-wall turbulence, a logarithmic velocity profile prevails, while $P \simeq \dot\varepsilon$ and the convection of $\dot\varepsilon$ is negligible. With this inserted, the $\dot\varepsilon$ equation reduces to

$$C_1 = C_2 - \frac{K^2}{\sigma_\varepsilon \sqrt{C_D}}, \tag{2.72}$$

from which, through computer optimization, $C_1 \simeq 1.4$. These computations lead to the following recommendations for the empirical constants:

C_1	C_2	C_D	σ_k	σ_ε
1.44	1.92	0.09	1.0	1.3

Two-equation models have gone a long way in reducing the arbitrary nature of specifying turbulence, and reducing a bagful of rules and constants associated with algebraic models to a single set of empirical constants. Universality of these constants is not complete, however, and alternative formulations not based on the specification of eddy viscosity/diffusivity concepts, but rather on stress relations *per se*, have also been developed. However, while involving higher-order stress terms, they ultimately require similar empirical constants for closure. The reader is referred to the texts of Launder and Spalding (1972) and Rodi (1980), for further details.

As noted previously, for inertially dominated flows, simple turbulence models, and even a constant (time- and space-independent) effective viscosity/diffusivity value, can give effective *flow* predictions (e.g., Sahai and Guthrie 1982). Consequently, the level of computational and intellectual effort to be expended by including a particular turbulence model must match the level of information required, and recognize the importance, or otherwise, of turbulence on the particular flow problem in question.

(h) Worked examples: Estimation of turbulent viscosity, eddy mixing length, and mean turbulence kinetic energy in a gas-stirred ladle

(i) Eddy viscosity

Sahai and Guthrie (1982) have proposed, on the basis of dimensional and physical arguments, that the mean turbulent, or eddy, viscosity μ_E in a gas-stirred ladle can be related to the flow rate Q of gas injected into the vessel (compensated for temperature and pressure), the mean diameter D of the ladle, and the filled height L, according to

$$\mu_t \simeq 5.5 \times 10^{-3} \rho_L L \left(\frac{gQ}{D}\right)^{1/3}.$$

Consider a typical ladle metallurgy station (as illustrated in Fig. 6.12) in which the ladle diameter is 3 m, the filled height is 2.8 m and a submerged gas flow-rate of 1 Nm^3 min^{-1} is metered; estimate the turbulent energy viscosity of steel at 1600°C. How does this compare with the molecular viscosity of liquid steel?

Solution. Referring the gas to mean height and temperature,

$$Q = 1 \times \frac{\theta}{273} \times \frac{1}{P_{L/2}} = 3.5 \text{ m}^3 \text{ min}^{-1}$$

$$\mu_t = 5.5 \times 10^{-3} \times 7000 \times 2.8 \left(\frac{9.81 \times 3.5/60}{3.0} \right)^{1/3}$$

$$= 62 \text{ kg m}^{-1} \text{ s}^{-1}$$

This compares with a molecular, or dynamic, viscosity of 7 $\times 10^{-3} \text{ kg m}^{-1} \text{s}^{-1}$ or 7 mPa s (see Table A3.11), the ratio μ_t/μ being approximately 10^4.

(ii) Eddy mixing length and kinetic energy of turbulence

On the basis of turbulence kinetic energy measurements in a one-third scale water model of a steelworks ladle, using laser Doppler anemometry, the average scale σ of turbulence was shown to be about 30–50% that of the mean speed $|\bar{U}|$ of the recirculating bath. Deduce the kinetic energy of turbulence and the eddy mixing length on the basis of the Prandtl–Kolmogorov formulation, given that the mean speed of liquid recirculation, and plume velocity are given (Sahai and Guthrie 1982) in SI units by

$$\bar{U} = \frac{0.18 U_p}{R^{1/3}}, \quad U_p = 4.4 \frac{L^{1/4} Q^{1/3}}{R^{1/3}}.$$

Solution. Inserting the appropriate steel data: $U_p = 1.9 \text{ m s}^{-1}$ and $|\bar{U}| = 0.30 \text{ m s}^{-1}$. Assuming isotropic turbulence, $\tilde{u} = \tilde{v} = \tilde{w}$, then

$$k = \tfrac{3}{2}\tilde{u}^2 = \tfrac{3}{2}(\sigma U)^2,$$

where σ represents the scale of turbulence. Taking an average fluctuating velocity equal to 0.4 that of the mean speed of recirculation,

$$k = \tfrac{3}{2}(0.4 \times 0.3)^2 = 0.022 \text{ m}^2 \text{ s}^{-2} \text{ kg}^{-1},$$

compared to a mean kinetic energy of flow $\bar{E}$ of 0.045 m² s⁻² kg⁻¹. Applying the Prandtl–Kolmogorov formulation,

$$\mu_t = \rho k^{1/2} l_k,$$

we have

$$l_k = \frac{62}{7000 \times (0.022)^{1/2}}$$

$$\approx 0.06 \text{ m}$$

Conclusion

It is predicted that the turbulent viscosity $\mu_t \approx 10^4 \mu$, that the kinetic energy of turbulence $k \approx 0.02\,\mathrm{m^2\,s^{-2}\,kg^{-1}}$ and that the mean eddy mixing length $l_k \approx 60\,\mathrm{mm}$. Naturally, there will be a much wider range of eddy sizes, but the energy-intensive eddies should be in this general size range.

2.7.   REAL, POTENTIAL, AND IRROTATIONAL FLOWS

The opposite extreme from a viscous fluid is an ideal, or perfect, fluid. Such a fluid has the property of zero viscosity. Many real fluids, flowing at high Reynolds numbers, in the absence of any flow divergences (or at least with only small ones) can closely approximate the ideal, or inviscid, fluid. From a metallurgical point of view, the flow of liquid metals through nozzles or channels, vortexing phenomena, the flow of fluids around the upper faces of submerged objects such as large rising bubbles, or of water jets impinging on hot strips for run-out cooling operations, and so on, can all approximate ideal flow behaviour. If one's objective is to predict a detailed flow field, there are a number of simplifications that can be made before a full frontal attack through numerical solutions to the turbulent Navier–Stokes, or Reynolds, equations, is launched.

The first simplification is to determine whether or not the bulk fluid flow is flowing in a convergent manner with zero angular velocity ω (i.e., that it is irrotational). If one imagines a z-directed flow and defines ω_z, the angular velocity about the z axis, as being the *average* rate of *counterclockwise* twisting of fluid in the x and y planes, then two fluid lines of length $\mathrm{d}y$ and $\mathrm{d}x$ will exhibit an average rotation of

$$\omega_z = \frac{1}{2}\left(\frac{\mathrm{d}\alpha}{\mathrm{d}t} - \frac{\mathrm{d}\beta}{\mathrm{d}t}\right) \tag{2.73}$$

during their translation through the fluid over a time interval of $\mathrm{d}t$. This is shown in Fig. 2.20, where we see that in the limit, as $\mathrm{d}t \to 0$, and $\Delta x \to \mathrm{d}x$,

$$\mathrm{d}\alpha = \lim_{\mathrm{d}t \to 0}\left(\tan^{-1}\frac{(\partial V/\partial x)\,\mathrm{d}x\,\mathrm{d}t}{\mathrm{d}x + (\partial U/\partial x)\,\mathrm{d}x\,\mathrm{d}t}\right) = \frac{\partial V}{\partial x}\,\mathrm{d}t, \tag{2.74}$$

and similarly, that

$$\mathrm{d}\beta = \frac{\partial U}{\partial y}\,\mathrm{d}t. \tag{2.75}$$

Combining the two, we have the condition

$$\omega_z = \frac{1}{2}\left(\frac{\partial V}{\partial x} - \frac{\partial U}{\partial y}\right) \tag{2.76}$$

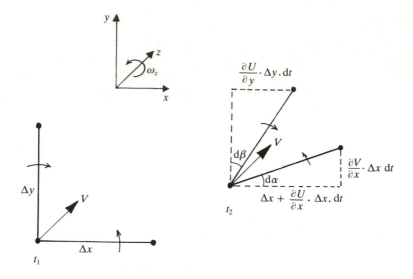

Fig. 2.20. Illustration of two fluid lines deforming in the x–y plane during their translation in the z direction at velocity V. Vorticity is related to the nett angular velocity in the anticlockwise direction ($\zeta = 2\omega = \operatorname{curl} V$ or $\nabla \times V$).

where $(\partial V/\partial x - \partial U/\partial y)$ is defined as the *vorticity*.

A simple example of an element behaving irrotationally, or with zero vorticity, is that of a passenger on a Ferris wheel at the fair, where his compartment (one hopes!) remains perpendicular to the horizon despite two rotational movements. You should also note that these angular vorticity terms are exclusive to the viscous parts of the laminar and turbulent N–S equations. Thus, outside a shearing boundary layer, many macroscopic flows exhibit little, if any, vorticity. They may therefore be treated as being irrotational. For *steady state*, two-dimensional *irrotational* flows we have, for zero vorticity that

$$\frac{\partial V}{\partial x} - \frac{\partial U}{\partial y} = 0. \tag{2.77}$$

Continuity requires, for steady, two-dimensional flows, that

$$\frac{\partial U}{\partial x} + \frac{\partial V}{\partial y} = 0. \tag{2.78}$$

(a)   The stream function

Seeking a higher-order function $\Psi(x, y)$ such that it may simultaneously replace U and V in the continuity equation, one requires that

$$\frac{\partial}{\partial x}\left(\frac{\partial \Psi}{\partial y}\right) + \frac{\partial}{\partial y}\left(\frac{\partial \Psi}{\partial x}\right) = 0. \tag{2.79}$$

We see that this can be managed, if one makes

$$U = -\frac{\partial \Psi}{\partial y}, \text{ and } V = +\frac{\partial \Psi}{\partial x},$$

since

$$-\frac{\partial^2 \Psi}{\partial x \partial y} + \frac{\partial^2 \Psi}{\partial y \partial x} = 0. \tag{2.80}$$

For the condition of irrotational flow,

$$\frac{\partial V}{\partial x} - \frac{\partial U}{\partial y} = 0 \text{ becomes } \frac{\partial}{\partial x}\left(\frac{\partial \Psi}{\partial x}\right) - \frac{\partial}{\partial y}\left(-\frac{\partial \Psi}{\partial y}\right) = 0,$$

or

$$\frac{\partial^2 \Psi}{\partial x^2} + \frac{\partial^2 \Psi}{\partial y^2} = 0. \tag{2.81}$$

This brilliant piece of mathematical trickery means that if we solve this differential equation for any particular steady, two-dimensional, irrotational flow, then continuity is also automatically satisfied. The function Ψ is termed the *stream function*. Unfortunately, this technique cannot generally be extended to three-dimensional flows.

Equation (2.81) is the celebrated Laplace equation, which requires two boundary conditions for solution.

Example: uniform stream
Let us consider a simple example, shown in Fig. 2.21, in which liquid is flowing through a slot towards a nozzle, at an average x velocity of U_0. The total flow rate per unit width into the paper Q' is given by

$$Q' = \int_{y_0}^{y_1} U_0 \, dy = \int -\frac{\partial \Psi}{\partial y} \, dy = -\int_{\Psi_0}^{\Psi_1} d\Psi = \Psi_0 - \Psi_1. \tag{2.82}$$

A second boundary condition is that the velocity component normal to a bounding surface must be zero, since there can be no penetration of the boundary by the fluid. Thus, for flow upstream of the convergence, $V_n = 0$, or

$$\frac{\partial \Psi}{\partial x} = 0 \text{ therefore } \Psi = \text{constant.} \tag{2.83}$$

Thus the boundary must represent a streamline of constant Ψ. The solution to $\Psi(x, y)$ is, of course

$$\Psi = U_0 y \tag{2.84}$$

for the upstream flow component, which the reader should confirm.

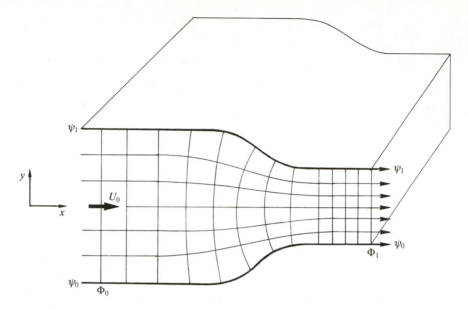

ψ_1

y

U_0

x

ψ_0 Φ_0

ψ_1

ψ_0

Φ_1

Fig. 2.21. Flow of an ideal fluid through a slot towards a nozzle (e.g., feed into a thin strip caster machine), illustrating the construction of an orthogonal flownet of streamlines (ψ), and velocity potential lines (Φ).

(b) The velocity potential

One can similarly define a function $\Phi(x,y)$ such that

$$\frac{\partial \Phi}{\partial x} = U \text{ and } \frac{\partial \Phi}{\partial y} = V, \tag{2.85}$$

which will also satisfy continuity but not zero vorticity. $\Phi(x,y)$ is termed the velocity potential, and is equivalent to temperature (i.e., heat flow potential) for heat flow, species concentration for diffusive mass transport, or voltage for flow of electrical current. As the Laplace equation describes all these phenomena for steady, two-dimensional flows, solutions to particular flows are interchangeable, i.e:

$$\Phi = (\text{velocity potential, temperature, concentration or voltage}) \tag{2.86}$$

$$\int_{\Psi_1}^{\Psi_2} d\Phi \equiv (\text{flowrate, heat flow, mass flow or electric current flow}) \tag{2.87}$$

One can therefore use simple electrical analogues employing conductive paper, or shallow baths of electrolyte constrained by suitable geometrically equivalent boundaries and impose a voltage potential between the 'inflow' and 'outflow' to deduce ideal fluid flow patterns.

In this respect, it is intriguing to note that the governing equations for potential two-dimensional, incompressible flow ($\nabla^2\Psi = 0$) contain no parameters, nor do their boundary conditions. Further, a line of constant Φ, implies

$$d\Phi = 0 = \frac{\partial\Phi}{\partial x}\,dx + \frac{\partial\Phi}{\partial y}\,dy, \tag{2.88}$$

or

$$0 = U\,dx + V\,dy. \tag{2.89}$$

Solving:

$$\left(\frac{dy}{dx}\right)_{\Phi\text{const}} = -\frac{U}{V} = \frac{\left(\dfrac{\partial\Psi}{\partial y}\right)}{\left(\dfrac{\partial\Psi}{\partial x}\right)} = \frac{-1}{\left(\dfrac{dy}{dx}\right)_{\Psi\text{const}}} \tag{2.90}$$

This is the mathematical condition that Φ and Ψ lines be mutually orthogonal. Thus, one can construct a geometrical *flow net* of orthogonal squares, almost by eye, to deduce the flow patterns illustrated in Fig. 2.21. Intuitively, one knows that temperature isotherms are also orthogonal to the conductive heat flux, as are isoconcentration, and isovoltage lines to molecular diffusive flux and electrical current vectors. The same holds true for the velocity potential for irrotational flows.

During the eighteenth century, mathematicians such as Daniel Bernoulli, Leonhard Euler, Pierre-Simon Laplace, Joseph-Louis Lagrange, Jean D'Alembert, produced many elegant solutions to problems involving the flow of inviscid liquids. While they achieved many successes, there were notable failures, the most infamous, or famous, depending on one's point of view, being D'Alembert's Paradox. He showed, in 1752, using potential/inviscid flow theory, that the nett drag force on an object travelling through an inviscid liquid, would be zero! As this was so far removed from reality, the subject fell into disfavour, and the empiricism of hydraulics gained pre-eminence.

The later works of Osborne Reynolds, L. Prandtl, W. Tollmien and Hermann Schlichting on boundary layers bridged the gap in comprehension. Potential flow theory has been very useful in modern aeronautical engineering, in aerofoil design and behaviour.

(c) Some useful potential flows

Several relevant potential flow problems can be constructed from three different types of elementary solutions. These are (i) uniform streams, (ii) sources or sinks, and (iii) potential, or free, vortices.

The potential and stream functions for these three flows can be usefully combined, since the principle of superposition, also introduced in Chapter 4, allows additive solutions to the Laplace equations: i.e., if,

$$\nabla^2 \Phi_1 = 0 \text{ and } \nabla^2 \Phi_2 = 0,$$

then
$$\nabla^2 (\Phi_1 + \Phi_2) = 0. \tag{2.91}$$

Line source. Imagine a point source along the z axis from which fluid issues radially (in the form of a fan). Then the total flow Q' crossing any cylindrical surface, of radius r and unit vertical length, is constant:

$$Q' = (2\pi r) V_r, \tag{2.92}$$

giving
$$V_{r,\text{source}} = \frac{A_s}{r}, \tag{2.93}$$

where $A_s = (Q/2\pi)$. The solutions are illustrated in Fig. 2.22(a). Thus, for the plane polar version of Ψ and Φ

$$V_r = \frac{A_s}{r} = \frac{1}{r}\frac{\partial \Psi}{\partial \theta} = \frac{\partial \Phi}{\partial r}. \tag{2.94}$$

Integrating, one obtains

$$\Psi = A_s \theta \text{ and } \Phi = A_s \ln(r). \tag{2.95}$$

Line sink. For a sink, the signs are reversed:

$$\Psi = -A_s \theta, \quad \Phi = -A_s \ln(r) \tag{2.96}$$

Line vortex. For a line vortex, one can reverse the roles of Ψ and Φ to give

$$\Phi = A_v \theta, \quad \Psi = A_v \ln(r). \tag{2.97}$$

This is shown in Fig. 2.22(b).

Sink plus vortex. An interesting flow pattern having metallurgical implications is the sink plus vortex. Adding eqns (2.96) and (2.97):

$$\Psi = A_s \theta - A_v \ln(r), \tag{2.98}$$

$$\Phi = A_s \ln(r) - A_v \theta. \tag{2.99}$$

Figure 2.22 (c) shows some results (White 1979) plotted by the graphical method. Equally spaced radial lines for the sink are combined with circles for a vortex whose diameter increases in the same ratio (here taken as 1.65). The streamlines cross the intersections of these two families of curves, and are seen to be spirals inward towards an origin, or common centre. These streamlines

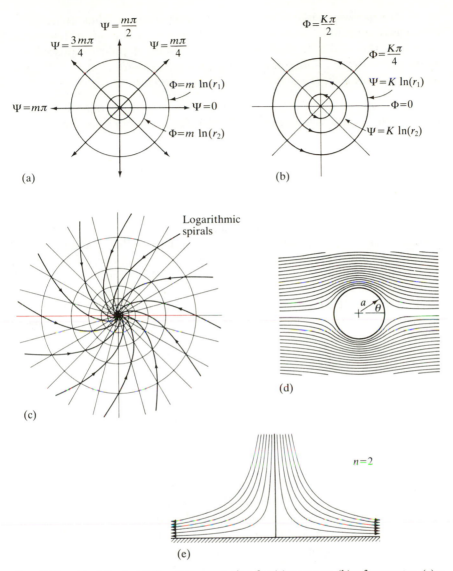

FIG. 2.22.   Ideal (inviscid) flow representation for (a) a source, (b) a free vortex, (c) a vortex plus sink, (d) flow past a sphere or cylinder, (e) normal flow towards a flat plate.

have the form

$$r = C_1 e^{As\theta/Av}, \text{ where } C_1 = e^{(\Psi/Av)} = \text{const.} \tag{2.100}$$

While being of relevance in the design of compressor vanes for turbo-machinery or diffusers for batch anneal furnaces, etc., vortexing flows are also of concern to many liquid metal/slag transfer operations. Thus, Byrne *et al.* (1985) have shown that steel-making slag, ladle slag, and tundish slag can all appear in the final solidified product as a result of pouring procedures, associated vortexing, and excessive stirring. This, of course, has an extremely detrimental effect on the quality of steel slab produced.

Flows around spheres and cylinders. Other potential flows of interest are those around spheres and cylinders (Fig. 2.22 (d)). Thus, one can show for a sphere,

$$V_r = U_\infty \cos(\theta)\left(1 - \frac{a^3}{r^3}\right), \tag{2.101}$$

$$V_\theta = -\frac{1}{2}U_\infty \sin(\theta)\left(2 + \frac{a^3}{r^3}\right), \tag{2.102}$$

where *a* is the sphere's radius, *r* the radial distance into the fluid flowing at an upstream velocity of U_∞. Note that on the surface of the sphere the radial velocity component reduces to zero, and the angular, or tangential velocity V_θ varies from zero at the front stagnation point, ($\theta = 0$, or 180°), to $1.5U_\infty$ at $\theta = 90°$. High-Re flows around spherical bodies closely approximate these solutions but breakaway at about 120° from the front stagnation point in real fluids, result in a turbulent wake and pressure loss, manifested as 'form' drag.

The equivalent results for a cylinder are:

$$V_r = U_\infty \cos(\theta)\left(1 - \frac{a^2}{r^2}\right), \tag{2.103}$$

$$V_\theta = -U_\infty \sin(\theta)\left(1 + \frac{a^2}{r^2}\right). \tag{2.104}$$

These solutions are of more relevance to aeronautical engineers interested in aerofoil sections, but are given for completeness.

Flows towards flat plates. For flows towards flat plates (shown in Figure 2.22(e)), or corners, the methods of combining sources and sinks is replaced by a method of complex variables, in which the real part is taken as the velocity potential Φ and the imaginary part, as the stream function. The solution for flat plates is

$$U = \frac{V_0 x}{L}, \quad W = -\frac{V_0 z}{L}, \tag{2.105}$$

where V_0 is a reference velocity and *L* is a reference length.

(d) The Euler (momentum) and Bernoulli equations for inviscid fluids

The discussion of inviscid fluids has so far been restricted to a consideration of vorticity (or rather a lack of it), and continuity. We should also address the question of the momentum equations (i.e. Navier–Stokes), given in Table 2.2 (a,b,c). The calculations are simplified when either viscosity, or irrotationality, or both, are zero. Let us consider steady, incompressible inviscid and/or irrotational flow along a streamline coinciding with the x component vector. This we will take to be directed vertically upwards. Then Table 2.2(a), for rectangular coordinates, reduces to

$$\rho U \frac{dU}{dx} = -\frac{dP}{dx} + \rho g, \qquad (2.106)$$

or

$$U dU = -\frac{1}{\rho}\frac{dP}{dx} + g. \qquad (2.107)$$

Taking gravity to act downwards, $g = -g$. Integrating with respect to x

$$\tfrac{1}{2}U^2 + \frac{P}{\rho} + gx = \text{constant} \qquad (2.108)$$

Equation (2.106) is the Euler equation, and its integrated form (2.108), is the famous Bernoulli equation. The latter expresses the fact that the total of the kinetic, pressure and potential energies per unit mass along a streamline remains constant for inviscid, and/or, irrotational flows.

(e)   Worked example: laminar cooling of hot strip

In the production of steel sheet in a hot strip mill (see Section 5.10, Exercise 7), thin steel strip passes at high velocity (~ 10–$20\,\text{m/s}^{-1}$) over a series of table rolls. A problem with 'cobbling' (i.e., buckling of strip, with pile-ups) was experienced, following the installation of a new runout cooling system incorporating laminar flow water curtains. Deflection of the sheet passing over the table rolls, as a result of water impact, was suspected. The minimum thickness of the water curtain (Fig. 2.23) was gauged to be 25 mm at a height of 3.0 mm above the strip, just before spreading commenced. The water issued at a velocity of $0.1\,\text{ms}^{-1}$ from a header box set 3 m above the strip. It is necessary to estimate the pressure distribution over the region of water impact on the strip as a first step towards a full stress analysis on the strip.

Solution

Recognizing that surface tension forces are negligible, and that the falling jet corresponds to a high Reynolds number, incompressible, two-dimensional flow, we will ignore the two-phase aspect of the problem, and make use of eqns

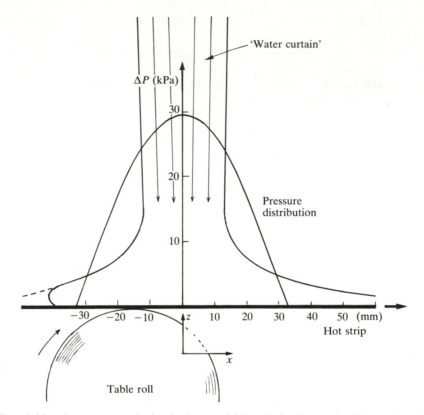

Fig. 2.23. A water curtain impinging on high-velocity hot strip, illustrating the gauge pressure distribution around the stagnation point at $x=0$, $z=0$.

(2.105) and the Bernoulli equation (2.108). Thus,

$$\frac{P}{\rho}+\frac{V^2}{2}+gz=\text{constant}.$$

Replacing V with x and z velocity components U and W, substituting for U and W according to eqns (2.105), and expressing the Bernoulli equation in terms of equivalent pressures:

$$P+\rho\frac{V_0^2(x^2+z^2)}{2L^2}+\rho gz=C.$$

Calculating the velocity of the water curtain at the reference point ($x=0$, $z=0.03$), following a fall of 3 m (less 0.03 m), we can use particle physics to conclude that the reference velocity $V_0=(W_i^2+2gs)^{1/2}=(0.1^2+2\times9.8\times2.97)^{1/2}=7.63\text{ m s}^{-1}$.

Inserting appropriate values:

$$C = 101 \times 10^3 + \frac{1000 \times 7.63^2(0+0.03^2)}{2 \times (0.03^2)} + 1000 \times 9.8 \times 0.03$$

$$= (101 + 29.17 + 0.29) \times 10^3$$

$$= 130.5 \text{ kPa.}$$

We may now calculate the distribution of nett gauge pressure on the strip (i.e. above atmospheric pressure) at $z=0$, $x=x$, since

$$(P - P_{At}) + \frac{\rho V_0^2(x^2+0^2)}{2 \times 0.03^2} = C - P_{At}.$$

Thus,

$$\Delta P_x = (130.5 \times 10^3 - 101 \times 10^3 - \frac{1000 \times 7.63^3}{2 \times 0.03^2}(x)^2 \qquad \text{Pa}$$

$$= 29.46 \times 10^3 - 32.34 \times 10^6(x^2).$$

Thus the pressure on the strip will fall off from 29 kPa at the stagnation point to zero at $x=30$ mm, falling off in proportion to the square of the distance from 'ground zero' ($x=0$, $z=0$).

2.8. MACROSCOPIC BALANCES FOR ISOTHERMAL SYSTEMS

(a) The overall mass, momentum, and energy relationships

Despite the foregoing, many engineering problems are best suited to a macroscopic approach, since a detailed description of flow in these transport operations is often not required. For instance, the mineral processor might be concerned with the pipeline transport of slurry from the mine site to a concentrator circuit where metal valuables are to be recovered. Similarly, the process metallurgist might be concerned with the pneumatic transport of solids (such as $CaSi_2$, CaC_2) for injection into liquid iron or steel ladles for desulphurization. Neither would require a knowledge of the precise paths followed by individual particles of solid or fluid during their transport from one part of the system to the other, provided of course that they did not settle out along the route! However, they would obviously need to know the power requirements of their pumping systems, bulk flow rates, pressure drops, etc. It is for these reasons that a *macroscopic* approach to many fluid flow operations is often more appropriate than any attempt at detailed microscopic or differential analysis of the flow.

This is accomplished by viewing the process from outside its physical enclosure. For instance, in developing the energy equation based on the first law of thermodynamics, changes within the enclosure, or control volume, are measured in terms of the properties of the inlet or outlet streams, together with any exchanges of energy (as heat or work) between the enclosure and its surroundings. Similar comments will apply for the mass continuity and momentum equations.

In the following derivation, therefore, the control volume concept is related to the fluid system that is in motion, in terms of a general fluid property N, corresponding to mass, momentum or energy. It will be shown that the time rate of increase of N within a *system* is equal to the time rate of increase of N within the *control volume* (fixed relative to x, y, z) plus the net rate of efflux of N across the bounding surfaces of the control volume. As the control volume is fixed in space, it corresponds to an Eulerian frame of reference, while the system itself, being in motion, corresponds to the Lagrangian frame of reference. Figure 1.9 demonstrates the approach in relation to Newton's second law of motion. Consider the more generalized flow situation illustrated in Fig. 2.24. This shows air flowing from the bustle pipe of a blast furnace towards the tuyère.

As seen, at time, t, a control volume exactly surrounds a certain mass of fluid (i.e. the system) which is moving from left to right. A short time, δt later,

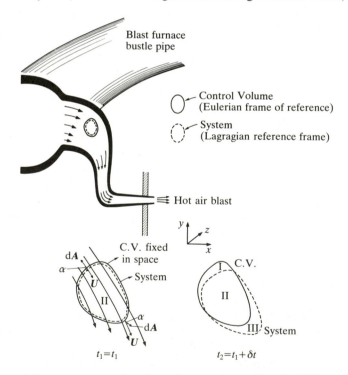

Fig. 2.24.   A flow system moving through space but enclosed within an identical control volume at time t_1 whose coordinates are fixed.

the system will have moved across somewhat, with each mass particle moving at a velocity associated with its location. In the figure, the particles in the upper region of the system are moving more rapidly so the system boundary has distorted slightly to reflect this. At $t_1 + \delta t$, the system occupies regions II and III, while at time t_1, it occupied region II. Therefore, the increase in N (if any) in time δt is given by:

$$N_{t+\delta t} - N_t = \left(\int_{II} n\rho \, dV + \int_{III} n\rho \, dV \right)_{t+\delta t} - \left(\int_{II} n\rho \, dV \right)_t. \qquad (2.109)$$

Here dV is a volume element, n is the amount of N per unit mass of fluid, and ρ is the density. If we consider the influx of material into region I during this time interval, we can introduce the term $\left(\int_I n\rho \, dV \right)_{t+\delta t}$ twice on the right-hand side of the above equation (by adding it and subtracting it), to yield

$$N_{t+\delta t} - N_t = \left(\int_{II} + \int_I \right)_{t+\delta t} - \left(\int_{II} \right)_t + \left\{ \left(\int_{III} \right)_{t+\delta t} - \left(\int_I \right)_{t+\delta t} \right\}, \qquad (2.110)$$

$$\underbrace{}_{\text{output}} \quad \underbrace{}_{\text{input}}$$

$$\frac{\text{Net accumulation}}{\text{in } system \text{ in } \delta t} = \frac{\text{net accumulation in}}{control \text{ } volume \text{ in } \delta t} + \frac{\text{net efflux from } control}{volume \text{ in } \delta t},$$

where the integrals have been written in short-hand notation. Dividing by δt,

$$\frac{N_{t+\delta t} - N_t}{\delta t} = \left(\frac{\left\{ \int_{II} + \int_I \right\}_{t+\delta t} - \left\{ \int_{II} \right\}_t}{\delta t} \right) + \left(\frac{\left\{ \int_{III} \right\}_{t+\delta t}}{\delta t} - \frac{\left\{ \int_I \right\}_{t+\delta t}}{\delta t} \right) \qquad (2.111)$$

and taking the limits as $\delta t \to 0$,

$$\lim_{\delta t \to 0} \left(\frac{N_{t+\delta t} - N_t}{\delta t} \right) = \frac{\partial N}{\partial t},$$

$$\lim_{\delta t \to 0} \left(\frac{\left(\int_{II} + \int_I \right)_{t+\delta t} - \left(\int_{II} \right)_t}{\delta t} \right) = \frac{\partial}{\partial t} \int_{\text{c.v.}} n\rho \, dV,$$

$$\lim_{\delta t \to 0} \left(\frac{\left(\int_{III} \right)_{t+\delta t}}{\delta t} - \frac{\left(\int_I \right)_{t+\delta t}}{\delta t} \right) = \int_{\substack{\text{outflow} \\ \text{c.s.}}} n\rho U \, dA + \int_{\substack{\text{inflow} \\ \text{c.s.}}} n\rho U \, dA = \oint_{\text{c.s.}} n\rho U \, dA.$$

Here subscripts c.v. and c.s. refer to the control volume and the control surface respectively.

We finally have

$$\frac{\partial N}{\partial t} = \frac{\partial}{\partial t} \int_{c.v.} n\rho \, dV + \oint_{c.s.} n\rho U \, dA. \qquad (2.112)$$

Note that a minus sign is required for the surface area of the inflow since $U \, dA$ is negative for inflow. In scalar quantities, $\oint_{c.s.} n\rho U \, dA = \oint_{c.s.} n\rho U \, dA \cos(\alpha)$, representing the integral $\oint$ taken over the whole control volume surface (c.s). Finally, the coordinate system for the control volume can be given a uniform velocity of translation without affecting the validity of the final equation.

(b)   The mass continuity equation

Taking the property of the system N to represent mass, $n = $ mass/unit mass $= 1$, so that on the basis of mass conservation, $dm/dt = 0$, and

$$0 = \frac{\partial}{\partial t} \int_{c.v.} \rho \, dV + \oint_{c.s.} \sigma U \, dA. \qquad (2.113)$$

This represents a general statement of the continuity or mass conservation equation. In words, the equation states that the rate of mass accumulation within the control volume is equal to the net influx of material into it.

Consider, for instance, the steady flow of fluid through a collection of stream tubes shown in Fig. 2.25. Taking the control volume to be made up of the walls of the stream tubes between sections 1 and 2 plus the end areas of sections 1 and 2, eqn (2.113) reduces to $0 = \oint_{c.s.} \rho U \, dA$, since there can be no accumulation. Similarly, and by definition, no flow occurs across the wall of a stream tube, so that for each stream tube of cross-sectional area dA perpendicular to the entering flow, and of dA_2 perpendicular to the discharging flow:

$$\rho_1 U_1 \, dA_1 + \rho_2 U_2 \, dA_2 = 0 \qquad (2.114(A))$$

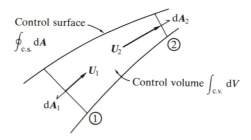

FIG. 2.25.   Steady flow through a stream tube.

or

$$\rho_1 U_1 dA_1 = \rho_2 U_2 dA_2, \qquad (2.114(B))$$

using scalar notation, and $dA_1 \equiv dA_1 \cos(180°)$.

For the collection of stream tubes shown in Fig. 2.25, with the boundaries 1 and 2 orthogonally placed, the average velocity, $\bar{U}$ is $(1/A)\int U \, dA$ and the discharge flowrate, $Q_2 = A_2 \bar{U}_2$.

(c) The linear momentum equation

Taking the property N of the system to represent linear momentum, mU, then $n = mU/m = U$. Substitution in eqn (2.112) then shows

$$\sum F = \frac{d(mU)}{dt} = \frac{\partial}{\partial t} \int_{c.v.} \rho U \, dV + \int_{c.s.} \rho U U \, dA. \qquad (2.115)$$

In words, eqn (2.115) states that the sum of forces acting on the system (which is equivalent to the rate of change of momentum of the system) is equal to the rate of accumulation of momentum within the control volume plus the net efflux of momentum from its surfaces. Consider, for instance, steady flow of fluid through a portion of the stream tube shown in Fig. 2.26, in which the entry and exit boundaries are orthogonally placed; we have

$$F = \rho U_2 U_2 A_2 - \rho U_1 U_1 A_1.$$

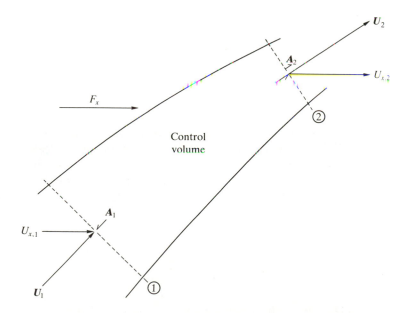

FIG. 2.26.   Control volume of fluid within a reducing band of stream tubes.

Averaging the fluid momentum entering at location 1 and leaving at 2, one may write

$$F_{\text{total}} = [\dot{m}U_2 - \dot{m}U_1], \tag{2.116}$$

since the mass entering or leaving per second is $\rho Q = \rho_1 Q_1 = \rho_2 Q_2 = \dot{m}$. The magnitude of the resolved force in the x direction can then be written as:

$$F_x = \dot{m}\{U_2 \cos(\alpha_2) - U_1 \cos(\alpha_1)\} \tag{2.117}$$

Consequently, a force F_x has to be exerted on the system in order to change the direction and magnitude of fluid momentum.

(d)   The general energy equation

The first law of thermodynamics states that the internal energy change ΔE of a system in changing from state E_1 to state E_2, is equal to the heat added to it, less the useful work performed by the system. E, the internal energy, is a property of state and can include intrinsic, chemical, kinetic, gravitational, electromagnetic, and surface energy effects unless otherwise restricted.

$$E_2 - E_1 = Q - W. \tag{2.118}$$

Denoting the internal energy/unit mass by e and applying eqn (2.112) with $N = E$ and $n = \rho e/\rho = e$, we have

$$\frac{dE}{dt} = \frac{\partial}{\partial t} \int_{\text{c.v.}} \rho e \, dV + \oint_{\text{c.s.}} \rho e \, U \, dA. \tag{2.119a}$$

From the first law of thermodynamics, we can then write

$$\frac{\delta Q}{\delta t} - \frac{\delta W}{\delta t} = \frac{\partial}{\partial t} \int_{\text{c.v.}} \rho e \, dV + \oint_{\text{c.s.}} \rho e \, U \, dA. \tag{2.119b}$$

The work done by the system is generally broken down into a component, W_S, W_τ and W_{pr}. W_S is known as the shaft work in which a torque force may be generated by the fluid to perform useful work (e.g. hydroelectric power), W_τ is the shearing work of the fluid at the boundaries and this is generally lost as frictional heat, W_{pr} the pressure work, is the result of pressure forces acting on the boundaries of the system and causing them to move, i.e.

$$\delta W = \delta W_S + \delta W_\tau + \delta t \oint P U \, dA. \tag{2.120}$$

Hence,

$$\delta \dot{Q} - \delta \dot{W}_S - \delta \dot{W}_\tau = \frac{\partial}{\partial t} \int_{\text{c.v.}} \rho e \, dV + \oint_{\text{c.s.}} \left(\frac{P}{\rho} + e \right) \rho U \, dA \tag{2.121}$$

where e, the energy/mass can be regarded as representing the sum of the potential, kinetic and internal energies, within the system, in the absence of the others just noted, i.e.

$$e = gz + \tfrac{1}{2}U^2 + i.$$

(e)   The steady-flow energy equation

In Fig. 2.27 a situation is depicted that lends itself directly to interpretation via the generalized macroscopic energy equation. At section 1, air is drawn in by turbo-blowers and then pumped through a heat-exchanger system (Cowper stoves) before being blown through the refractory-lined main blast line to the iron blast furnace. Of critical interest is the condition (i.e. pressure, temperature, velocity) of the hot air blast as it leaves section 2. As seen section 2 represents the total cross-sectional area of the tuyères. (This presumes that there is an equal distribution of flow to the approximately twenty tuyères.)

It will readily be appreciated that any variations in U, P, ρ, θ, and i across the surfaces of the control volume areas of interest (i.e. 2 and 1) can be ignored, at least, to a first approximation. This follows since the significant changes in fluid properties occur along the system, which is essentially linear.

We may therefore write through integration of eqn (2.121) that for steady state

$$\dot{Q} - \dot{W}_s - \dot{W}_\tau = [e + P/\rho]_1^2 \rho Q. \qquad (2.122)$$

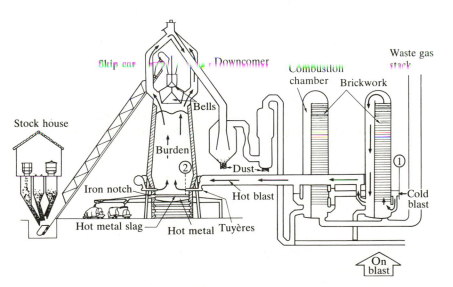

FIG. 2.27.   Hot blast equipment for iron blast furnace, illustrating application of the steady-flow energy equation.

Dividing by ρQ, and expanding e into its components parts, we obtain

$$Q'' - W_S'' - W_\tau'' = \left[i_2 + \frac{P_2}{\rho_2} + \frac{\bar{U}_2^2}{2\alpha} + gz_2 \right] - \left[i_1 + \frac{P_1}{\rho_1} + \frac{\bar{U}_1^2}{2\alpha} + gz_1 \right] \quad (2.123)$$

where Q'' represents the heat energy added to unit mass of fluid;
W_S'' represents the work performed by unit mass of fluid;
W_τ'' represents the frictional work (low-grade thermal energy losses) performed by unit mass of fluid;
$\alpha \simeq 1$ for turbulent flows, $+0.5$ for laminar flow in pipes;
$\bar{U}$ represents the mean velocity at the location in question.

(f)   The Bernoulli equation

When all the terms on the left-hand side of eqn (2.123) are negligible (i.e., when friction losses are negligible, no work is being done by the system, and no heat is transferred into it), the total energy of the system must remain constant:

$$\left[\frac{P}{\rho} + \frac{U^2}{2\alpha} + gz \right]_1 = \left[\frac{P}{\rho} + \frac{U^2}{2\alpha} + gz \right]_2 \quad (2.124)$$

This macroscopic relationship is equivalent to that deduced through integration of Eulers differential equation (2.108). In words, while the pressure, kinetic and potential energy are interconvertible and may change from one place to another, the nett energy/mass of fluid flowing through the system remains constant.

One can usefully write Bernoulli's equation in terms of pressure head, velocity head and potential (or elevation) head, such that:

$$\left[\frac{P}{\rho g} + \frac{U^2}{2g\alpha} + z \right] = \text{constant.} \quad (2.125)$$

As seen, the pressure head $p/\rho g$, kinetic head $u^2/2g\alpha$, and potential head z all have the dimensions of length, or height, above a datum line.

(g)   Friction losses in flow systems

When fluids are pumped through pipes or conduits, turbulent friction losses inevitably lead to energy or head losses. Consider, for instance, incompressible flow through a pipe of length L, diameter d, shown in Fig. 2.28(a). Under isothermal conditions, with $U_1 = U_2$, $z_1 = z_2$, $i_2 = i_1$, $\rho_2 = \rho_1$, $Q'' = 0$, $W_s'' = 0$, the energy equation yields

$$F \text{ or } -W_\tau'' = \frac{P_2 - P_1}{\rho}. \quad (2.126)$$

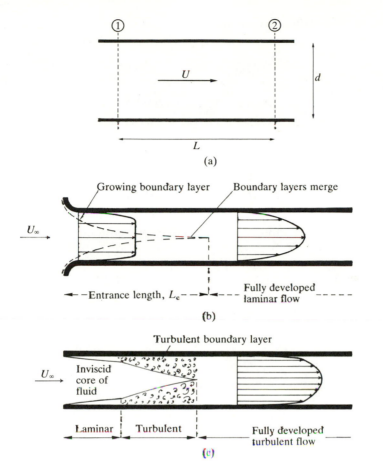

Fig. 2.28. (a) Incompressible flow through a pipe of length L and diameter d. (b) Entry flow of fluid into a pipe, illustrating the growth of a laminar boundary layer and the charcteristic parabolic velocity profile. Note $\bar{U} \approx 0.5\, U_{max, \ell}$. (c) Turbulent flow in a pipe, illustrating development of laminar, then turbulent, boundary layers, and finally the velocity profile in the fully developed turbulent flow region. Note $\bar{U} \approx 0.82\, U_{max, \ell}$.

Since W''_τ is positive, $P_2 < P_1$, W''_τ is often used to denote these frictional energy losses per unit mass resulting from the shearing work done by the fluid. (Some texts use the symbol, F, the frictional energy losses per unit mass. This is synonymous with $-W''_\tau$.)

We may also carry out a simple force balance on the system. Noting that no momentum changes occur in passing between (1) and (2), the macroscopic momentum equation reduces to

$$\sum F = 0,$$

or

$$(P_2 - P_1)\pi R^2 + \tau_0 2\pi R L = 0, \qquad (2.127)$$

where τ_0 is the shear stress acting on the fluid at the periphery of the pipe,

$$\Delta P_f = (P_1 - P_2) = \tau_0 \frac{2L}{R}, \qquad (2.128)$$

and

$$\frac{\Delta P_f}{\frac{1}{2}\rho U^2} = \left(\frac{\tau_0}{\frac{1}{2}\rho U^2} \frac{4L}{D} \right) = f \frac{4L}{D}. \qquad (2.129)$$

Here, the customary friction factor f has been introduced to replace $\tau_0/\frac{1}{2}\rho u^2$. It is a dimensionless grouping representing the ratio of the friction head loss/velocity head of fluid.

For a given Reynolds number and internal pipe roughness, f is constant. In general, it is empirically determined. It is shown in Fig. 2.29, which is often referred to as the Moody chart. This is probably the most famous, and useful, figure in fluid mechanics, at least for plumbers. For laminar flow it can be

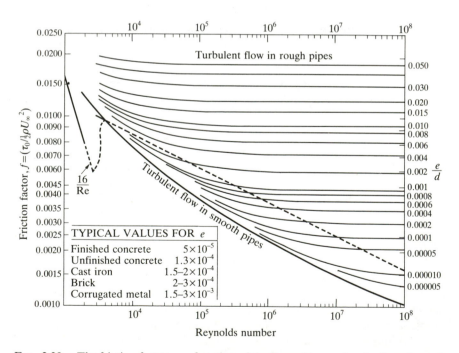

FIG. 2.29.   The friction factor as a function of the Reynolds number for flow through pipes (the Moody chart).

shown that $f = 16/\text{Re}$. In older texts the friction factor defined can be four times that used here, so caution is advised in using these charts. Writing frictional energy losses in terms of a head loss h_L, we have

$$h_L = \frac{\Delta P_f}{\rho g} = \frac{2 f u_L^2}{g D}. \tag{2.130}$$

This is known as Darcy's formula, which Darcy developed while a student with Prandtl. As we might have anticipated for developed flow in a pipe, the head losses increase in proportion to the length of piping, and are inversely proportional to pipe diameter. For rough pipes, at high Reynolds numbers, f is approximately constant and is typically 0.01–0.02.

Friction losses through pipe fittings, bends, opening, etc., are generally handled in an equivalent manner, and expressed in terms of equivalent pipe diameters. Thus it is found by experiment that the head loss associated with flow through a valve can be expressed in the form

$$h_L = K \frac{U^2}{2g}, \tag{2.131}$$

whereas flow through a pipe gives

$$h_L = \left(\frac{4 f L}{D} \right) \frac{U^2}{2g}. \tag{2.132}$$

This has led to the preparation of tables providing information on the equivalent length of pipe of diameter D to give the same head loss. One can then conveniently sum up equivalent pipe lengths for the various valves and fittings, in arriving at system pressure losses. Some typical equivalent lengths of pipe fittings are given in Table 2.4.

TABLE 2.4.   *Equivalent lengths of pipe fittings*

Fitting	Pipe diameters
45° elbows	15
90° elbows (standard radius)	30–40
90° square elbows	60
Entry from leg of T-piece	60
Entry into leg of T-piece	90
Unions and couplings	Generally very small
Globe valves fully open	60–300
Gate valve fully open	7
$\frac{3}{4}$ open	40
$\frac{1}{2}$ open	200
$\frac{1}{4}$ open	800

One can readily show that the laminar flow of fluid through a pipe results in a parabolic velocity profile once the flow has become fully developed. This corresponds to the growth and penetration of boundary layers towards the centreline, where they merge as illustrated in Fig. 2.28(b). The accepted correlation for entry length, L_e, is:

$$\frac{L_e}{d} \simeq 0.06\,\text{Re}, \tag{2.133}$$

while for turbulent conditions (i.e., $\text{Re} > \text{Re}_{d,\text{crit}} = 2300$)

$$\frac{L_e}{d} \simeq 4.4\,\text{Re}_d^{1/6}. \tag{2.134}$$

This result suggests that a turbulent boundary layer thickens more rapidly than its laminar counterpart, and that entry lengths are correspondingly shorter (e.g., 30 pipe diameters at $\text{Re} \simeq 10^5$). Finally, Prandtl showed that the velocity profile for fully developed turbulent flow in a pipe ($\text{Re} \sim 10^5$), corresponds closely to the relationship

$$\frac{u}{U_\ell} = \left(\frac{y}{R}\right)^{1/7}. \tag{2.135}$$

This is sketched in Fig. 2.28(c). The mean, or average axial velocity there, equals $0.82\,U_{\text{max}}$, whereas for laminar flow, shown in Fig. 2.28(b), $U = 0.5\,U_{\text{max}}$.

(h) Flow through nozzles: emptying rates for BOF vessels

In demonstrating the use of macroscopic energy balances, let us consider the example of a 150-tonne BOF vessel whose contents are being tapped into a teeming ladle through a 15-cm diameter nozzle. It is required to estimate (see Fig. 2.30) the initial mass flow rate of liquid steel out of the vessel, given the

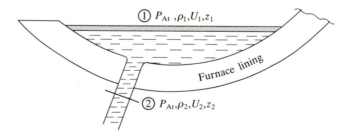

F$_{IG}$. 2.30. Tapping rates of BOF vessels.

fact that the effective initial head of metal is 1 m. The extra slag is converted into an equivalent head of metal for these purposes.

Applying the steady flow energy equation between (1) and (2), we should note that both surfaces of the control volume which are of interest to us (i.e., 1 and 2) are at atmospheric pressure. Ignoring any temperature effects, noting that no energy is added to, or shaft work done by, the system, the relevant form of the SFEE is the Bernoulli equation, so that

$$\frac{P_1}{\rho_1} + \frac{U_1^2}{2} + gz_1 = \frac{P_2}{\rho_2} + \frac{U_2^2}{2} + gz_2. \tag{2.136}$$

Taking $z_2 = 0$ as the datum line, we have $P_1 = P_2 = P_{At}$, $U_1 \simeq 0$, so that

$$U_2 = \sqrt{2gz_1}, \tag{2.137}$$

$$Q_{th} = A_2\sqrt{2gz_1}, \text{ and } Q_{act} = C_D A_2 \sqrt{2gz_1} \tag{2.138}$$

Here C_D, an empirical nozzle discharge coefficient has been introduced to allow for a 'vena contracta' effect which leads to an overestimate of A_2 (the cross-sectional area of the nozzle) and frictional losses within the nozzle. Taking $C_D = 0.9$ as typical, we can substitute appropriate numerical valves, to obtain

$$\dot{m} = \rho C_D A_2 \sqrt{2gz_1}$$

$$= 7000 \times 0.9 \times \frac{\pi}{4} \times 0.15^2 \sqrt{2 \times 9.81 \times 1} \text{ kg s}^{-1}$$

$$\simeq 500 \text{ kg s}^{-1} \text{ or } 30 \text{ tonnes min}^{-1}.$$

(i) Flow measurement by Pitot tubes, Venturi meters, and orifice plates

One of the most important class of flowmeters is that in which the fluid is either accelerated or retarded at the measuring section and the change in the kinetic energy of the fluid is measured by the pressure difference produced. Included in this class are Pitot tubes, orifice meters, nozzles, Venturi meters and notches or weirs.

Consider, under ideal conditions, the dynamic pressure that would be registered by a Pitot tube (in mm of mercury) facing an air blast velocity of 100 m s^{-1}. Ignoring compressibility effects and referring to Fig. 2.31, we can assume ideal flow conditions with no loss of total energy between sections (1) and (2). Thus, applying Bernoulli's equation over the differential control volume between (1) and (2):

$$\frac{P_1}{\rho_1} + \frac{U_1^2}{2} + gz_1 = \frac{P_2}{\rho_2} + \frac{U_2^2}{2} + gz_2.$$

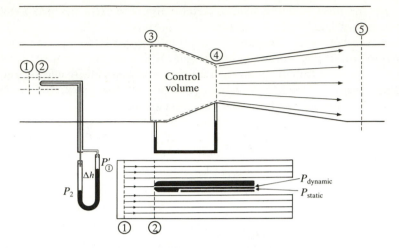

F<small>IG</small>. 2.31.   Principle of Pitot tube and Venturi meter for measuring local velocities
and bulk flow rates respectively.

We can note the gas velocity at the nose of the Pitot tube at (2) must reduce to
zero, as it is a point of stagnation. The kinetic energy of the stream is thus
converted into equivalent pressure energy, or dynamic head. Noting that
$z_1 = z_2$, $\rho =$ constant and that the pressure P_1 represents the static pressure
just upstream of the probe, (conveniently measured by equivalent static
pressure taps essentially in the pilot head), we have:

$$\frac{P_2 - P_1}{\rho} = \frac{U_1^2}{2}, \quad \Delta P = 0.5\,\rho U_1^2.$$

Assuming ideal gas behaviour, we can deduce gas density by

$$PV = nRT$$

$$\rho = \frac{nM}{V} = \frac{PM}{RT} = \frac{1.01 \times 10^5 \times 28}{8.3 \times 10^3 \times 273} \simeq 1.25 \text{ kg m}^{-3}$$

$$\Delta P = 0.5 \times 1.25 \times 10^4 \simeq 6240 = 13600 \times 9.81 \times h_{\text{Hg}}$$

Hence, the dynamic head of mercury, h_{Hg}, that would be registered by a Pitot
tube is

$$h_{\text{Hg}} = 46.8 \text{ mm Hg}$$

Also shown in Fig. 2.31 is the Venturi meter, designed to measure overall fluid
flow through a tube or pipe. In the converging section between (3) and (4), the
boundaries of the control volume, energy losses are negligible, so that once
more Bernoulli's equation can be applied. In this case, contrary to the Pitot

tube situation, the static pressure at the exit to the control volume (i.e. the throat) is lower than that registered at (3). The explanation is, of course, that the flow, in accelerating from (3) to 4, must increase its kinetic energy, and this is only possible through a loss in static, or pressure energy. Consequently, we write

$$\frac{P_3}{\rho_3} + \frac{U_3^2}{2} = \frac{P_4}{\rho_4} + \frac{U_4^2}{2}.$$

From continuity, we have

$$\rho_3 A_3 U_3 = \rho_4 A_4 U_4.$$

Substitution for U_3, and taking $\rho \simeq$ constant, yields

$$U_4 = \left(\frac{2(P_3 - P_4)}{\rho\{1 - (A_4/A_3)^2\}}\right)^{1/2},$$

or

$$\dot{Q}_{th} = A_4 \left(\frac{2(P_3 - P_4)}{\rho\{1 - (A_4/A_3)^2\}}\right)^{1/2}. \tag{2.139}$$

Allowing for any energy losses in practice, one can write more generally

$$\dot{Q}_{act} = \frac{C_D A_4}{\{1 - (d_4/d_3)^4\}^{1/2}} \left(\frac{2(P_3 - P_4)}{\rho}\right)^{1/2} \equiv K A_4 \left(\frac{2(P_3 - P_4)}{\rho}\right)^{1/2}. \tag{2.140}$$

K is termed the *flow coefficient*. Typical values for C_D, the discharge coefficient, are 0.98 to 1.0 for Venturi meters operating under turbulent flow conditions.

The same principles can be applied to sharp-edged orifices, shown in Fig. 2.32. While these are the most convenient in terms of fabrication, they suffer from significant pressure energy losses downstream of the constriction. This is avoided in the Venturi meter by limiting the angle of divergence to 7° or less

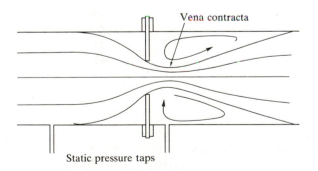

Vena contracta

Static pressure taps

Fig. 2.32.   An orifice meter, showing a square-edged circular plate constricting flow.

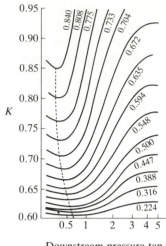

Downstream pressure tap
location in pipe diameters

FIG. 2.33.   Flow coefficient for square-edged, circular orifices at orifice Reynold's numbers greater than 3×10^4. The effect of downstream tap location, as well as $D_{\text{orifice}}/D_{\text{pipe}}$, is illustrated.

from the horizontal. This avoids flow separation, and the recirculating vortex formed behind the sharp-edged orifice. For such situations, K, the flow coefficient, needs to be measured before flow rates can be established precisely.

Figure 2.33 shows how this correction factor varies with orifice diameter/pipe diameter at Reynolds numbers above 3×10^4, for square-edged orifices. As seen, K is at a minimum when the downstream pressure tap is located below the vena contracta shown in Fig. 2.32.

2.9.   EXERCISES ON THE MACROSCOPIC MASS, MOMENTUM, AND ENERGY BALANCES

1. Show from first principles that a sudden enlargement in pipe diameter from d_1 to d_2 leads to a head loss coefficient or resistance coefficient, K, given by $K = (1 - d_1^2/d_2^2)^2$. (*Hint*: combine SFEE and momentum equations.)
2. A simple open ended U-tube containing mercury is used to measure the pressure in a 50-mm pipe carrying carbon dioxide gas at 30°C. The mercury level with no gas flow is 0.7 m below the pressure tap.

(a) If the reading in the manometer is 25 mm Hg, what is the pressure in the line?

(b) If water were flowing in the line under the same pressure as the CO_2, what would the manometer reading be?

The molecular weight of CO_2 is 44.

3. An oil-and-mercury manometer has the dimensions shown in the figure. The end A is permanently open to the atmosphere and the end B is connected to a vessel containing gas. The level of mercury in the left-hand limb is observed to rise 2 cm. What is the pressure of the gas? The density of the gas may be neglected and the relative weights of oil, mercury, and water are 0.85, 13.6, and 1.0. The surface of the oil remains in the enlarged portion of the tube. The oil–mercury interface is on the right–hand side of the 4 mm diameter tube (i.e. below the enlarged portion).

4. Two circular jets of the same velocity meet head-on, giving the idealized flow pattern shown below. Derive an expression for θ in terms of d_1 and d_2. Ignore cavity and surface tension forces.

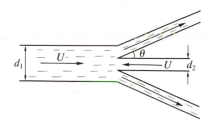

5. A pipe 3000 m long and 0.3 m diameter ($e/D = 0.001$) connects two water-storage tanks whose water levels differ by 6 m. It is necessary to increase the flow by 50%. Compare the *running costs* of an electric pump at 4 cents per kW h with the alternative of paying 7 per cent interest on a second

pipe (0.3 m diameter) which extends for an appropriate distance before joining the original pipe. The cost of the pipe plus installation is $500 per metre. The overall pump efficiency is 80 per cent.

6. Two reservoirs are connected by 120 m of horizontal pipe, 0.25 m in diameter, $e/D = 0.004$. The difference in level of the reservoirs in 10 m. Find the quantity of water flowing if there is (a) a turbine working at a head of 7 m in the pipe, (b) a pump generating 17 m head in the pipe line. In this case the water is being pumped uphill.

7. A 0.3 m pipe is joined to a 0.15 m pipe by a 60° reducing bend. The flow is 0.3 $m^3 s^{-1}$ and the pressure in the larger pipe is 2×10^5 N m^{-2}. Find the force acting on the bend and express as components parallel and normal to the 0.3 m diameter pipe.

8. Waste gases from a roasting furnace and a reverberatory smelting furnace from flues A and B respectively (see figure) pass into a common flue C at plane (1) and are discharged to the stack. In order to calculate the height of the stack necessary to provide adequate draught, it is essential to calculate loss of head in the flue owing to wall friction, bends, enlargements, etc. Loss of energy is also caused by mixing two fluids which are travelling at different velocities. Assume that:

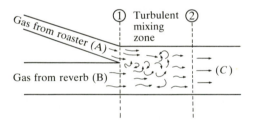

(a) frictional forces due to the action of the wall, over the short length of flue considered, are negligible;

(b) the density of the gases in all flues is constant and is 0.9 kg m^{-3};

(c) complete mixing has occurred and the fluid is travelling at a uniform velocity at plane (2);

(d) all flues are horizontal.

Calculate:

(a) the reading of a differential water manometer connected between planes (1) and (2);

(b) the net loss of mechanical energy between planes (1) and (2) due to the mixing of two gases at different velocities (see data below).

Data

	Area (m^2)	Flow rate (m^3 min^{-1})
Flue A plane (1)	1.0	40
Flue B plane (1)	1.5	160
Flue C plane (2)	2.5	200

9. A steel-works ladle, in the shape of a vertical cylinder (3 m diameter, 3 m high) with a closed base, is filled to a depth of 2.8 m with molten steel. The liquid steel is teemed through a 6 cm-diameter nozzle, in the botton of the ladle, into ingot moulds of 2- tonne capacity. Teeming is carried out continuously until the slag–metal interface reaches the ladle bottom.

   (a) Calculate the time required for teeming.
   (b) Compare the time required to teem the first 2 tonnes of molten *metal* from the ladle, with that required for the last 2 tonnes of *metal*.
   (c) To what extent does the slag layer affect the teaming time?

Data

Average density of molten steel	7000 kg m^{-3}
Average density of molten slag	3000 kg m^{-3}
Coefficient of discharge of the nozzle	0.92
Thickness of slag layer	0.15 m

10. Molten metal is continuously poured by the ladle operator from a ladle into a tundish of wide cross-section. The metal in the tundish issues from a 3 cm-diameter orifice, with a coefficient of discharge of orifice of 0.955, set in the bottom of the tundish, into the mould of a continuous casting machine. If the height of molten metal above the orifice in the tundish is 0.3 m, calculate the flow rate of metal (m^3 s^{-1}) from the tundish, stating any assumptions that you have made. If the ladle operator can only maintain the level of metal in the tundish to ± 20 mm, calculate the variation in flow rate to the mould (i.e., compute standard deviation in flow rate).

11. In the continuous casting of steel, it is current practice to protect the jet of molten iron issuing from the ladle nozzle from contamination with air, by inserting a refractory tube between the ladle and the tundish, as shown in the diagram below. One of the difficulties of this method is ensuring that a

good seal is obtained between the ladle bottom and the top of the refractory tube. Show why this is a problem, by obtaining an expression for the static pressure in the steel at the junction between the two. Also, if the residual gas content of dissolved carbon and oxygen in the vacuum-degassed metal corresponds to an equilibrium CO pressure of 10^{-4} atmospheres, what is the maximum length L of tube that can be safely used if cavitation is to be avoided. Calculate the associated exit velocity $(m s^{-1})$ of the liquid steel jet under these conditions.

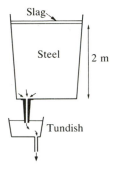

12. Experimental work on the design of a casting wheel for copper wire bar necessitated calculations of the trajectory of the jet of molten copper issuing from the tap hole of a surge vessel into the mould. The surface of the mould was 150 mm below the horizontal plane of the centreline of the tap hole. The level of copper in the vessel varied between 450 mm and 600 mm above the tap hole centreline. Calculate:

(a) the maximum distances in the horizontal direction between the edge of the mould and the front of the tap hole;

(b) the minimum width of the mould opening in order to contain the jet whether the surface of copper is at the higher or lower levels specified.

Assume the tap hole is a round orifice with a coefficient of discharge $=0.95$. Neglect the resistance of air on the jet.

13. A furnace system consists of a large, wide furnace with a 3 m diameter 10 m long horizontal flue connected to a 3 m diameter, 50 m high chimney. The mean temperature of the exhausting hot gases is 510 °C; ambient conditions are normally 25 °C and 1 atm. If waste gas production is to be increased to 6 000 Nm³ min⁻¹ (assume physical properties of air apply), what extra pumping capacity would you recommend so as to maintain the draft (i.e., gauge pressure) in the furnace adjoining the flue entrance at zero.

Take $e/D = 0.001$ for calculating friction factors and K values of 0.5 and 1.0 for the flue entrance and exit losses, and a blower efficiency of 80%. Note that draft (i.e. gauge pressure) at the top of the stack is zero.

14. What pressure (cm Hg) is developed by a Pitot tube facing an airstream (0 °C and 1 atm static pressure) for air velocities of 250, 25 and 2.5 m s^{-1}, respectively. Calculate associated temperatures at the tip of the Pitot tube for adiabatic flow at 250 m s^{-1} ($\gamma = 1.4$). Assume isothermal flow for other velocities.

15. A Venturi meter with a throat diameter of 0.2 m is in a 0.5 m main, supplying a reservoir in which the water level is kept 3 m above the centreline of the meter. What is a maximum flow through the meter if the flow pressure must not fall below 2 m of water absolute. Under these conditions what will be the gauge pressure in the main upstream of the flow? Neglect frictional loss of energy. If the minimum flow is 1/10 of this, what range of pressure differences, between the main and throat, is to be measured?

16. Air flows steadily from a 150-mm-diameter convergent nozzle in a large tank out into the atmosphere. The velocity in the tank far from the nozzle is negligible and the pressure there is 50 per cent greater than atmospheric pressure. The relation between pressure and density anywhere in the flow is $(P/\rho)^{1.4} = $ constant. At the nozzle outlet, the pressure is 1 atm, with a density of 1.2 kg m^3.

Find the mass rate of flow through the nozzle.

2.10. CONCLUSIONS

This chapter has attempted to strike a practical balance between the theory of fluids and the industrial applications of fluids, with specific reference to systems of metallurgical interest. While the principles have been underscored, and some important areas for process metallurgists fleshed out, the depth and range of subject matter presented remains restricted. Some recommended texts for further study are listed below.

FURTHER READING

1. **Inviscid flow**
   Lamb, H. (1945). *Hydrodynamics* (6th edn). Dover, New York.
2. **Turbulence**
   Davies, J. T. (1972). *Turbulence phenomena*. Academic Press, New York.
3. **Experimental methods**
   Bradshaw, P. (1964). *Experimental fluid mechanics*. Pergamon Press, London.

4. **Boundary layers**
   Schlichting, H. (1979). *Boundary layer theory* (7th edn). McGraw-Hill, New York.
5. **Flow visualization**
   Taylor (1968). *Flow visualization.* McGraw-Hill, New York.
6. **For undergraduate and graduate mechanical engineers**
   Daily, J. W. and Harleman, D. R. F. (1973). *Fluid dynamics.* Addison-Wesley, Reading, MA.
   Batchelor, G. K. (1967). *An introduction to fluid dynamics.* Cambridge University Press, London.
   Frank M. White (1979). *Fluid mechanics.* McGraw-Hill, New York.
7. **For research chemical and metallurgical engineers**
   Clift, R., Grace, J. R. and Weber, M. E. (1978). *Bubbles, drops and particles.* Academic Press New York.
   Szekely, J. (1979). *Fluid flow phenomena in metals processing.* Academic Press, London.
8. **Review of metallurgy**
   Sahai, Y. and Guthrie, R. I. L. (1982). *Advances in transport processes* (ed. A. Majumdar), Wiley Eastern.

3

DIMENSIONAL ANALYSIS AND REACTOR DESIGN

3.0. INTRODUCTION

It becomes apparent at an early stage in one's career as an engineer that purely theoretical analyses rarely produce completely correct solutions to the problems set by real systems. For example, the detailed performance of many metallurgical and chemical reactors of the type mentioned in Chapter 1 would defy analytical solution. While it may be desirable to have a complete mathematical solution describing gas, steel and slag flows in a basic oxygen furnace, together with associated chemical reactions, the task would be difficult for even the most sophisticated of mathematical models. The problem is certainly not amenable to analytical solution. It is clear that alternative approaches are needed in the analysis of such problems.

This chapter is concerned with two techniques that can be useful in representing the overall effect of complex transport phenomena on typical metallurgical processing operations. The first of these is dimensional analysis. It is often used by experimenters to summarize the effects of heat, mass, or fluid flow in, or around, specific objects, vessels or models thereof. The second method involves elementary reactor analysis for characterizing fluid flows and associated chemical reactions within the metallurgical or chemical reactor vessel of interest.

Both techniques avoid detailed descriptions of flow patterns, etc., that may be associated with the reactor processing vessel under investigation. Rather, they attempt to describe the overall effect of these phenomena through simple mathematical expressions between the physical variables deemed to be of particular importance. These reactor analysis methods will, to some extent, be superseded in the coming years. Thanks to the advent of high-speed digital computers, it is now possible to describe complex multidimensional flow systems with some confidence through numerical solutions to the governing equations.

Mathematical models are therefore opening the way towards more accurate and sophisticated techniques for metallurgical process design and optimization. A description of such analyses is reserved for the final chapter.

3.1. DIMENSIONAL ANALYSIS

Dimensional analysis, or the principle of similitude, is an important tool for the experimenter in thermofluid dynamics. Its purpose is to give certain information about the relations which hold between measurable quantities associated with various phenomena. The advantage of the method is that it is rapid, and guides the experimenter towards efficient and correct experimental procedures and modelling criteria. It also enables one to dispense with writing down transport equations, etc., and attempting to solve them. However, its main disadvantage is that it does tend to require previous experience on the part of the investigator, so that he can identify those factors which are likely to be important to his analysis, and those which can be ignored.

Furthermore, while the mechanics of the technique are simple to master, they can be deceptive. Three important methods of dimensional analysis are illustrated here: the differential equation approach and those developed by Rayleigh and Buckingham. The various 'tricks' that can be used to extract a maximum of information from the technique of dimensional analysis are illustrated by way of a number of worked examples. For those readers unfamiliar with heat and/or mass flows, the heat flow examples can be reserved until Chapters 4 and 5 have been covered.

(a)   Dimensional formulae

It is well known that all satisfactory mathematical formulae of the form

$$x_1 = f(x_2, x_3, x_4, \dots) \tag{3.1}$$

must be dimensionally coherent. Considering a simple case, such as a body in free fall towards the Earth's surface, x_1 might represent the body's velocity, while x_2 and x_3 would represent possible factors affecting this motion. These might, for instance, be the distance travelled by the body, the air resistance encountered, the gravitational constant, and so on. Saying that x_1 represents the body's velocity is, in fact, a lazy or shorthand way of stating that x_1 is a number which is a measure of velocity. Similarly, x_2 would be a number which is a measure of distance travelled, and x_3 is a number which is representative of air resistance. All these numbers are arbitrary, in that they depend on the system of measurement peculiar to the nationality of the engineers or scientists making the observations. Consequently, to write

$$U = U_0 + gt \tag{3.2}$$

and take measurements of U will result in different numerical values depending on the preferred system of measurement. By contrast, one can appreciate that the *ratio* of any two quantities must have an absolute significance that is independent of the measuring system. For instance, if one object is travelling twice as fast as another, the ratio of speeds would be 2, no matter whether m.p.h., $m\ s^{-1}$ or ft min^{-1} were used as a measure of speed.

Turning to another aspect of eqns (3.1) and (3.2), one sees that the arguments appearing in these equations may either be of a primary, or 'fundamental' nature (i.e., mass, length and time) or of a secondary nature (e.g., velocity and acceleration). The latter are said to be secondary quantities, since they are expressed in terms of ratios of powers of the primary units. For example, velocity is measured in units of length/time or LT^{-1} and acceleration as LT^{-2}. Returning to eqn (3.1), we can rewrite it in the general form

$$f(\alpha, \beta, \gamma, \delta, \dots) = 0. \qquad (3.3)$$

The equation, as written, is dimensional. The *dependent variable of interest*, α, has been incorporated into a general expression containing a number of *independent variables* as quantities β, γ, δ, etc. Whatever the unknown form of this relationship, it can be empirically represented or fitted by a power series expansion or polynomial expression of the type

$$f(\alpha, \beta, \gamma, \delta, \dots) = A\alpha^{a_1}\beta^{b_1}\gamma^{c_1}\delta^{d_1} \dots + B\alpha^{a_2}\beta^{b_2}\gamma^{c_2}\delta^{d_2} \dots, \text{ etc.,} \qquad (3.4)$$

provided a continuous relationship between α and the other parameters exists. For such an expression to be a *complete* representation of the phenomenon of interest, however, it must be cast in such a form as to be independent of the system of units by which the various quantities are measured. This can be accomplished by dividing throughout by the first term in the power series, $A\alpha^{a_1}\beta^{b_1}\gamma^{c_1}\delta^{d_1}$:

$$f'(\alpha, \beta, \gamma, \delta, \dots) = 1 + \frac{B}{A}\alpha^{a_2-a_1}\beta^{b_2-b_1}\gamma^{c_2-c_1}\delta^{d_2-d_1} \dots$$

$$+ \frac{C}{A}\alpha^{a_3-a_1}\beta^{b_3-b_1}\gamma^{c_3-c_1}\delta^{d_3-d_1} \dots + \dots \qquad (3.5)$$

Note that the first term has been reduced to unity and is therefore dimensionless. It follows that the successive terms containing the products of powers of the quantities must also be dimensionless. For this to hold, each one of the products must become dimensionless with respect to the primary quantities by which the secondary quantities are measured. The result can be summarized in the statement:

$$f'(\alpha, \beta, \gamma, \delta, \dots) = f(\alpha^a \beta^b \gamma^c \delta^d) \qquad (3.6)$$

3.2.   RAYLEIGH'S METHOD OF INDICES

Having laid the groundwork to the subject of dimensional analysis, it is appropriate to consider Rayleigh's method of indices, which he first used in 1900. There is a rather famous analysis (Rayleigh 1915) of liquid flow past a heated sphere, in which he sought to deduce the type or form of relationship that one should expect between the rate of heat loss $\dot{Q}$ from the sphere and other pertinent system factors. Table 3.1 lists the dependent variable $\dot{Q}$,

TABLE 3.1.   *Rayleigh's method of indices: assembly of variables and dimensions for heated sphere in cold stream of liquid*

Name of quantity	Symbol	Dimensional formula
Rate of heat transfer (dependent variable)	$\dot{Q}$	QT^{-1}
Diameter of sphere	d	L
Velocity of stream	U	LT^{-1}
Temperature difference	$\Delta\theta$	θ
Volumetric heat capacity of liquid	C_v	$QL^{-3}\theta^{-1}$
Thermal conductivity of liquid	k	$QL^{-1}\theta^{-1}T^{-1}$

together with the independent variables d, U, $\Delta\theta$, C_v and k on which he
thought the phenomenon depended.

Referring to the result obtained in eqn (3.6), we see that the unknown
relationship can be represented by a power series containing product
grouping of the dependent and independent variables or quantities. A typical
product of those quantities can be written:

$$\dot{Q}^a d^b U^c \Delta\theta^d C_v^e k^f. \tag{3.7}$$

This must be dimensionless in terms of the four primary quantities Q, θ, L and
T in which the arguments are expressed. Writing eqn (3.7) in dimensional
terms, we have

$$Q^0 T^0 L^0 \theta^0 = (QT^{-1})^a (L)^b (LT^{-1})^c (\theta)^d (QL^{-3}\theta^{-1})^e (QL^{-1}\theta^{-1}T^{-1})^f. \tag{3.8}$$

This allows us to write down the following exponent relationships:

for Q^0 $\qquad\qquad\qquad 0 = a + e + f,$ $\qquad\qquad\qquad\qquad$ (3.9)

for T^0 $\qquad\qquad\qquad 0 = -a - c - f,$ $\qquad\qquad\qquad\qquad$ (3.10)

for L^0 $\qquad\qquad\qquad 0 = b + c - 3e - f,$ $\qquad\qquad\qquad\qquad$ (3.11)

for θ^0 $\qquad\qquad\qquad 0 = d - e - f.$ $\qquad\qquad\qquad\qquad$ (3.12)

We have six unknown exponents and only four equations derived from this
dimensional analysis. Since two of the exponents will therefore have to remain
arbitrary, we will choose a, which contains the dependent variable, $\dot{Q}$, and b.
After some manipulation of eqns (3.9)–(3.12), we obtain $c = a + b$, $d = -a$,
$e = a + b$, $f = -b - 2a$, whence

$$f\left\{\left[\frac{\dot{Q}UC_v}{k^2\Delta\theta}\right]^a \left[\frac{dUC_v}{k}\right]^b\right\} = 0, \tag{3.13a}$$

which can also be cast in the form

$$f\left\{\left[\frac{\dot{Q}}{kd\Delta\theta}\right]^a\left[\frac{dUC_v}{k}\right]^{a+b}\right\}=0.\tag{3.13b}$$

This allowed Rayleigh to conclude that

$$\dot{Q}=kd\,\Delta\theta\,f'\left\{\frac{dUC_v}{k}\right\}.\tag{3.14}$$

The equation shows that the rate of heat loss from the sphere must be directly proportional to the liquid's thermal conductivity, to the diameter of the sphere to the first power, and to the temperature difference $\Delta\theta$, between the sphere and the bulk of the liquid. The analysis goes on to show that its rate of heat loss is dependent on the liquid's velocity past it and the liquid's volumetric heat capacity. This analysis would therefore help in setting up an appropriate experimental investigation of the phenomenon. However, on reflection, the example leaves many unanswered questions. For instance, why was Rayleigh justified in choosing heat and temperature as independent or primary quantities? If temperature is defined as the mean kinetic energy of molecules, we would lose θ as a primary quantity, and as a consequence also lose eqn (3.12). Similarly taking heat as equivalent to work energy, we would lose eqn (3.9). The net result would be a more general, and therefore less useful relationship than that given by eqn (3.14), in which only mass, length and time were used as "truly" fundamental units. We may similarly ask why Rayleigh was able to neglect liquid viscosity, compressibility, density, thermal expansivity, absolute temperature or the gravitational constant? Clearly his previous experience was almost an essential prerequisite to a successful analysis.

3.3.   BUCKINGHAM'S π THEOREM

Rayleigh's method tends to become unwieldy when a large number of quantities are involved. A more rapid approach that involves less algebra was presented by Buckingham (1914). Buckingham's π theorem states:

If the equation $\phi(\alpha, \beta, \gamma, \dots)=0$ is to be a complete equation, the solution has the form

$$F(\pi_1, \pi_2, \dots)=0,$$

where the π are the $n-m$ independent products of the arguments α, β, $\gamma \dots$, which are dimensionless in the fundamental units.

We will demonstrate the method by re-running the last example, and using the following set of rules based upon the above statement.

Rule 1. Choose the maximum number m, of primary quantities that is consistent with the problem.

Rule 2. Choose a maximum number n of independent variables, or arguments, such that they themselves cannot be grouped into a dimensionless product and such that all the primary units appear in at least one of the n variables chosen.

Rule 3. Successively combine the remaining dependent and independent variables with the group of n variables chosen according to Rule 2.

Solution

From eqn (3.3), we have

$$f(\dot{Q}, d, U, \Delta\theta, C_v, k) = 0. \tag{3.15}$$

Taking Q, θ, L, T as primary quantities, $m = 4$; while from eqn (3.15), we see that $n = 6$. Consequently the number of π groups is two: $n(\pi) = 6 - 4 = 2$. Choosing m variables; $d, \Delta\theta, k, U$ and successively combining the remainder, our π assemblies are:

$$\pi_1 = \dot{Q} d^{a_1} \Delta\theta^{b_1} k^{c_1} U^{d_1}, \tag{3.16}$$

$$\pi_2 = C_v d^{a_2} \Delta\theta^{b_2} k^{c_2} U^{d_2}. \tag{3.17}$$

Writing π_1 in dimensional terms, we have the condition:

$$Q^0 \theta^0 L^0 T^0 = \pi_1 = (QT^{-1})(L)^{a_1} \theta^{b_1} (QL^{-1} T^{-1} \theta^{-1})^{c_1} (LT^{-1})^{d_1}. \tag{3.18}$$

Equating exponents:

for Q	$0 = 1 + c,$	$\therefore c = -1,$
for θ	$0 = b - c,$	$\therefore b = -1,$
for L	$0 = -1 - c - d,$	$\therefore d = 0,$
for T	$0 = a - c + d,$	$\therefore a = -1,$

so that

$$\pi_1 = \frac{\dot{Q}}{d\Delta\theta k}.$$

Note that π_1 is the dependent π group containing $\dot{Q}$, the dependent variable. Similarly,

$$\pi_2 = \frac{dUC_v}{k}$$

Consequently, we finally write:

$$\pi_1 = f(\pi_2) \tag{3.19}$$

or

$$\dot{Q} = d\Delta\theta k \, f\left(\frac{dUC_v}{k}\right). \tag{3.20}$$

This is the same result as that obtained previously (eqn 3.14), but obtained more rapidly. It is worth remarking that the π groups produced can look very different, depending on the choice of initial variables. However, all the expressions will be consistent, and one may be obtained from an other through appropriate manipulation of π groups. For example, if we divide both sides of eqn (3.19) by π_2, then an equally valid and equivalent expression (eqn 3.13a) results.

3.4. DIMENSIONAL ANALYSIS VIA DIFFERENTIAL EQUATIONS

This technique is perhaps the most satisfactory of the three presented, provided the governing differential equations describing the process under investigation are known. By this means, the major difficulty of ensuring that no significant variables have been omitted from the analysis is obviated. Another advantage of the approach is that it shows up the physical significance of dimensionless criteria as ratios of flows, forces or analogous quantities.

For instance, we can recall from Chapter 2 that for isothermal flow of a Newtonian viscous liquid of constant viscosity and density, the Navier–Stokes equation applies:

$$\rho\frac{DU}{Dt} = -\nabla P + \mu\nabla^2 U + \rho g. \tag{3.21}$$

In terms of a rectangular coordinate system, the x coordinate component becomes

$$\underbrace{\rho\frac{\partial U}{\partial t}}_{\text{I}} + \underbrace{\rho\left(U\frac{\partial U}{\partial x} + V\frac{\partial U}{\partial y} + W\frac{\partial U}{\partial z}\right)}_{\text{II}} = \underbrace{-\frac{\partial P}{\partial x}}_{\text{III}} + \underbrace{\mu\left\{\frac{\partial^2 U}{\partial x^2} + \frac{\partial^2 U}{\partial y^2} + \frac{\partial^2 U}{\partial z^2}\right\}}_{\text{IV}} + \underbrace{\rho g_x}_{\text{V}}$$

$$\tag{3.22}$$

Equation (3.22) may be written in the dimensional form:

$$\underbrace{\rho\frac{U}{T}}_{\text{I}} + \underbrace{\rho\frac{U^2}{L}}_{\text{II}} = \underbrace{-\frac{\Delta P}{L}}_{\text{III}} + \underbrace{\frac{\mu U}{L^2}}_{\text{IV}} + \underbrace{(\rho g)}_{\text{V}} \tag{3.23}$$

where we see that I and II are dimensionally equivalent. Consequently, we can make eqn (3.23) dimensionless by dividing throughout by expression V, for example, to obtain *three* independent groupings

$$\frac{U}{gt} \quad \text{or} \quad \frac{U^2}{gL}, \quad \frac{\Delta P}{\rho g L} \quad \text{and} \quad \frac{\mu U}{\rho g L^2},$$

as based on the M, L, T system of primary quantities. Alternatively, we may wish to divide the pressure force term $\overline{\overline{III}}$, by the inertial force term $\overline{\overline{II}}$, to obtain $\Delta P/\rho U^2$, or to take the ratio of inertial forces to viscous forces by dividing term $\overline{II}$ by term $\overline{IV}$ to give us $\rho U L/\mu$.

We are therefore at liberty to write down the following dimensionless function which is representative of the Navier–Stokes equation for incompressible isothermal flow of a viscous liquid.

$$\phi\left(\frac{\rho U L}{\mu}, \frac{U^2}{gL}, \frac{\Delta P}{\rho U^2}\right) = 0, \tag{3.24}$$

$$\phi(\text{Re, Fr, N}) = 0 \tag{3.25}$$

The reason for choosing the particular π groupings will become clearer in the next section. While those written in eqn (3.24) may appear to be arbitrary, they do in fact tend to be more useful and common than other dimensionless arrangements.

The first grouping is known as the Reynolds number, Re, named in honour of the famous pioneer of fluid mechanics. It represents the ratio of inertial and viscous forces. The second is the Froude number, Fr, again named after a pioneer, representing the ratio of inertial and potential forces. Sometimes Fr is represented as $U/\sqrt{gL}$. Finally $\Delta P/\rho U^2$, sometimes known as the Newton number, or force coefficient, represents the ratio of pressure and inertial forces within the flow system.

There are a number of such preferred groups which regularly appear, and therefore well worth the effort of remembering. The reader is referred to Table 3.2 for a selection of those groups which are particularly relevant to metallurgical operations.

3.5. MODELLING AND SIMILARITY CRITERIA

It is now appropriate to discuss the general problem of how to model or characterize metallurgical processes. As a foreword, it is perhaps self-evident, but important to note, that if the same forms of dimensionless differential equations apply to two or more such metallurgical operations, and if an equivalence of dimensionless velocity, temperature, pressure, or concentration (etc.) fields also exists between the two, then either becomes a faithful representation of the other; i.e. one can be used as a model of the other. This is a general statement of the need for *similarity* between a model and a prototype which requires that there be constant ratios between corresponding quantities. The states of similarity would normally include geometrical, mechanical, thermal, or chemical similarity. Mechanical similarity is further subdivided into requirements of static, kinematic and dynamic similarity between a model and its prototype.

(a) Geometric similarity

A basic requirement of any effort to model a full-scale system or prototype is that the two be geometrically similar. Obviously, both must derive from the same generic species — a small feathered bird could certainly not replace its full-scale human equivalent. Two bodies are said to be geometrically similar when for every point in one body, there exists a corresponding point in the other. Such point-to-point geometrical correspondence normally allows a single characteristic linear dimension to be used in representing the sizes of a model and a prototype. For instance, the model ladle shown in Fig. 3.1 can be represented by its diameter D, and compared to its equivalent full-scale counterpart by noting its relative size or scale according to

$$\lambda = \frac{D_m}{D_p} \tag{3.26}$$

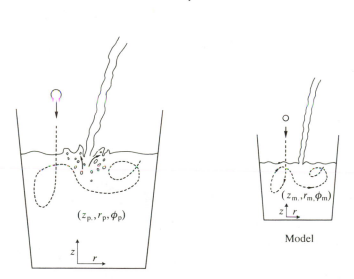

FIG. 3.1. Prototype and model ladle. Demonstration of requirement of kinematic similarity criteria, viz. corresponding particles in a geometrically similar system trace out geometrically similar paths in corresponding intervals of time.

TABLE 3.2. *Some dimensionless groups useful in fluid mechanics and related heat/mass flow phenomena in metallurgical processing operations*

Group	Formula	Nomenclature	Quantities represented	Typical applications
Reynolds, Re	$\dfrac{\rho U L}{\mu}$	L characteristic length dimension of system ρ fluid density μ fluid viscosity	inertial force/viscous force	All fluid flows (e.g. flow through pipes, around submerged bodies, laminar/turbulent transitions)
Froude, Fr	$\dfrac{U^2}{gL}$	L characteristic length dimension of system	inertial force/gravitational force	Free surface flows (e.g., tapping operations, gas driven flows, casting streams)
Friction factor, f	$\dfrac{\tau_0}{\frac{1}{2}\rho U^2}$ $\dfrac{\Delta P_f D}{\frac{1}{2}\rho U^2 L}$	U mean velocity through pipe or conduit ρ density of fluid τ_0 shear stress on wetted perimeter ΔP_f pressure drop due to friction losses over length L of pipe	shear stress/velocity head	Pressure drops in pipes and conduits caused by frictional wall losses, diverging flows (e.g., off-gas handling systems, flows through nozzles and orifices)
Drag coefficient, C_D	$\dfrac{F_D}{\frac{1}{2}\rho U^2 A}$	F_D drag force on object A characteristic area of object $\perp$ to flow U relative velocity	drag force/inertial force	Free settling or rising of submerged bodies in steady translation

Criterion	Formula	Symbols	Symbol meaning	Ratio	Application
Pressure coefficient or Newton, N	$\dfrac{\Delta P}{\frac{1}{2}\rho U^2}$	ΔP	pressure difference between two points in flow of interest	pressure force/inertial force	All fluid flow phenomena which involve pressure gradients
Froude modified, Fr_m	$\dfrac{\rho U^2}{(\rho_L - \rho)gL}$	ρ ρ_L	gas density or solid density liquid density	inertial force/buoyancy force	Fluid behaviour of sub-merged gas–liquid or solid–liquid systems (e.g., jets, air entrainment, buoyant additions, and inclusions)
Added mass coefficient, C_A	$\dfrac{m_a}{m_s}$	m_a m_s	added mass of liquid mass of particle	inertial force of liquid/inertial force of solid	Accelerations of submerged solids in fluids
Mach, M	$\dfrac{U}{U_s}$	U U_s	fluid velocity velocity of sound	local velocity/velocity of sound in fluid	High-speed flow (oxygen steel-making top and bottom blown jets)
Morton, M_0	$\dfrac{g\mu^4}{\rho_L \sigma^3}$	μ	liquid viscosity	(gravitational × viscous)/(surface tension) forces	Velocity of bubbles in liquids; classification of bubble type
Power, P	$\dfrac{P}{\rho\omega^3 L^5}$	P L ω	power input to agitator characteristic dimension of agitator paddle angular speed of rotation	drag force on paddle/inertial force	Power consumption in agitated vessels

TABLE 3.2. (*continued*)

Group	Formula	Nomenclature	Quantities represented	Typical applications
Weber, We	$\dfrac{\rho U_0^2 L}{\sigma}$	σ surface tension	inertial force/surface tension force	Free surface flows, atomization of liquids, bubble/droplet formation, interfacial turbulence
Roughness	ε/L	ε roughness (length-scale)	wall roughness/body length or diameter	
Heat and Mass Transfer Phenomena				
Nusselt, Nu	$\dfrac{hL}{k}$	h heat transfer coefficient L characteristic length k thermal conductivity of fluid	heat flow by convection/heat flow by conduction	A dependent π group giving values for h
Sherwood, Sh	$\dfrac{k_L L}{D}$	k_L mass transfer coefficient D diffusivity of solute	mass flow by convection/mass flow by conduction	A dependent π group giving values for k_L
Biot, Bi	$\dfrac{hL}{k_s}$	h heat transfer coefficient at solid–fluid interface L characteristic length k_s thermal conductivity of solid	fluid's heat conductance/solid's heat conductance	Heat flow between a fluid and solid
Thermal Grashof, Gr_θ	$\dfrac{\rho^2 \beta_\theta g \Delta\theta L^3}{\mu^2}$	β_θ coefficient of thermal volume expansion $\Delta\theta$ temperature difference across thermal boundary layer adjacent to solid μ viscosity of fluid	buoyancy force/viscous force	Free or natural thermal convection phenomena

Name	Formula	Symbols	Physical interpretation	Application
Mass transfer Grashof, Gr_m	$\dfrac{\rho^2 \beta_X g \Delta X L^3}{\mu^2}$	β_X coefficient of density change with concentration ΔX concentration difference of solute across mass boundary layer adjacent to solid	buoyancy force/viscous force	Mass transfer induced natural convection (e.g., dissolution of alloys in liquid metals, scrap melting in steel, electrolysis cells)
Prandtl, Pr	$\dfrac{\nu}{\alpha}$ or $\dfrac{\mu C_p}{k}$	α thermal diffusivity C_p specific heat of fluid μ viscosity of fluid k thermal conductivity	momentum diffusivity/thermal diffusivity	Forced and free thermal convection
Schmidt, Sc	$\dfrac{\nu}{D}$	ν kinematic viscosity D solute diffusivity	momentum diffusivity/molecular diffusivity	Forced and free mass convection
Rayleigh, Ra	$(Gr)(Pr)$	$\dfrac{\rho^2 \beta g \Delta \theta L^3 C_p}{\mu k}$		Natural convection (e.g. Bénard type convection)
Fourier, Fo	$\dfrac{\alpha t}{L^2}$	α thermal diffusivity of solid t time of heating L characteristic length	time elapsed/thermal equilibrium time	Transient heat conduction
Peclet, Pe	$\dfrac{UL}{\alpha}$ or $\dfrac{\rho C_p U L}{k}$	U fluid velocity C_p specific heat α thermal diffusivity	heat flow by bulk motion/conductive heat transfer	Forced convection

We see that the model ladle is a six-tenth-scale replica.

Liquid steel at position r_p, z_p, θ_p would then correspond to liquid in the model at r_m, z_m, θ_m, its equivalent location, according to

$$\theta_m = \theta_p \quad \text{and} \quad \frac{r_m}{r_p} = \frac{z_m}{z_p} = \frac{D_m}{D_p} = \lambda. \tag{3.27}$$

Sometimes, distorted models or model elements are used. The above criteria would then only apply along given axes of the model and prototype. For instance, in modelling heat flow in a sinter strand (Fig. 3.2), one would employ an element or model element of the packed bed operation in which a vertical slice of the coke and ore bed would be built, and combustion experiments performed. Similarly, in scaling up a blast furnace, a distorted model would prevail in which the furnace diameter would increase proportionately more rapidly than its height so as to maintain chemical similarity at the expense of other criteria.

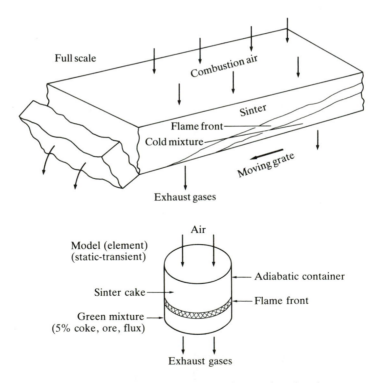

FIG. 3.2.   Schematic of heat flow in a sinter strand operation showing prototype and full scale model element.

(b)   Mechanical similarity

For systems of interest to process metallurgists, mechanical similarity is a common requirement. As previously mentioned, mechanical similarity comprises static similarity, kinematic similarity and dynamic similarity.

(i)   Static similarity

Static similarity is concerned with solid bodies or structures which are subject to loads and is chiefly of interest to mechanical and structural engineers. For instance, in modelling the torques that would be developed in rotating a BOF vessel on its trunnions during tapping procedures, an equivalent low temperature small-scale model in which mercury replaces liquid steel and water replaces liquid slag might be used. For static loads and torques to be quantitatively related, a scale factor for the static body force $(\rho L^3 g)$ would be introduced as a force ratio, i.e.

$$\frac{\text{Torque force}_p}{\text{Torque force}_m} = \frac{(\rho L^3)_p}{(\rho L^3)_m} =$$

$$\frac{\text{weight of prototype} + \text{contents}}{\text{weight of model} + \text{contents}} = C_{SF}. \tag{3.28}$$

Torque characteristics on the model could then be translated into equivalent torques on the full-scale vessel.

(ii)   Kinematic similarity

The requirement of kinematic similarity may be stated as follows: geometrically similar moving systems are kinematically similar when corresponding particles trace out geometrically similar paths in corresponding intervals of time. The concept is illustrated diagramatically in Fig. 3.1. This shows an alloy addition being dropped into a filling ladle of steel during a tapping operation. In order that the low-temperature analogue on the right be a true representation, it is necessary that the model alloy addition (a wooden sphere in water having the same density ratio as the ferro-alloy addition in steel) move through the water such that

$$\frac{r_{m,t}}{r_{p,t}} = \frac{z_{m,t}}{z_{p,t}} = \frac{\phi_{m,t}}{\phi_{p,t}} = \frac{t_m}{t_p} = C_t. \tag{3.29}$$

We shall return to this example later.

(iii) Dynamic similarity

Dynamic similarity is concerned with the forces which accelerate or retard moving masses in dynamic systems. It requires that the corresponding forces acting at corresponding times at corresponding locations in the model should also correspond. It is worthwhile considering the typical forces met in fluid flow systems, viz. inertial, viscous, gravitational and pressure, and their dimensional representation. Let us consider Fig. 3.3, in which we wish to model the flow of molten iron and/or slag out of a blast furnace tap hole. In this model we can use glycerol and mercury to see how a maximum amount of metal can be tapped before slag appears. Evidently, too high an internal blast pressure acting over the hearth region can exhaust pig iron from the hearth sump before liquid metal drainage through the coke bed has been allowed for. Consequently, the furnace will not be completely emptied of metal before slag starts running out, and this can be modelled with a low-temperature analogue. Figure 3.4 shows a differential volume element of fluid (iron) moving up through the furnace tap hole. The dominant differential pressure,

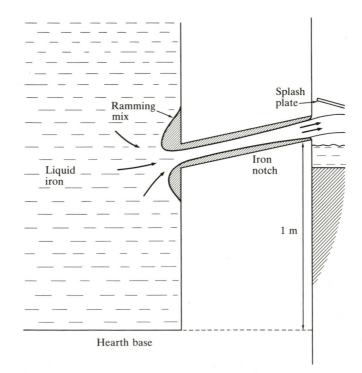

Fig. 3.3.   Iron flow through blast furnace taphole during casting operations.

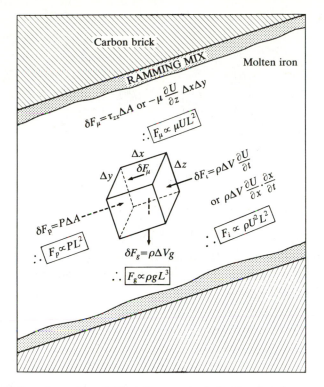

FIG. 3.4. Enlarged section of blast furnace taphole showing forces acting on a differential volume element of liquid iron.

viscous and inertial forces acting on appropriate surfaces of the moving volume element are shown, together with the body force term. We see that:

Pressure force, F_p = pressure × cross-sectional area   or PL^2
Inertial force, F_i = mass × acceleration   or $\rho U^2 L^2$
Gravity force, F_g = weight of fluid   or $\rho g L^3$,
Viscous force, F_μ = tangential shearing stress × area   or $\mu U L$.

Dynamic similarity requires that, at a given point,

$$\frac{F_{p,m}}{F_{p,p}} = \frac{F_{i,m}}{F_{i,p}} = \frac{F_{g,m}}{F_{g,p}} = \frac{F_{\mu,m}}{F_{\mu,p}} = C_F, \tag{3.30}$$

i. e., the polygons of forces for corresponding particles or fluid elements must be geometrically similar between model and full-scale.

As a further consequence, it follows that corresponding ratios of different forces in model and prototype should be equal; i. e., for inertial/frictional force equivalence,

$$\frac{F_{i,m}}{F_{\mu,m}} = \frac{F_{i,p}}{F_{\mu,p}},$$

or

$$\left(\frac{\rho U^2 L^2}{\mu U L}\right)_m = \left(\frac{\rho U^2 L^2}{\mu U L}\right)_p,$$

or

$$Re_m = Re_p. \tag{3.31}$$

For inertial/gravity force equivalence,

$$Fr_m = Fr_p \tag{3.32}$$

Under such conditions, the (normally) dependent grouping, telling us what the overall pressure drop along the flow system is, will again bear a constant ratio such that

$$N_m = N_p. \tag{3.33}$$

These statements are equivalent to saying that if all equivalent π groupings in an unknown functional relationship are numerically equal in both model and prototype, one has achieved corresponding conditions. The practical difficulty is how to reach these objectives through appropriate selection of the model's physical characteristics and properties.

(c)　Thermal similarity

In modelling heat transfer operations, thermally similar systems are those in which corresponding temperature differences bear a constant ratio to one another at corresponding positions. When the systems are moving, kinematic similarity is a pre-requisite to any thermal similarity. Heat transfer rates by conduction, convection and/or radiation to a certain location in the model must bear a fixed ratio to the corresponding rates in the prototype. This situation can be expressed algebraically as:

$$\frac{\dot{Q}_{r,m}}{\dot{Q}_{r,p}} = \frac{\dot{Q}_{k,m}}{\dot{Q}_{k,p}} = \frac{\dot{Q}_{c,m}}{\dot{Q}_{c,p}} = C_Q \tag{3.34}$$

where subscripts r, k and c refer to heat transfer by radiation, conduction, convection respectively, and $\dot{Q}$ represents quantities of heat transferred per unit of time.

(d)   Chemical similarity

For adequate chemical similarity to be achieved between a model and a prototype, it is normal that dynamic and thermal similarity be first satisfied: the former since mass transfer and chemical reaction usually occur by convective and diffusive processes during motion of reacting material through the system, and the latter since chemical kinetics are normally temperature sensitive. Consequently, chemically similar systems may be defined as those in which corresponding concentration differences bear a constant ratio to one another at corresponding points within the geometrical system of interest.

(e)   Successful modelling

The objective of the physical modeller is to achieve geometrical, mechanical, thermal and chemical states of similarity between a model and the full-scale system (prototype). His objective is achievable provided certain criteria are met. These criteria are that ratios of like quantitities (forces, heat flows, mass flows) should correspond on a point-to-point basis within the physical domains of interest. As we have seen, any one of the dimensional analysis techniques presented at the start of this chapter can be used to deduce what corresponding conditions are needed for a model. The differential equation approach is probably the safest.

Having carried out such an analysis, one usually finds that only one ratio of corresponding quantities (or π groups) can be satisfied in practice. It is then up to the skill of the physical modeller to decide which is the most important criteria requiring his attention. Fortunately, some latitude is generally allowable in the less sensitive force ratios, since they will (or may) have little bearing on modelling the phenomenon. The trick is to decide which factors can be ignored in setting up the model.

3.6.   WORKED EXAMPLE: FLUID FLOW AND FERRO-ALLOY HYDRODYNAMICS IN LADLE TAPPING OPERATIONS

Objective

Recommend the essential parameters and experimentation needed to provide a low-temperature simulation of alloy addition practices during furnace tapping operations into a 250-tonne, 4 m-diameter cylindrical ladle. Both fluid flow patterns and associated alloy trajectories are to be simulated, and the analogue is to be a 0.15-scale model. The ferro-alloys range in size from 20–200 mm and are between 0.3 and 0.9 times the density of liquid steel.

Solution

(i)   *Fluid flow behaviour*

Referring to Fig. 3.5, we see that a jet of liquid steel issues from the tap hole of a basic oxygen furnace and plunges a height H into a filling bath of steel. The most important factors governing the resulting flow field are, by reference to the turbulent form of the Navier–Stokes equation, forces due to inertia, to gravity, to friction and possibly to surface tension. We therefore require equivalence of forces between model and prototype, such that

$$\mathrm{Re}_m = \mathrm{Re}_p \quad \text{or} \quad \left(\frac{\rho U L}{\mu}\right)_m = \left(\frac{\rho U L}{\mu}\right)_p$$

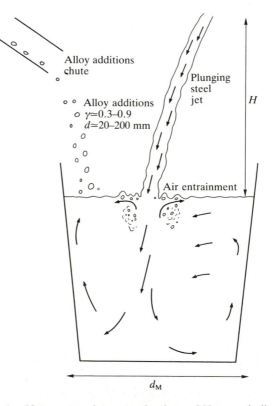

FIG. 3.5.   Schematic of furnace tapping operation into a 250-tonne ladle of steel, mean diameter 4.0 m.

and

$$\mathrm{Fr_m} = \mathrm{Fr}_{p} \quad \text{or} \quad \left(\frac{U^2}{gL}\right)_m = \left(\frac{U^2}{gL}\right)_{p},$$

and perhaps equality of Weber numbers, We $\left(\dfrac{\rho U^2 L}{\sigma}\right)$, according to:

$$\mathrm{We_m} = \mathrm{We_p} \quad \text{or} \quad \left(\frac{\rho U^2 L}{\sigma}\right)_m = \left(\frac{\rho U^2 L}{\sigma}\right)_{p}.$$

Choosing water as the low-temperature analogue fluid, we have the condition that $(\rho/\mu)_{\text{water}} = (\rho/\mu)_{\text{steel}} = 10^6 \, \text{s m}^{-2}$. We see therefore that a water model satisfying Reynolds criteria would need a characteristic velocity (i.e., jet entry velocity into the ladle) such that

$$U_m = \frac{U_p}{\lambda} = \frac{U_p}{0.15}.$$

However, if we are to satisfy the Froudé criteria, we have

$$\left(\frac{U^2}{L}\right)_m = \left(\frac{U^2}{L}\right)_p \quad \text{or} \quad U_m = \lambda^{1/2} \, U_p.$$

Obviously, both conditions cannot be met with $\lambda = 0.15$. The question is which one should we adopt? If we consider the numerical value of the Reynolds number for steel entering a 4 m-diameter ladle at $10 \, \text{m.s}^{-1}$,

$$\mathrm{Re}_p = \frac{7000 \times 10 \times 4}{0.007} = 4 \times 10^7,$$

where $\rho = 7000 \, \text{kg m}^{-3}$, $\mu = 7 \times 10^{-3} \, \text{kg m}^{-1} \text{s}^{-1}$ (or mPa s). This indicates that the ratio of inertial to molecular viscous forces is enormous, and that the flow is therefore dominated by inertial forces rather than by viscous forces.

By contrast, the numerical value of the Froude relationship,

$$\mathrm{Fr}_p = \frac{U_0^2}{gL} = \frac{10^2}{9.81 \times 4} = 2.5,$$

is close to unity, where the characteristic length scale is taken to be the vessel diameter. The jet entry velocity U_0 is governed by the height of drop from the furnace and the ferrostatic head in the emptying BOF furnace. The result indicates that inertial and potential forces are of the same order of magnitude and therefore of equal importance.

Finally, we should consider surface tension forces which appear in the Weber number as a ratio of inertial/surface tension forces. Again,

$$\mathrm{We} = \frac{\rho U_0^2}{\sigma} = \frac{7000 \times 100 \times 4}{1.63} = 1.7 \times 10^6,$$

where the surface tension of steel, at 1600 dynes cm^{-1}, translates to 1.63 kg s^{-2}. Evidently, with inertial forces almost two million times greater than surface tension forces, we can neglect the latter, at least in terms of bulk movement of steel.

One difficulty with this analysis is the phenomenon of air entrainment. Air is entrained into liquid steel surrounding the impact zone of the plunging jet. In fact, significant quantities can be drawn in to (a) create very dirty reoxidized steel, and (b) considerably modify flow patterns within the ladle. However, this entrainment phenomenon is not likely to be surface-tension-related at full-scale, but rather Froudé-based. Definitive studies have yet to be performed. Summarizing these arguments, we conclude that the operation of tapping a furnace into a ladle should be modelled on a Froude basis.

For modellers with lingering doubts, a full-scale aqueous model would be needed: under such conditions, equality of both Fr and Re groups between model and prototype would be satisfied, since both $\lambda^{1/2} = 1$ and $\lambda^{-1} = 1$.

Recommendation

A 0.15-scale model of the tapping operation should be built, wherein lengths L, velocities U_o, entry flow-rates Q_o, and filling times t, should be modelled such that:

$$L_m = \lambda L_p \qquad \therefore \; L_m = 0.15 \times 4.0 \qquad = 0.60\,\text{m}$$
$$V_m = \lambda^3 V_p \qquad \therefore \; V_m = 0.15^3 \times 37.7 \qquad = 0.127\,\text{m}^3$$
$$U_m = \lambda^{1/2} U_p \quad \therefore \; U_m = 0.15^{1/2} \times 10 \qquad = 3.88\,\text{m s}^{-1}$$
$$Q_m = \lambda^{5/2} Q_p \quad \therefore \; Q_m = 0.15^{5/2} \times 0.11 \quad = 9.6 \times 10^{-4}\,\text{m}^3\,\text{s}^{-1}$$
$$t_m = \lambda^{1/2} t_p \quad \therefore \; t_m = 0.15^{1/2} \times 360 \qquad = 140\,\text{s}$$

(ii)   Ferro-alloy hydrodynamics

To model particle trajectories of alloy additions in liquid steel, a little background information on heat transfer to alloy additions is needed. One finds that all alloys dropped into steel baths, no matter what their actual thermal properties and melting points, freeze a shell of steel around them. This leads to the phenomenon of a molten ferro-alloy core contained within a frozen steel shell, as shown in Fig. 3.6. Consequently, the hydrodynamic simulation involves treating the motion of a buoyant lump (a sphere, for simplicity) of relatively constant diameter as it moves through a swirling flow of liquid steel.

We will use the *differential equation technique* to determine what important parameters need to be considered, and therefore how we should set about modelling particle behaviour.

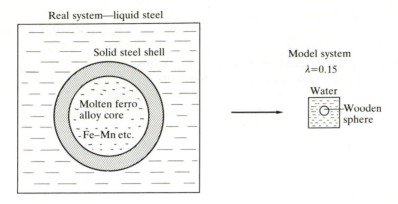

FIG. 3.6. Schematic of thermal phenomena associated with alloy additions and possible hydrodynamic simulation by a 0.15-scale aqueous model.

Applying Newton's second law of motion, therefore, which states that the sum of the applied forces acting on a body is equal to the product of the body's mass and acceleration, we have, taking z positive downwards (see Fig. 3.7),

$$m_s \frac{dU}{dt} = F_B + F_A + F_D + F_g$$

where

$$F_g = \rho_s V_s g,$$

$$F_B = -\rho V_s g,$$

$$F_D = -C_D \times \tfrac{1}{2} \rho \, U_r |U_r| \left(\frac{\pi d_s^2}{4} \right)$$

$$F_A = +C_A \rho V_s \frac{dU_p}{dt}$$

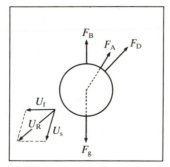

FIG. 3.7. Forces acting on a sphere moving through a liquid in motion, together with associated velocity vectors.

and U_r is the relative velocity.

The only force worthy of extra comment is the added mass force F_A. This force is a measure of the resistance experienced by an accelerating or decelerating submerged body during its translation through a fluid. Evidently, acceleration of a body through a high-density liquid such as molten steel is considerably more difficult than the same body's acceleration through a gas, for instance. For a sphere, the coefficient of proportionality C_A, the added mass coefficient, is 0.5, and

$$F_A = 0.5 \rho V_s \frac{dU}{dt}.$$

Substituting for the various forces, and collecting like terms, we can write the following differential equations:

$$(M_s + M_A)\frac{dU}{dt} = F_g + F_B + F_D$$

or

$$(\rho_s + C_A \rho)V_s \frac{dU_p}{dt} = \rho_s g V_s - \rho g V_s - \tfrac{1}{2}\rho |U_r| U_r C_D \frac{\pi d_s^2}{4}$$

or

$$\frac{dU}{dt} = \frac{(\rho_s - \rho)g}{(\rho_s + C_A \rho)} - \frac{3\rho |U_r| U_r C_D}{4 d_s(\rho_s + C_A \rho)}.$$

Defining $\gamma = \rho_s / \rho$,

$$\frac{dU}{dt} = \frac{(\gamma - 1)g}{(\gamma + C_A)} - \frac{3|U_r| U_r C_D}{4 d_s(\gamma + C_A)},$$

or, simplifying, and reducing to dimensionless form,

$$\frac{d_s}{U_r^2}\frac{dU}{dt} = \frac{(\gamma - 1)g d_s}{(\gamma + C_A)U_r^2} - \frac{3 C_D}{4(\gamma + C_A)}.$$

For the ferro-alloy in liquid steel to be modelled by a particle travelling through water, this general equation must apply to both systems and there must be a correspondence of like quantities or forces. Denoting the real system by subscript 1, and the model by subscript 2 (and dropping the subscript for solid, we require a correspondence of quantities such that

$$d_2 = K_d d_1$$
$$U_{r,2} = K_{U_r} U_{r,1}$$
$$\gamma_2 = K_\gamma \gamma_1,$$
$$C_{A,2} = K_A C_{A,1}$$
$$C_{D,2} = K_C C_{D,1}$$
$$t_2 = K_t t_1.$$

We can write for System 1:

$$\frac{d_1}{U_{r,1}^2}\frac{dU_{p,1}}{dt_1}=\frac{(\gamma_1-1)}{(\gamma_1+C_{A,1})}\frac{gd_1}{U_{r,1}^2}-\frac{3C_{D1}}{4(\gamma_1+C_{A,1})},$$

and for System 2 in terms of 1:

$$\frac{K_d}{K_{U_r}^2 K_t}\frac{d_1}{U_{r,1}^2}\frac{dU_{p,1}}{dt_1}=\frac{(K_\gamma\gamma_1-1)}{(K_\gamma\gamma_1+K_A C_{A,1})}\frac{gK_d d_{s,1}}{K_{U_r}^2 U_{r,1}^2}-\frac{3K_C C_{D,1}}{4(K_\gamma\gamma_1+K_A C_{A,1})}.$$

For these two differential equations to be numerically identical as required for a perfect model, we need $K_A=1$, $K_C=1$, $K_\gamma=1$, $K_t=1$, and $K_d/K_{U_r}^2=1$. Consequently, provided the added mass coefficient is identical for both large and small spheres, the drag coefficient C_D is numerically equivalent (true provided Reynolds number $\sim 10^4$ or more), the solid/liquid density ratio is the same, and Froude modelling is adopted, $(K_d/K_{U_r}^2)=\{(d_2/d_1)/(U_2^2/U_1^2)\}=1$, we can achieve a perfect correspondence between model and prototype, in which times are equivalent, as well as particle trajectories.

Recommendation

A 0.15-scale model requires that the particles entering the aqueous analogue be scaled, together with entry velocities and density ratios. Hence,

$$d_m=\lambda D_p \quad d_m=(0.15\times 20)\text{ to }(0.15\times 200)\text{ mm}=3\text{–}30\text{ mm},$$

$$\gamma_m=\gamma_p,\quad \rho_m=0.3\rho_{H_2O}\text{ to }0.9\rho_{H_2O}=300\text{–}900\text{ kg m}^{-3},$$

and entry velocities should be scaled according to Froude criteria, as for the plunging jet.

As a final note (Guthrie *et al.* 1975), one can treat the impact and penetration of a sphere with a liquid as a case of momentum conservation, such that $M_s U_0=(M_s+M_a)U_0'$ where M_s is the mass of the sphere, M_a is the added mass, U_0 is the pre-impact velocity, and U_0' the post-interface velocity.

Alternative approach

If we had used the Buckingham π theorem, and again considered the acceleration/deceleration of the alloy addition (dependent variable) to be governed by the remaining listed independent variables according to

$$f(a_s, U_r, \rho, \rho_s, g, d_s, \mu)=0,$$

we would have, taking mass, length, and time as primary quantities, and choosing U_r, ρ and D to assemble our π groups:

$$\frac{d_s a_s}{U_r^2}=f\left(\frac{\rho_s}{\rho}, \frac{U_r^2}{gd_s}, \frac{\rho U_r d_s}{\mu}\right).$$

This result is essentially the same as before, but with a little less information on the structure of the interrelationships.

3.7.   REDUCING THE NUMBER OF SIMILARITY CRITERIA

There is an obvious advantage if the criteria of similarity can be legitimately reduced in number. There are several possibilities:

1. A particular parameter, or quantity, can be dropped on physical grounds.
2. Parameters may be combined into a single parameter.
3. The number of basic or primary units may be increased.

Listing a few examples:

(a) For creeping flow around a sphere, the drag force F depends on viscous rather than inertial forces, so that the liquid density parameter can be dropped, giving $F_D = f(\mu, d_s, U)$ or $F_D/\mu U d_s = f(0) = 3\pi$.
(b) For gases bubbling into liquids at very low flow rates, bubble sizes are governed by buoyancy and surface tension forces. Inertial forces and viscous forces can therefore be eliminated. Similarly, at high flow rates they are inertially governed.
(c) For a natural-convection-dominated phenomenon, the density difference between different elements of fluid(s), together with the coefficient of volumetric expansion and gravity, can all be lumped as a single buoyancy force parameter, e.g. $\beta g \Delta\theta$, $\beta g \Delta X$, or $\Delta\rho g$.
(d) For heat flow problems in which mechanical work and thermal energy are not interlinked, we can use five primary Q's (M, L, T, θ, Q) rather than four (M, L, T, θ) or three (M, L, T). However, when these quantities are linked, then Joule's mechanical equivalent of heat must be included as a dimensionless independent constant when using the M, L, T, θ, Q system. Under these circumstances, then, we do not gain an advantage in reducing the number of π groups.

(a)   Example: planar turbulent jets

As an example of combining parameters, consider incompressible flow of a planar turbulent jet into still surroundings. We may expect that U, the mean velocity (time-averaged) at any downstream location x, will depend on the exit velocity U_0 of fluid leaving the jet, the nozzle's width b, the distance downstream x, the centreline displacement y, and the density and viscosity of the fluid:

$$U = f(U_0, \rho, \mu, x, y, b).$$

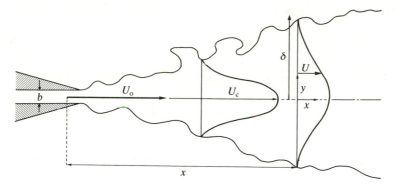

FIG. 3.8.   Fluid exiting from a nozzle into stagnant surroundings in the form of a submerged planar jet. Note coordinate system and velocity profiles across jet.

Consequently, by inspection, and referring to Fig. 3.8, we see that

$$\frac{U}{U_0}=f\left(\frac{x}{b},\frac{y}{b},\frac{\rho U_0 x}{\mu\cdot}\right).$$

We may first drop the fluid's viscosity from the analysis on the grounds that viscous effects are confined to the smallest eddies in the jet. Such eddies have little effect on the transport of momentum by the larger turbulent eddies associated with moderate and high Reynolds number flows. Consequently, dropping Re_x,

$$\frac{U}{U_0}=f\left(\frac{x}{b},\frac{y}{b}\right).$$

One can go beyond this result by considering another relationship linking U and U_0, as embodied in the principle of momentum conservation. Thus, the principle of momentum requires that the momentum input $\dot{M}_0$ ($=\rho U_0^2 b$), of the jet should conserved downstream since there are no net external forces acting on the system. Consequently, the flow downstream is determined by $\dot{M}_0$, with a small change in origin from the jet's exit (Fig. 3.8). This allows us to combine U_0 and b, reducing the number of π groups to two:

$$U=f\{\dot{M}_0, x, y, \rho),$$

giving

$$\frac{\rho U^2 x}{\dot{M}_0}=f\left(\frac{y}{x}\right).$$

Clearly, where $y=0$, $f(y/x)=f(0)=$ constant, and

$$\frac{\rho U_c^2 x}{\dot{M}_0}=f(0)\quad\text{or}\quad U_c\propto\frac{1}{\sqrt{x}}.$$

At the edge of the jet boundary, $y = \delta$, $U = 0$, so we also have

$$0 = f\left(\frac{\delta}{x}\right) \quad \text{or} \quad \delta \propto x.$$

Finally, at distance x from the jet source, $U = f(\delta, y, U_c)$ so that

$$\frac{U}{U_c} = f\left(\frac{y}{\delta}\right).$$

This analysis shows the benefits of reducing the number of π groupings. It shows that the slot jet's centreline velocity should drop off inversely to the square root of the distance from the origin, that the jet's width increases in proportion to distance from its source, and that velocity profiles are similar.

This example concludes remarks on techniques for dimensional analysis and modelling. Some of the more useful correlations derived for convective heat and mass transfer rates in which the dimensional groups given in Table 3.1 appear are reserved for Chapter 5. The following section provides some typical metallurgical examples.

3.8.   EXERCISES IN DIMENSIONAL ANALYSIS AND MODELLING

(1)   Hydrodynamics of non-metallic inclusions in liquid metals

The settling (or rising) velocity of a spherical particle in a fluid depends upon the weight of the particle *in* the fluid, the diameter of the particle and the viscosity and density of the fluid. By dimensional analysis, show how the weight of the particle in the fluid must be related to its rising velocity.

Model experiments to ascertain the rising velocity of spherical slag particles in molten copper were carried out by allowing geometrically similarly shaped particles of bakelite to rise in bromoform. One result showed that a particle of bakelite 1.3 mm in diameter rises at a speed of 5.4 cm s^{-1}. Estimate the rising speed and the diameter of a slag particle that can be predicted from this result.

Data:

	Density (g cm^{-3})	Viscosity (P)
Bromoform	1.55	0.011
Molten copper	8.25	0.033
Bakelite	1.27	—
Slag particle	2.35	—

(2) Soaking pits: models

A model of a soaking pit (or reheat) furnace has been proposed so that gas flow patterns generated in a new one-way fired pit design can be optimized for uniform convective heat flow to the reheating ingots.

Develop an expression showing how the gas velocity at any given point within the system is related to the length L, and width W, of the pit, the combusted gas (i.e. burnt oxygen fuel mixture) flow rate G_0, and the mean density and viscosity of these combusted gases.

In order to model the system at the one-fifth scale, it is proposed that water, rather than cold air, be used as the fluid medium. Comment on this proposal and recommend the inlet gas and water flows and other dimensions that would be required to simulate the following set of conditions expected for the prototype.

Data:

Viscosity of air at 700°C	$4.3 \times 10^{-5} \, \mathrm{kg\,m^{-1}\,s^{-1}}$
Density of air at 700°C	$0.36 \, \mathrm{kg\,m^{-3}}$
Viscosity of water at 20°C	$1.01 \times 10^{-3} \, \mathrm{kg\,m^{-1}\,s^{-1}}$
Density of water at 20°C	$1000 \, \mathrm{kg\,m^{-3}}$
Inlet combusted gas + air flow	$300 \, \mathrm{m^3\,s^{-1}}$
Length of soaking pit	15 m
Width of soaking pit	3 m
Height of soaking pit	4 m
Height of ingots	2.5 m
Width of ingots	1.2 m
Number of ingots	24

(3) Oil atomization in blast furnace operations

In order to reduce the coke requirements per NTHM (net tonne hot metal) in modern blast furnace operations and to improve operational characteristics, Bunker C Oil may be used as a partial replacement for coke. The oil is commonly fed into the blowpipe region some 0.5 m before the end of the tuyère's nose, via a simple oil lance (see figure). Assuming a liquid jet of oil is fed into the high velocity air blast, estimate on the basis of dimensional analysis, and the data below, the equilibrium size of oil droplets that should result (i.e. diameter). On the basis of these calculations would you recommend primary atomization of the fuel oil exiting the lance?

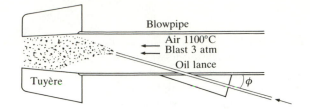

Data:

Air blast temperature	1100 °C
Air blast pressure	3×10^5 Pa
Air blast flow rate	$4 \text{ m}^3 \text{ s}^{-1}$
Air density	0.9 kg m^{-3}
Oil density	970 kg m^{-3}
Oil viscosity	$0.1 \text{ kg m}^{-1} \text{ s}^{-1}$
Oil surface tension	$2.2 \times 10^{-5} \text{ kg s}^{-2}$
Tuyère/blowpipe I.D.	150 mm
Oil flow rate	$1.5 \times 10^{-4} \text{ m}^3 \text{ s}^{-1}$

(4) Tight coil batch annealing operations

It is necessary to increase the rate of heating and cooling of steel coils in batch annealing operations so as to obtain a 10 per cent productivity increase. This can be achieved by a 13 per cent increase in the flow rate of recirculating gases $(5\% \text{H}_2 + 95\% \text{N}_2)$ within the inner hood. Using methods of dimensional analysis, obtain a relationship between gas flow and system pressure drop.

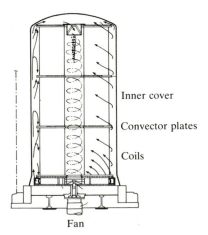

Use this to estimate the percentage increase in fan input power requirements needed to obtain the 13 per cent flow increase. Assume incompressible flow and that the efficiency for the recirculating fan remains constant.

(5) Thermal stresses developed in continuously cast ingots: Example

The thermal stress σ developed in the centre of a continuously cast cooling ingot is deemed to be a function of its shape and the following independent variables:

- E   modulus of elasticity;
- β coefficient of linear thermal expansion;
- $\Delta\theta$ initial temperature difference between cooling medium and melting point of metal;
- L   characteristic linear dimension of ingot;
- h   heat transfer coefficient of cooling medium at surface of ingot;
- k   thermal conductivity of ingot;
- C   thermal capacity of ingot;
- ρ density of metal;
- t   time from start of cooling (i.e. distance from mould/downward velocity of ingot);
- LH   Latent heat of metal.

Using Buckingham's π theorem, or otherwise, show by dimensional analysis what form the unknown physical equation relating σ to the pertinent variables must take.

   In an industrial process, a 6-inch diameter steel ingot was found to develop an acceptable maximum stress after 4 min. What process variables would you adjust to ensure that this stress would not be exceeded in a similar steel, whose thermal conductivity was 9/10 that of the steel investigated? In particular, what diameter must the mould have, and at what time will maximum thermal stress be developed?

(6) Regenerator stoves: scale-up

A regenerator filled with hot metal spheres is used to heat incoming cold gas. Using the π theorem, or otherwise, derive the form of the expression relating $\Delta\theta_t$ (the total rise in temperature of the gas as it passes the regenerator at time t) to the following independent variables.

- $\Delta\theta_0$ initial temperature difference between gas and spheres (time zero);
- t   time after the gas started to flow;
- C_g *volume* specific heat of gas;
- C_s volume specific heat of metal spheres;

U_0 velocity of gas at point of entry into regenerator;
k thermal conductivity of the gas;
μ_0 viscosity of the gas;
ρ density of gas;
L characteristic linear dimension of regenerator (width);
d diameter of metal spheres.

Do you think that the variables considered above are sufficient to describe the system if the metal spheres in the regenerator are replaced by ceramic spheres?

Comment on the possibility of building a geometrically similar regenerator of double the linear dimensions, to raise the temperature of twice the quantity of gas (i.e., $U_2 L_2^2 = 2U_1 L_1^2$) by the same amount $\Delta\theta$.

Calculate the relative times taken in the two regenerators for the same $\Delta\theta$ to be recorded.

(7)   Continuous copper-making: Noranda process

You are required to initiate a study of the important process variables in continuous copper-making. By the method of dimensional analysis, suggest the likely form of the unknown physical relationship relating the rate of oxidation of FeS to the geometry of the system, the diameter of the tuyères, the angle of tuyères, and so on. Try to simplify the problem, where possible, by reasonable assumptions concerning the mechanisms of oxidation, and suggest where model experiments could usefully be employed.

(8)   Killing steel with aluminium bullets

In an effort to improve the recovery of aluminium in ladle steel deoxidation processes, a Japanese steelworks company considered the possibility of firing bullet-shaped pieces into the melt, in the hope that penetration depths and immersion times would be sufficient to enable subsurface melting to occur. It was hoped that this would diminish Al/slag and Al/air interactions resulting from the buoyancy forces, causing such additions to float before melting into the steel. Some preliminary experiments were carried out by firing wooden bullets vertically into water using an air gun. It was found that an immersion time of 5 s could be achieved by firing 3 cm diameter wooden bullets into water at a velocity of 30 m s^{-1}. Based on this one experiment, provide a set of predictions for an aluminium bullet entering steel (i.e. specify size, required entry velocity and resulting immersion time) using the methods of dimensional analysis; i.e. develop a relationship between t (the immersion time and dependent variable), and the relevant independent variables for which data is provided below.

	Viscosity $(\mathrm{kg\,m^{-1}\,s^{-1}})$	Density $(\mathrm{kg\,m^{-3}})$
Water	1×10^{-3}	1000
Steel	6.7×10^{-3}	7000

Wooden bullet	{	Diameter of cylindrical section	$= 30$ mm
		Density	$= 380 \mathrm{\,kg\,m^{-3}}$
		Entry velocity	$= 30 \mathrm{\,m\,s^{-1}}$

Aluminium bullet	Density	$= 2700 \mathrm{\,kg\,m^{-3}}$

(9)   Thermal model of blast-furnace hearth

Williams and Burton (1955) constructed a 1/55-scale model of a blast furnace. They used exactly the same raw materials as the prototype in constructing the furnace hearth region. Their objective was to determine the way the temperatures at any location would change during the course of a typical furnace campaign. In one model experiment, the temperature of a probe climbed from 0 °C to 6 °C at the location marked, after a period of 10 h. In the tests, the carbon powder (simulating the molten pig iron in the hearth) was maintained at 100 °C by passing steam through the heater container (i.e. at $t > 0$, $\theta_\mathrm{H} = 100$ °C).

Using the Buckingham's π method of dimensional analysis, and recognizing the following independent variables as being of key importance, show the form of expression relating $\theta_{x,t}$ to:

(1) temperature of molten pig iron/or carbon powder, θ_H;

(2) ambient temperature of surrounding earth foundation and air, θ_AMB;

(3) densities of carbon, fire-brick and concrete layers, ρ;

(4) heat capacities of fire-brick and concrete layers, C;

(5) thermal conductivities of fire-brick and concrete layers, k;

(6) thickness of fire-brick and concrete layers, l;

(7) time from start of campaign (or experiment), t.

Calculate the corresponding time during the furnace campaign which is equivalent to 10 h of model time. What temperature would be observed at that time in the furnace hearth at the geometrically equivalent location?

Data:

	Furnace	Model
Materials	same	same
Scale	1	1/55
Iron temperature θ_H	2740°F	100°C
Ambient temperature, θ_{AMB}	40°F	0°C

3.9. BATCH REACTORS AND CONTINUOUS FLOW SYSTEMS

(a) Introduction

The importance of good reactor design for efficient processing operations cannot be over-emphasized. For instance, only recently has it been realized by steel-makers just how inefficient top-blown oxygen steel-making furnaces are with respect to adequate stirring of the underlying steel bath for carbon removal at low concentrations ($\sim 0.05\%$ C). To avoid unnecessarily high yield losses of iron through over-oxidation during the final stages of steel decarburation, retrofit solutions involving submerged gas injectors placed in the bottom of the furnace are now common, twenty years after the process was first developed!

Such vessels are known as *batch reactors* for obvious reasons. The other class of chemical or metallurgical processing vessels commonly met with are the continuous reactors. Perhaps the most impressive of all such reactors are today's blast furnaces. These are capable of producing up to 5000 tonnes of iron daily. Again, as recently as 1975, the importance of layering and proper burden distribution on gas flow patterns and associated chemical reactions had not been appreciated. Figure 3.9 provides a schematic of some typical metallurgical reactor vessels of the batch and continuous type: items shown are ladles, torpedo cars and oxygen furnaces, (batch); and clarifiers, pachuka tanks and a twin strand billet caster (continuous).

(b) Characterization of reactors

Apart from classifying reactor vessels into batch and continuous categories, subdivisions of continuous reactors are made according to the flow characteristics exhibited by these vessels. Referring to Fig. 3.10, two extreme flow possibilities are considered together with an intermediate case. The first type depicts a flow in which each element of the entering fluid moves through the vessel with a uniform velocity, and exits having spent the same time inside.

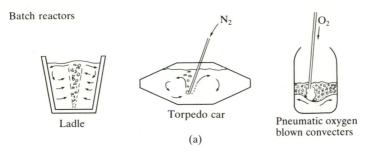

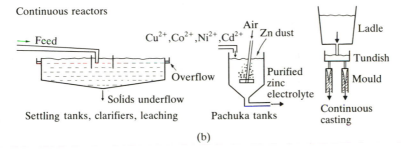

FIG. 3.9. Some typical metallurgical reactors of the batch (a) and continuous (b) variety.

Such a reactor is known as a *plug flow, piston flow* or *linear* reactor.

At the opposite end of the spectrum of possible flow characteristics is the *well-mixed* or *backmix* reactor. Here all fluid elements entering the reactor are mixed instantaneously with fluid already inside. Consequently, some of the entering elements immediately exit while others can remain within the system longer.

Evidently, if V is the volume of the reactor vessel, and v is the volumetric flow rate of fluid into it, then for incompressible steady-state flow conditions, $\bar{t}$, the average time spent by fluid within a continuous reactor is logically defined as

$$\bar{t} = \frac{V}{v};$$
(3.35)

$\bar{t}$ is also known as the nominal holding time of fluid within the reactor.

Referring to the last flow situation that might be expected of a real reactor vessel (Fig. 3.10(c)), one sees that much of the flow entering the vessel will *short-circuit* in a plug flow mode, going straight towards the exit. Some of the remainder will be caught up in a recirculatory mode of flow space more characteristic of perhaps a well-mixed reactor. In the corners of the vessel, the

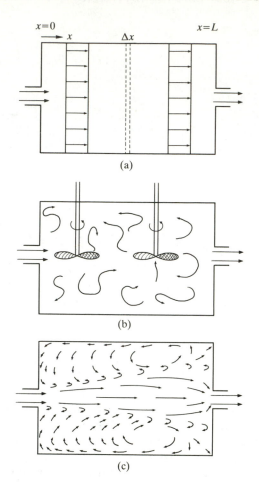

FIG. 3.10.   Continuous flow reactors exhibiting (a) plug flow, (b) well-mixed and (c) intermediate characteristics.

fluid elements will hardly be moving, so fluid components in those regions are essentially 'dead', and their residence times are very prolonged.

Using these analogies, one can characterize an actual reactor in terms of how well it fits well-mixed or piston flow criteria, by the proposition that the reactor's volume can be split up into three component parts:

$$V \equiv V_m + V_p + V_d \qquad (3.36)$$

where V_m is the well-mixed volume within vessel, V_p is the plug flow volume within reactor, and V_d is the dead volume of vessel. The definition of dead volume is again rather arbitrary, and can be regarded as that volume of fluid which stays within the reactor longer than twice the nominal residence time.

In order to characterize a reactor for design or diagnostic purposes by this approach, we need to know more about the incoming fluid's *residence time distribution* characteristics.

(c) Tracer additions

This information has to be obtained experimentally in one of two ways. In the first case, we can introduce a continuous tracer source or addition into the inlet fluid (e.g., a solution of potassium permanganate added to water entering the model reactor), and then observe how its concentration varies in the outlet flow. The second way is to add a pulse tracer addition (i.e., an instantaneous point or line source), and again mark its output concentration from the vessel. By studying the output characteristics of the tracer concentration, one can then diagnose the reactor's mode of behaviour and likely chemical efficiency.

Case 1.   Continuous tracer addition into plug flow reactor

Consider an experiment in which water, flowing through a reactor, is continuously fed at $t \geqslant 0$ with a potassium permanganate addition, such that the inlet feed then contains a permangantate concentration equal to C_0 (kg m^{-3}). Given the behaviour of a plug flow reactor, the exit water will remain clear until the nominal residence time is reached. At that point, there will be a sudden increase in the outlet concentration of dye from zero to C_0. We may plot the result in dimensionless form as in Fig. 3.11(a), where $\mathscr{F}$, the fractional concentration of permanganate ($\mathscr{F} = C_t/C_0$) is plotted against the fractional holding time ratio $t/\bar{t}$, or θ.

Case 2. Continuous addition to well-mixed reactor

Consider now the continuous introduction of tracer into a well-mixed reactor. We can carry out a mass balance, by stating:

Input of tracer	−	output of tracer	=	rate of accumulation of tracer

$$v C_0 \quad - \quad v C_t \quad = \quad \frac{\mathrm{d}}{\mathrm{d}t} V C_t. \qquad (3.37)$$

Note that C_t represents the instantaneous concentration of tracer material inside the reactor *and* in the exit. This follows from the idea that for a well-mixed reactor with $D_E \simeq \infty$, no concentration gradients can exist. Dividing

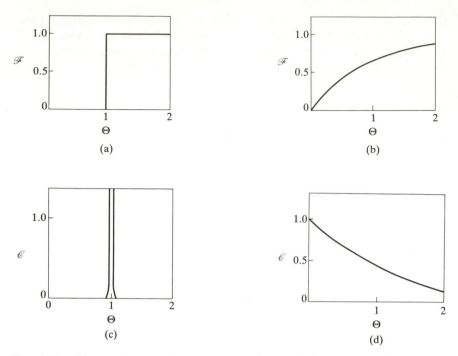

FIG. 3.11.   Diagnostic curves (tracer response characteristics) of plug and well-mixed reactors for continuous and pulse tracer additions to input feed.

by C_0,

$$\frac{v}{V}(1 - \mathscr{F}) = \frac{d\mathscr{F}}{dt}. \tag{3.38}$$

Integrating with boundary conditions that $\mathscr{F} = 0$ at $t = 0$ and $\mathscr{F} = \mathscr{F}$ at $t = t$;

$$\left[-\ln(1 - \mathscr{F}) \right]_{\mathscr{F}=0}^{\mathscr{F}=\mathscr{F}} = \frac{vt}{V}, \tag{3.39}$$

giving

$$\mathscr{F} = 1 - e^{-\Theta}. \tag{3.40}$$

The result shows that the output concentration of tracer from a well-mixed reactor would immediately start to rise, following its initial introduction, and would slowly approach unity in an asymptotic manner. The diagnostic $\mathscr{F}$ curve is shown in Fig. 3.11(b). The advantage of presenting the data in dimensionless form is obvious in that the results are thereby generalized rather than being specific to a given vessel's dimensions and input feed conditions.

Case 3.   Pulse addition to plug flow reactor

In industrial practice, it is usually easier to make a pulse, rather than continuous, addition for making diagnostic tests on reactor performance. Radioactive gold, or copper or nickel additions could be used for liquid steel, or fluorescent dye for aqueous systems.

Again, in order to express the exit concentration of tracer from the reactor in a dimensionless form, we can present the results in terms of dimensionless concentration $\mathscr{C}$ curves, where $\mathscr{C} = C_{\text{exit},t}/\bar{C}$. $\bar{C}$ is defined as that bulk concentration of tracer that would be observed in the vessel, if an amount equal to mass, δM, of tracer were uniformly mixed into a volume, V, of fluid corresponding to the fluid volume of the reactor, i.e. $\bar{C} = \delta M/V$.

For the plug flow reactor then, we would expect the concentration to remain zero at the reactor's outlet until the nominal residence time were reached. At this point, the tracer material would come out at the same concentration as it entered, (assuming equal inflow and outflow velocities). Referring to Fig. 3.11(c) therefore, where $\mathscr{C} = C/(\delta M/V)$ or $CV/\delta M$, we see that $\mathscr{C} \to \infty$ at $\theta = 1$ for a line source addition.

Case 4. Pulse addition to well-mixed reactor

Considering our final case of a pulse addition to a well-mixed continuous reactor, we can carry out a mass balance as before. Thus, for a tracer addition of mass, δM to the reactor at $t = 0$, the situation for all times beyond the instant of injection is:

$$\text{Input} - \text{output} \qquad = \text{rate of accumulation},$$

$$v(0) - vC \qquad = \frac{d}{dt}VC. \tag{3.41}$$

Noting that the exit concentration at time t is equal to C, the reactor's concentration at time t (which is everywhere uniform) is

$$\frac{dC}{C} = -\frac{v}{V}dt, \tag{3.42}$$

which may be integrated with the boundary condition $C = \delta M/V$ at $t = 0$, to yield

$$\frac{C}{\delta M/V} = e^{-\Theta} \quad \text{or} \quad \mathscr{C} = e^{-\Theta}. \tag{3.43}$$

Case 5. Intermediate situations

Clearly, ideal plug flow or well-mixed reactors are rarely encountered in practice. Very often, there will be intermixing of material in a plug flow reactor (i.e., some back-mixing) as a result of turbulent eddy diffusion processes. For instance, considering Fig. 3.12, which depicts a tubular reactor of length L, through which fluid is flowing at a volumetric flow rate v, some intermixing between marked and unmarked fluid is to be expected as a result of random turbulent fluctuations in the mean velocity, $\bar{U}$.

For an observer moving at a steady velocity $\bar{U}$ through the reactor, at $x = 0$, the moving origin (Fig. 3.12), the equivalent of Fick's second law for unsteady state molecular diffusion applies:

$$\frac{\partial C}{\partial t} = D_E \frac{\partial^2 C}{\partial x^2}, \qquad (3.44)$$

where D_E is an eddy diffusivity (units of $L^2 T^{-1}$). Chapter 4 gives further information.

The initial conditions are

$$C = 0, \quad x > 0,$$
$$C = C_0, \quad x < 0;$$

The boundary conditions are

$$C = C_0, \quad x = -\infty, t = t,$$
$$C = 0, \quad x = +\infty, t = t,$$

and the particular solution is (Chapter 4)

$$C_{x,t} = \tfrac{1}{2} C_0 \left(1 - \operatorname{erf} \frac{x}{\sqrt{4 D_E t}} \right). \qquad (3.45)$$

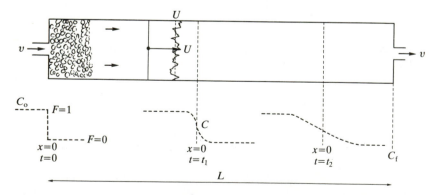

FIG. 3.12.   Plug flow reactor with back-mixing characteristics.

However, we are only interested in the concentration of tracer material appearing in the exit from the reactor, i.e. at L. Clearly the distance x_L between the moving origin and the exit of the reactor diminishes with time according to

$$x_L' = L - \bar{U}t, \quad \text{where} \quad \bar{U} \equiv vL/V \equiv L/\bar{t}.$$

Consequently, we have the result

$$C_{L,t} = \tfrac{1}{2}C_0\left(1 - \text{erf}\frac{L(1-\Theta)}{\sqrt{4D_E t}}\right), \tag{3.46}$$

which can be rearranged to yield

$$\mathscr{F}_{L,t} = \tfrac{1}{2}\left(1 - \text{erf}\frac{(1-\Theta)}{\sqrt{\dfrac{4D_E}{L\bar{U}}}}\right). \tag{3.47}$$

Here D_E/UL is known as the dispersion number, and is the inverse of the well known Peclet number. An equivalent analysis can be carried out for the condition of a line source tracer addition to the feed. The results of these analyses are therefore summarized in the $\mathscr{F}$ and $\mathscr{C}$ diagrams shown in Figs 3.13 (a,b,c).

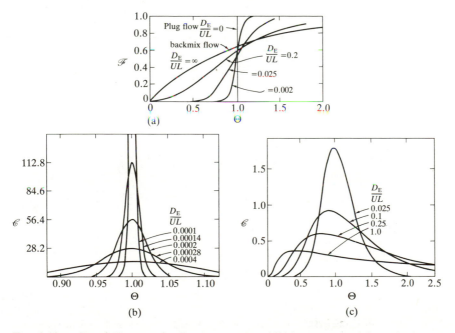

FIG. 3.13. $\mathscr{F}$ and $\mathscr{C}$ curves for linear reactors exhibiting some backflow characterized by D_E/UL.

3.10. DISPERSION NUMBERS

The type of analysis just carried out is of particular relevance to reactors in which longitudinal/width aspect ratios are high. It would, for instance, be relevant to the leaching of a soluble component from a packed bed of solid particles contained in a long tubular reactor, or to the flow of gases through a packed bed to effect oxidation, desulphurization or deoxidation reactions.

Figures 3.14(a) and (b) present observed values of *local* dispersion numbers summarized by Levenspiel (1967) for gas and liquid flow through tubes. For flow through pipes, the characteristic length is taken as the tube diameter d_t,

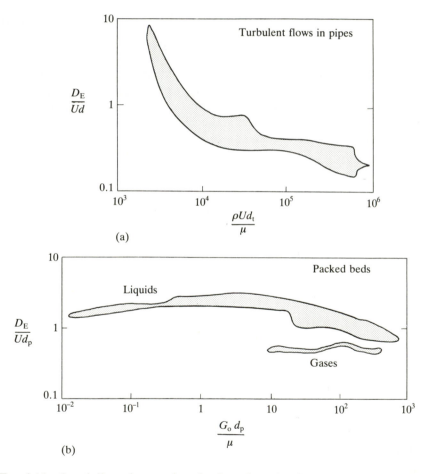

FIG. 3.14.   Local dispersion numbers for flow through tubes and packed beds as a function of local Reynolds numbers (d_t is the tube diameter and d_p the particle diameter).

while for flow through packed beds, the particle diameter d_p is taken as the characteristic length. Provided both are independent of reactor length, the reactor dispersion numbers are then simply estimated according to

pipe flow:
$$\frac{D_E}{UL} = \frac{D_E}{Ud_t}\frac{d_t}{L},$$

Packed beds:
$$\frac{D_E}{UL} = \frac{D_E}{Ud_p}\frac{d_p}{L},$$

at the relevant Reynolds number of interest to the designers.

3.11. REAL SYSTEMS AND MIXED MODELS

In many instances, none of the diagnostic curves provided in Fig. 3.13 may be relevant to a given situation. For instance, if the experimenter is faced with a curve of the type shown in Fig. 3.14 and compares it with the closest fitting C curve for a linear reactor with some backmixing, $(D_E/UL=0.1)$, he must conclude that a mixed model approach is needed. Using this tack, he must determine the relative volumes of mixed, plug and dead regions of the reactor.

Since the curve exhibits a sharp rise, characteristic of a plug flow system, but at a dimensionless time $\Theta = 0.25$ rather than 1, the plug volume can be taken to be given by

$$\Theta_p = \frac{V_p}{V} \quad \text{or} \quad V_p = V\Theta_p, \tag{3.48}$$

or 25% the reactor volume.

While all curves shown in Fig. 3.15 must have the same area under the curves, equal to $\delta M/V$, their mean residence times can be quite different, since the mean residence time t_{mean}, is defined as

$$t_{mean} = \frac{\int_0^\infty Ct\,dt}{\int_0^\infty C\,dt}. \tag{3.49}$$

Levenspiel (1967) suggests that in practical terms all fluid remaining in a reactor at times greater than $\Theta = 2$ can be regarded as dead volume. Clearly,

$$t_{mean} = \frac{V_{used}}{v} = \frac{V - V_d}{v},$$

where V is the nominal filled volume of a reactor. Recalling that the *nominal*

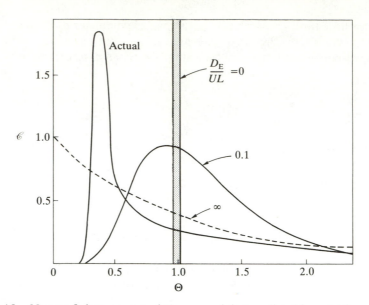

Fɪɢ. 3.15.   Nearest fit between actual $\mathscr{C}$ curve and that predicted for axial flow with a dispersion, or backmix number of 0, 0.1 and ∞.

holding time $\bar{t}$ is given by

$$\bar{t} = \frac{V}{u}$$

$$\Theta_{\mathrm{mean}} = \frac{t_{\mathrm{mean}}}{t} = \frac{V - V_{\mathrm{d}}}{V} = 1 - \frac{V_{\mathrm{d}}}{V}$$

where $\Theta_{\mathrm{mean}} \leqslant 1 (\simeq 0.6$ in this case).

Consequently, we can estimate a value for the dead volume within the reactor:

$$V_{\mathrm{d}} = V(1 - \theta_{\mathrm{mean}}) \quad (V_{\mathrm{d}} \simeq 40\% \text{ of } V). \tag{3.50}$$

The well-mixed portion of the reactor, V_{m}, then follows, since

$$V_{\mathrm{m}} = V - V_{\mathrm{p}} - V_{\mathrm{d}}. \tag{3.51}$$

Having decided what the active and inactive components of a reactor are, these data may then be used to predict a distribution curve for comparison with Fig. 3.15. If satisfactory, the reactor's anticipated chemical performance characteristics can then be predicted for any process design or scale-up work.

3.12.   CHEMICAL PERFORMANCE OF REACTOR TYPES

(a)   Introduction

Many metallurgical processing operations can be characterized by first-order chemical/transport kinetics. Pratically all our operations involve heterogeneous reactions between gas/solid, gas/liquid, liquid/liquid, liquid/solid, solid/solid, gas/gas and gas/liquid/solid phases. Examples, in the same order, would be reduction of iron oxides, oxidation of copper matte ($Cu_2S.2FeS$), organic/aqueous solvent extraction, aluminothermic reactions, fuming of gas vapour in the presence of oxygen, and finally copper-making in the Pierce-Smith converter.

Let us consider a simple reaction therefore, in which the reaction of substance A to substance B follows a first-order kinetics equation of the form

$$A \rightarrow B \qquad\qquad\qquad (3.52)$$
$$\dot{R}_A = kC_A$$

where $\dot{R}_A$ represents the rate of production of A per unit volume of the system, or consumption (when sign is negative);

k is an overall rate constant, having units of inverse time;

C_A represents the number of moles, or mass, of A per unit volume.

Let us suppose the initial concentration of A is $C_{A,0}$ at time $t=0$. We want to know what the concentration of A will be at the time fluid exits from a continuous reactor. Similarly, for a batch reactor, we want to know what the final concentration would be after the specified processing time t_f.

Case 1. Batch reactor. First-order kinetics

Referring to Fig. 3.10(b) and applying the basic equation for continuity of species A,

| Inflow of A | − | outflow of A | + | rate of chemical production of A | = | Rate of accumulation of A |

we have

$$0 \quad - \quad 0 \quad + \quad (-kCV) \quad = \quad \frac{d}{dt}(VC),$$

which may be integrated with the initial condition $C=C_0$ at $t=0$, to give

$$\frac{C}{C_0} = e^{-kt}. \qquad\qquad\qquad (3.53)$$

We see that the concentration of A drops off in an exponential manner with time.

Case 2. Plug flow reactor. First-order kinetics

Applying the same continuity equation for species A to a differential volume element δx of a plug flow tubular reactor of the type shown in Fig. 3.10(a):

$$vC_{A,x} - vC_{A,x+\delta x} - kC_{A,x}\delta V = \frac{d}{dt}(\delta V C)$$

Dropping the species A subscript, and noting that, $v/\delta V = U/A\delta x$, one has, for steady-state conditions:

$$-\frac{U}{\delta x}\frac{dC_x}{dx}\delta x - kC_x = 0.$$

Separating variables, and integrating between the entrance of the reactor and its exit at distance L, we have

$$\int_{C_0}^{C_f}\frac{dC}{C} = -\frac{k}{U}\int_0^L dx,$$

or

$$\frac{C_f}{C_0} = e^{-kL/U} \quad \text{or} \quad e^{-kV/v} \quad \text{or} \quad e^{-kt_R} \tag{3.54}$$

where t_R is the residence time of an element in the reactor.

Comparison with the previous result leads to a most important conclusion. *A plug flow continuous reactor behaves in an equivalent manner to a well-mixed batch reactor.* Thus, the same degree of chemical conversion of A to B will take place in the same time period t_R, provided the rate constant k remains unchanged. This result gives a clue to how bench-scale laboratory tests may be used for the design of full-scale commercial units.

Case 3. Well-mixed reactor. First-order kinetics

Again, applying the criteria of continuity to species A,

$$C_{A,0}v - C_A v - kVC_A = 0,$$

so that the exit concentration of A from a well-mixed reactor will be

$$C_{A,f} = \left(\frac{v}{v+kV}\right)C_{A,0} \tag{3.55}$$

or

$$C_{A,f} = \left(\frac{1}{1+(kV/v)}\right)C_{A,0}.$$

It is easily shown, and useful to verify, that the exit concentration from N

stirred tanks connected in series is given by

$$C_{A,J,N} = \left(\frac{1}{1+k\bar{t}}\right)^N C_{A,o}. \tag{3.56}$$

(b)   Efficiency of plug-flow versus well-mixed reactors

In order to compare the efficiency of well-mixed and plug flow reactors for carrying out a reaction involving first-order kinetics, we can compare the volume V_∞ of well-stirred reactor, that is needed to obtain the same fractional conversion of A to that volume, V_p, needed to accomplish the same result with a plug flow reactor.

We see that the ratio of the exit to entry concentrations of A is related to its conversion to B by

$$\frac{C_A}{C_0} = 1 - \frac{C_0 - C_A}{C_0} = 1 - f, \tag{3.57}$$

where f represents the fractional conversion of reactant to product.

Recalling eqns (3.54) and (3.55) and recasting them as follows,

$$V_p = \frac{-v\ln(1-f)}{k}, \tag{3.58}$$

$$V_\infty = \frac{fv}{(1-f)k}, \tag{3.59}$$

we may take the ratio, to give our result:

$$\frac{V_\infty}{V_p} = \frac{-f}{(1-f)\ln(1-f)}. \tag{3.60}$$

Generalizing, it is readily shown that for a reaction following nth-order kinetics ($n > 1$);

$$\frac{V_\infty}{V_p} = \frac{f}{(1-f)^n}\left(\frac{n-1}{\{1/(1-f)^{n-1}\}-1}\right) \tag{3.61}$$

We see from eqn (3.60) that at $f = 0.9$,

$$\frac{V_\infty}{V_p} \simeq \frac{-0.9}{0.1\ln(0.1)} \simeq 4$$

but at $f = 0.99$, $$\frac{V_\infty}{V_p} \simeq 20.$$

In other words, a well-mixed reactor is always less efficient than a plug flow reactor for any first-, or higher-order rate equation and rapidly becomes less efficient when fractional conversions in excess of 90% are needed.

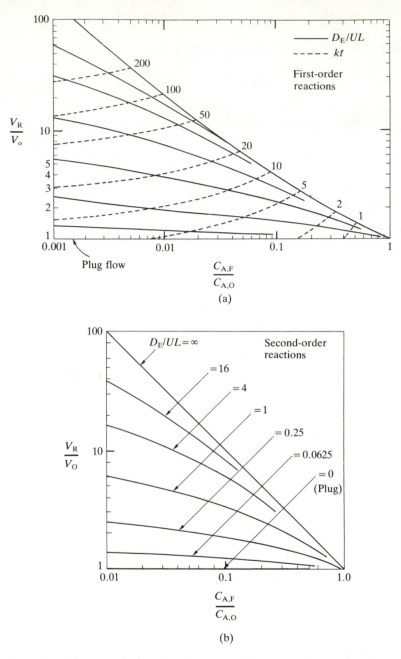

Fig. 3.16. (a) The volume of an actual reactor with a given dispersion number to that of an ideal plug flow reactor needed for the same duty under conditions of first-order kinetics. (b) The volume of an actual reactor with a given dispersion number to that of an ideal plug flow reactor needed for the same duty under conditions of second-order kinetics.

Figures 3.16(a) and (b) provide useful summaries of these types of analyses. The ordinate gives the volume ratio of an actual reactor to an ideal plug flow reactor, while the abscissa shows the ratio of final or exit concentration to the entry concentration. The curve corresponding to eqn (3.60) is the uppermost line in Fig. 3.16(a). The other curves are based on the axial dispersion model presented earlier, and show the effect of the backflow or dispersion number, D_E/UL, on results for $V_{reactor}/V_p$.

While most metallurgical reactions commonly encountered exhibit quasi-first-order kinetics, Fig. 3.16(b) has been included for the sake of completeness and shows the results of an equivalent analysis in which second-order kinetics are involved. It is also instructive to note that zero-order chemical kinetics (e.g., isotope decay) are independent of the reactor's flow characteristics.

In conclusion, we see that ideal plug flow reactors are generally more desirable than well-mixed continuous flow reactors for effecting chemical reactions, since they can provide the same fractional conversion of material in a substantially smaller volume.

3.13. HETEROGENEOUS REACTIONS IN REACTOR VESSELS

(a) Introduction

As we have noted, most metallurgical operations involve heterogeneous reactions, whereas the previous analyses of reactor performance have perhaps implied that homogeneous chemical reactions are taking place within the fluid. We therefore have to see how relevant this previous treatment of chemical reactors is, and exactly what factors might be involved when measuring a global first-order rate constant, according to the equation

$$\dot{R}_A = -kC_A.$$

(b) Example: cementation of Cu^{2+}, Co^{2+}, Ni^{2+}, on zinc powder in electro-winning

Let us consider a simple example of a cementation reaction carried out in the electrolytic zinc process (Fig. 3.9, Case 5). Zinc dust is fed into a Pachuka tank through which impure zinc electrolyte is being passed. Cementation reactions of the type

$$Cu^{2+} \quad + \ Zn \ \rightarrow \ \ Zn^{2+} \quad + \quad Cu$$

electrolyte   dust   electrolyte   on dust

remove copper, cobalt and nickel impurity ions during the electrolyte's passage through the tank and on to the electro-winning step for zinc.

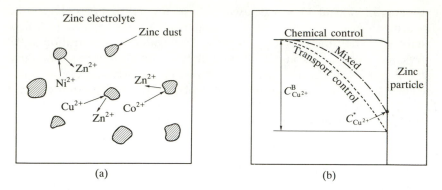

FIG. 3.17. Purification of zinc electrolyte by use of zinc dust to remove (cement) any dissolved copper, cobalt and nickel. (a) Physical schematic; (b) concentration profiles of cupric ion adjacent to a particle of zinc dust.

Clearly, referring to Fig. 3.17, this reaction takes place as a result of transport of copper ions to zinc particle interfaces, followed by Cu^{2+} adsorption and chemical reaction with the substrate. Desorbtion of Zn^{2+} product, together with subsequent diffusion and transport away into the bulk of the purified electrolyte then completes the steps involved.

Step 1: Transport

Writing the Cu^{2+} transport step, we know from the preliminaries in Chapter 1 that the flux of cupric ions can be characterized by a mass transfer coefficient k_L, in the equation

$$\dot{N}_{Cu^{2+}} = k_L A (C^B_{Cu^{2+}} - C^*_{Cu^{2+}}).$$

Here A is defined as the net exposed surface area of zinc dust available for cementation/unit mass of zinc dust and the flux of Cu^{2+} ions is $\dot{N}/A$ per unit mass of zinc.

Step 2: Adsorption of Cu^{2+}, chemical reaction and desorbtion of Zn^{2+}

$$\dot{R}_{Zn^{2+}} = k_c A C^*_{Cu^{2+}}$$

Here the rate of chemical reaction is expressed as the product of a chemical rate constant k_c (units of LT^{-1}), the area of zinc surface/unit mass of zinc dust and the driving force (i.e., the interfacial concentration of cupric ions at the zinc dust interfaces). Under steady-state conditions, the output of Zn^{2+} must equal the input of Cu^{2+}, so that

$$\dot{R}_{Zn^{2+}} = +\dot{N}_{Cu^{2+}}$$

Combining, eliminating $C^*_{Cu^{2+}}$ and sorting

$$\dot{R}_{Zn^{2+}} = + \left(\frac{k_c k_l}{k_c + k_l} \right) A C^B_{Cu^{2+}}$$

$$= + \frac{A C^B_{Cu^{2+}}}{(1/k_l + 1/k_c)},$$

or

$$\dot{R}_{Cu^{2+}} = -k_G A C^B_{Cu^{2+}}.$$

Where $\dot{R}$ is the number of moles of copper reacting per unit time and mass of zinc dust k_G is the overall rate constant and equal to the geometric mean of the intrinsic chemical rate constant, k_c, and the mass transfer coefficient k_L, i.e. $k_G = k_c k_L / (k_c + k_L)$ and $C^B_{Cu^{2+}}$ is the bulk concentration of cupric ions (moles/unit volume). If ρ is the mass of zinc powder/unit volume of electrolyte, we have

$$\dot{R}_{Cu^{2+}} \rho = -k_G \rho A C^B_{Cu^{2+}},$$

or

$$\dot{R}_{Cu^{2+}} = -K_G C^B_{Cu^{2+}},$$

where K_G is an overall, or global, rate, constant with units of inverse time, equivalent to the first-order rate constant introduced earlier in this chapter. The treatment supposes that the adsorption and desorbtion steps and product transport rates are all rapid compared with those considered. This will be appreciated following a reading of Chapter 5.

This example serves to illustrate the care one must take in any attempts to scale-up heterogeneous reaction systems where a global rate constant K_G has been obtained from batch- or laboratory-scale tests. It is most important to be aware of the intrinsic factors on which a measured value of K_G rests.

Consequently, where heterogeneous reactions are involved, one must consider all the various steps making up K_G and assess their relative importance. For instance, we see that for very fast cementation kinetics, $k_c \gg k_L$ and $K_G = \rho k_G = \rho k_L$ (i.e., transport control). Alternatively, for slow chemical kinetics, $k_c \ll k_L$, and $K_G = \rho k_G = \rho k_c$ (i.e., chemical control).

Another factor requiring caution is temperature. Activation energies for chemical reactions tend to be at least one order of magnitude greater than those of transport- (diffusion-) controlled reactions (i.e., the reactions are more sensitive). Consequently, for a mixed control reaction, falsely low activation energies obtained from an experimental analysis carried out between 20 and 40°C, say, would lead to unrealistically low global coefficients K_G if extrapolated to 80°C via an Arrhenius-type equation,

$$K_G = K_o e^{-\Delta E/RT}.$$

Similarly, there are many complex metallurgical reactions for which this treatment would prove to be inadequate and simplistic. In such cases, one must describe the kinetic phenomena taking place on, or within, particles or lumps involved. These particles would, for instance, be reacting within packed beds, fluidized beds, kilns and furnaces. Their individual behaviour must then be integrated into a global description of the process of specific interest.

3.14. REACTOR DESIGN AND SCALE-UP TECHNIQUES

The main objectives of the type of work described in the second part of this chapter are the improvement, scale-up, and/or design of efficient continuous flow reactors. It is therefore appropriate to briefly consider how a commercial-scale reactor might be designed from first principles using the approaches described.

We will take a specific system, so that we can focus on the steps needed. Consider therefore, how a continuous flow reactor might be built for liberating a metal from a crushed ore by an acid leaching operation in a tubular packed bed.

Step 1. Ascertain a global rate constant K, and reaction order n using batch, laboratory-scale leaching tests. Determine the contibution of transport kinetics (stirring) by carrying out tests under various conditions of relative velocity and temperature. Usually, transport kinetics can be predicted with some degree of confidence. Consequently, if an observed rate constant coincides with that expected for a transport-controlled process, one can anticipate $k_c \gg k_L$. On the other hand, if the observed rate is less, then chemical control is indicated.

Step 2. Decide on the degree of conversion desired, and therefore the leaching time needed. This time will be based on those of the batch tests.

Step 3. Knowing that the time indicated from the batch tests should be identical to the time that would be required in a plug flow reactor, the volume (provided it is modelled correctly), of a plug flow reactor needed can be calculated according to the equation

$$C_f/C_o = e^{-kV/v},$$

where C_f/C_o, k and v are all specified.

Step 4. Choose a suitable (reasonable) aspect ratio (say, 10) to ensure that the tubular reactor approaches ideal plug flow conditions, and use Fig. 3.14 to estimate the magnitude of the dispersion number D_E/UL.

Step 5. Use Fig. 3.16 to determine V/V_p, i.e. how much bigger the volume of a commercial unit should be to provide the same conversion when some backmixing is present.

(a) Example: Design a reactor for recovering tellurium from anode slimes

The recovery of tellurium from copper anode slimes involves pressure-leaching of the slimes, followed by precipitation of tellurium on copper shot within a reactor vessel by the reaction

$$\text{Te(OH)}_6 + 5\text{Cu} + 3\text{H}_2\text{SO}_4 \leftrightarrows \text{Cu}_2\text{Te} + 3\text{CuSO}_4 + 6\text{H}_2\text{O}$$

| liquid | solid | liquid | solid | liquid |

A continuous flow tubular reactor is to be designed to handle a flow rate of $3 \text{ m}^3 \text{ h}^{-1}$ of tellurium-rich solution and to recover 90% of the tellurium. To determine a chemical rate constant, the same reaction was carried out in a laboratory-scale batch reactor (Fig. 3.18) in which a rotating drum of 1 m diameter was half-filled with the tellurium-rich liquid and copper shot. The copper shot loading in the liquid was 35 per cent on a volume basis. By plotting $\ln(C)$ against time in Fig. 3.19, using semi-logarithimic graph paper, a

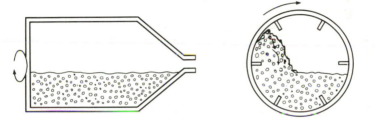

FIG. 3.18.   Batch-scale tests to determine rate constant for scale-up and design of a continuous-flow reactor.

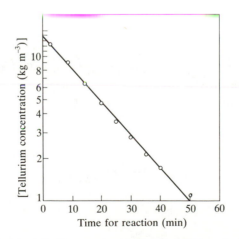

FIG. 3.19.   Variation in tellurium concentration in solution during batch cementation steps.

straight line of slope -3.1 h^{-1} was obtained, indicating an exponential decay in the concentration of dissolved tellurium with time, and first-order kinetics.

It is to be noted that the Cu_2Te product layer was continuously and mechanically abraded from the surfaces of the copper shot, so that any surface poisoning owing to a product layer of Cu_2Te was absent.

Solution:

Step 1. The plots of $\ln(C)$ vs t yielded a value for K, the global rate constant of 3.1 hr^{-1}.

Step 2. Knowing that 3 m^3 h^{-1} of solution is to be treated, for 90 per cent removal, we have

$$\frac{C_f}{C_o} = 0.1 = e^{-3.1t},$$

giving

$$t \simeq 0.75 \text{ h.}$$

Step 3. The volume of plug flow reactor's liquid contents = flowrate × reaction time needed for 90 per cent removal $= 0.75 \times 3$, or 2.25 m^3 of liquid.
  Volume of liquid + shot $= 2.25 \times 1.35 = 3.04$ m^3.
  Required geometric volume of plug flow reactor at 50 per cent loading $= 6.08$ m^3.

Step 4. Estimating the dispersion number, for a reactor of 1.0 m diameter:

$$\text{Re} = \frac{\rho U d}{\mu} \simeq \frac{\rho[\dot{Q}/(\pi d^2/8)]d_H}{\mu} \simeq \frac{8\rho\dot{Q}d_H}{\pi\mu d^2} = \frac{8 \times 1000 \times 3/3600 \times 0.61}{1 \times 10^{-3} \times 1^2 \times \pi},$$

i.e. $\text{Re} \simeq 0.15 \times 10^4$.

The diameter here is defined as the hydraulic diameter d_H†, of the vessel's liquid contents. Assuming the relevance of Fig. 3.14(a) to the problem at hand, a dispersion number of about 2 is indicated for turbulent flow through a pipe at Reynolds number of 0.5×10^4.
  The length of the plug flow reactor is such that

$$\frac{\pi d^2}{4} \times L_p = 3.04 \text{ m}^3,$$

$$L_p = \frac{4 \times 3.04}{\pi} = 3.87 \text{ m.}$$

The reactor dispersion number can be estimated by guessing the reactor will

---

† $d_H = 4 \times$ cross-sectional area of flow/wetted perimeter.

be about 1.5 times longer than the plug flow length:

$$\frac{D_E}{UL} = \frac{D_E}{Ud_h} \times \frac{d_h}{L_p} \simeq 2 \times \frac{0.61}{(3.87 \times 1.5)} \simeq 0.21.$$

While the use of Fig. 3.14 is an extrapolation to the situation described in this case, in which tumbling of the shot and rotation of the vessel are super-imposed on an axial flow situation, it is interesting to note that actual experiments gave a dispersion number of 0.18.

Step 5. The volume of reactor needed can now be obtained with the use of Fig. 3.16, where we see $V/V_p \simeq 1.4$ for $C_f/C_o = 0.1$.

Conclusion

In view of the approximate nature of estimating the dispersion number, further iterations on D_E/UL (i.e. repeating steps 4 and 5) would not be fruitful. Consequently, one would recommend that a cylindrical reactor be built such that:

Diameter of reactor	1 m
Length of reactor	5.4 m
Feed rate of Te solution	3.0 m^3 h^{-1}
Fractional conversion	90 per cent.

3.15. EXERCISES ON REACTOR DESIGN, DIAGNOSTICS, AND BEHAVIOUR

(1) First- and second-order reactions in plug flow and well stirred systems

Dissolved substance A flows into a reactor of volume V litres, at a rate of v litres min^{-1}. Derive expressions for the percent conversion of material A to inert material B for the cases of:

(1) First-order homogeneous chemical reaction, linear reactor;
(2) First-order homogeneous chemical reaction, well-mixed reactor;
(3) Second-order homogeneous chemical reaction, well-mixed reactor;
(4) Second-order homogeneous chemical reaction, linear reactor.

Calculate conversions to be expected from the data below.

Volume of reactor, V	1000 litres
Volumetric flow rate, v	10 litres min^{-1}
Inlet concentration, C	1 g mole litre^{-1}
Rate constant (1st-order)	6.7×10^{-3} min^{-1}
Rate constant (2nd-order)	2.0×10^{-2} litres gmole^{-1} min^{-1}.

On the basis of these results and any other necessary computations, what sort of reactor would you choose to achieve:

(1) maximum conversion per unit volume of reactor for equal inputs;
(2) maximum conversion of material from A to B following a first-order reaction, in the face of a competing second-order reaction of A to C.

(2)   Reactor design—precipitation of iron from solution

In the development of a hydrometallurgical process, part of the circuit involved the oxidation of ferrous chloride (soluble in water) to insoluble ferric hydroxide. It is known that the oxidation involves a first-order homogeneous chemical reaction, the rate constant from laboratory experiments having been determined as 0.05 h^{-1}. It was decided that a continuous process would be most suitable for the job. However, for reasons of available space, etc., the total volume V_T of the reactor(s) was not allowed to exceed 1 million litres of solution, whilst the hourly throughput of solution to be treated had to be 1000 litres.

Assuming the liquid in the reactor(s) to be *perfectly mixed* (owing to the passage of air bubbles up the reactor tanks), show whether it would be more efficient to use one large reactor of volume V_T, or four smaller reactors in series, each of volume $V = V_T/4$. Calculate the percentage transformation of A to B for the two cases and comment on the results. Finally, compare this result with what would be obtained by a plug flow reactor

Data:

Reaction rate constant, k	0.05 h^{-1}
Volumetric flowrate of solution, v	$1000 \text{ litres h}^{-1}$
Total permissible reactor volume, V_T	10^6 litres
Reactor volume of smaller units, V	$10^6/4$ litres

(3)   Electrowinning of zinc-purification of anolyte

In the treatment of impure zinc anolyte solution, iron is removed from the circuit by feeding anolyte through a series of five pachuka tanks. The anolyte feed rate into the system is $2000 \text{ litres/h}^{-1}$, each tank having a filled capacity of 500 litres of solution.

As shown in the figure, air is fed into the tanks by bubbling through a gas distributor plate. Zinc oxide calcine is fed as a finely dispersed powder into the top of the first tank. The net reactions involved are:

(1) oxidation of ferrous sulphate to ferric sulphate;

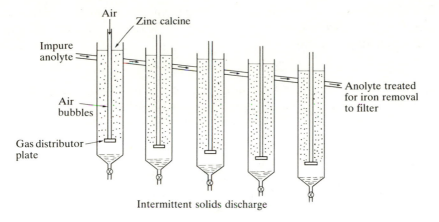

(2) reaction of calcine with ferric sulphate to precipitate iron hydroxide, i.e.,

$$Fe_2(SO_4)_3 + 3ZnO + 3H_2O \rightarrow 3ZnSO_4 + 2Fe(OH)_3.$$

In a study of the hydrodynamic characteristics of these tanks, 10 kg of inert finely divided material (simulating zinc calcine) was fed into the reactor as an instantaneous tracer addition. The outflow of material from the first tank was measured and the following results were recorded:

Time (min)	0	3	9	12	21
Concentration of material $(kg \ litre^{-1}) \times 10^4$	1000	368	50	18	0.1

Using the data and graph paper, construct a C–θ diagram for these results and compare the curve obtained with equivalent C curves for a well-mixed and a plug-flow reactor. Based on mixed model theory, show how you can model the hydrodynamic performance of these particular pachuka tanks with reference to the flow of zinc calcine. In particular, calculate the proportions of dead, plug and mixed volumes. Interpret the results in a physical sense and suggest how reactor performance can be improved.

(4) Electrowinning of nickel—purification of electrolyte

Impure nickel sulphate anolyte containing various cationic impurities is pumped at a rate of 300 000 litres h^{-1} to the purification section, where the first step is to remove copper ions by cementation with active sponge-nickel powder in two mechanically agitated pachuka tanks connected in series. The slurry of anolyte and nickel sponge is pumped into the first tank and discharged below the surface, and the copper ion concentration is reduced from 0.5 g l^{-1} down to 0.001 g l^{-1} by the exit of the second tank according to the cementation reaction:

$$Cu^{2+} + Ni \rightarrow Ni^{2+} + Cu.$$

In a planned extension of 200 electrolytic cells to the refinery tank house, 345 000 l h^{-1} of equivalent anolyte solution required treatment, and three options (shown diagramatically below) were proposed.

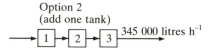

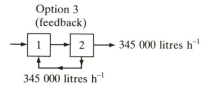

Calculate the percentage impurity removals that will be achieved and state which of the three you would recommend to management, based on a minimum impurity removal of 99.8 per cent Cu^{2+}.

The cementation reaction follows first-order chemical kinetics with practically identical rate constants, while the pachuka tanks can be regarded as well-stirred (or 100% backmix) chemical reactors.

Similarly, conditions are far from equilibrium so that one can assume an irreversible type of reaction and also, since a large excess of nickel powder is added to reactor 1 over that required stoichiometrically, one can consider the reaction surface area as being independent of time.

(5) Off-gas systems for BOFs.—explosion characteristics

A new torpedo-shaped dry precipitator has been designed for cleaning partially burned waste gases $(CO + H_2)$ from oxygen steel-making convertors. However, the occurrence of explosive gas mixtures is possible when critical

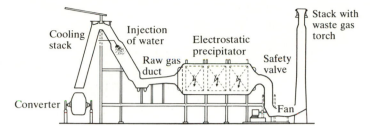

levels of carbon monoxide and oxygen are present (see the graph). These critical mixtures are most likely to occur at the start of the blow. Thus, at the beginning of a blow, oxygen is blown onto a mixture of scrap and molten iron and a finite time is required before ignition takes places after which CO begins to be evolved. Ideally (and ignoring any air entrainment around the hood), the concentration of oxygen in the gas from the convertor drops from 100% to zero as soon as the steelmaking reactions begin, while the CO concentration instantaneously rises from 0 to 100 per cent. However, since flow in the cleaning system is not completely plug flow, some intermixing of the two gas mixtures will occur as they flow together through the precipitator.

Knowing the length, diameter, and eddy diffusion coefficient of the gases through the precipitator, determine the minimum gas flow so as to prevent spontaneous explosions during the initial stages of the blow. Treat the problem as plug flow with some back mixing in an ideal cylindrical pipe with no dead spots.

Data:

Length of 'reactor'	$= 30$ m
Diameter of reactor	$= 6$ m
Eddy Diffusion Coefficient	$= 0.07 \, \text{m}^2 \, \text{s}^{-1}$
Hydrogen content	$= 0\%$
Ignition limits + erf table.	
Operating pressure	$\simeq$ Ambient
Operating temperature	$\simeq$ Ambient.

Adopt the criterion that 5.6 m³ of an explosive mixture containing back-diffused oxygen in carbon monoxide is needed to sustain a serious explosion. This problem requires familiarity with material presented in Chapter 4 on the error function relationship and relies on eqn (3.46) in which the $x = 0$ axis is about 5 metres downstream of the exit at the time the gas pocket is about to exhaust.

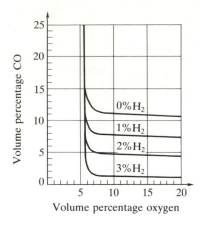

Volume percentage oxygen

(6) Global versus intrinsic kinetics—hydrofluorination of uranium oxide

Explain the meaning of global and intrinsic rates of reaction and the need for their careful distinction in reactor design. The hydrofluorination of UO_2 pellets with $HF(g)$ proceeds according to the reaction:

$$UO_2(s) + 4HF(g) \rightarrow UF_4(s) + 2H_2O(g)$$

This reaction was studied by suspending spherical pellets, 2 cm in diameter, in a stirred-tank reactor. In one run, at a bulk gas temperature of 377 °C, the surface temperature of the pellet was 462 °C and the observed rate was $-\dot{r}''_{UO_2} = 6.9 \times 10^{-2}$ mole UO_2 s^{-1} m^{-2} reaction surface. The bulk concentration of HF was 13.8 mole m^{-3} and the gas phase mass transfer coefficient was 106 mm s^{-1}.

The intrinsic rate of reaction was found to be given by

$$\dot{r}''_{UO_2} = 40\, C_{HF} \exp\left(-\frac{6070}{RT}\right) \text{g moles/cm}^2\text{s}.$$

Calculate the error involved in assuming bulk phase conditions at the interface in determining reaction rates. Estimate the percentage errors involved by neglecting (a) the thermal and (b) the mass transfer resistance terms.

N.B. The product layer does not affect the initial surface kinetics.

3.16. CONCLUSIONS

This chapter has attempted to show some of the skills and judgement needed by the experimental modeller of high-temperature metallurgical operations, and the pragmatic, order of magnitude approach needed to set up reliable, instructive models. The section on metallurgical reactors and associated tracer techniques was included because such methods can be used as an effective diagnostic tool by the experimentalist.

FURTHER READING

1. Johnstone, R. E. and Thring, M. E. (1957). *Pilot plants, models and scale-up methods.* McGraw-Hill, New York.
2. Levenspiel, O. (1967). *Chemical reaction engineering.* Wiley, New York.
3. Bridgman, P. W. (1922). *Dimensional analysis.* Yale University Press, New Haven.
4. Iron and Steel Institute (1973). *Mathematical process models in iron and steelmaking.* Iron and Steel Institute, London.
5. AIME (1982). *Mathematical process models in iron and steelmaking, Process Technology* (1982) Division of Iron and Steel Soc. of AIME, Chicago.
6. Smith, J. M. (1970). *Chemical engineering kinetics* (2nd edn). McGraw-Hill, New York.
7. Szekely, J., Evans, J. W., and Sohn, H. M. (1976). *Gas solids reactions.* Academic Press, New York.
8. Process Technology Division of the Iron and Steel Society of AIME. *Application of Mathematical and Physical Models in the Iron and Steel Industry.* Proceedings of the 3rd Process Technology Conference, Vol. 3. Pittsburgh meeting (1982). Edward Brothers, Ann Arbor, MI.

4

HEAT AND MASS TRANSFER THROUGH MOTIONLESS MEDIA

4.0. INTRODUCTION

A wide range of heat and mass transfer phenomena in metallurgical operations relate to transfer within solids or equivalent stationary media. For instance, while the steel bar being carburized, considered in section 1.4, was stationary, a movement of interstitial carbon atoms took place by diffusion mechanisms. Other movements at the atomic level manifest themselves through a broad range of metallurgical events (e.g., crystallization, re-crystallization, grain orientation, precipitation of new phases, phase and crystalline reconstitution and dislocation movements). Although these atomic-level movements lead to drastic changes in the physical properties of a metal or alloy, they lie beyond the subject matter of this book, except in a macroscopic phenomenological sense.

However, with respect to diffusion and thermal conduction phenomena through solids, it will be appreciated that there is no coupling of the equations describing the flow of heat through a solid, or of atoms diffusing through a solid, with the fluid flow equations presented in Chapter 2. Diffusion or heat conduction in solids therefore represents a logical starting point for the mathematical description of heat and mass transport phenomena. The category can be conveniently subdivided into steady-state, and unsteady-state processes.

4.1. HEAT AND MASS CONSERVATION EQUATIONS

Every process analysis dealing with the interchange of heat or mass between various parts of a system—in our example, the transfer of heat from the steel bar to the quenching medium and the transfer of carbon from the furnace atmosphere to the steel rod — should involve the equations of heat and mass conservation. Hence, the solution to a heat or mass transfer problem must combine the respective expressions of transfer (Fourier or Fick) with an appropriate mathematical statement of heat or mass conservation. The resulting differential equation is then solved by means of appropriate boundary conditions.

(a) Heat conservation equation

Consider the differential volume ΔV shown in Fig. 4.1. Conservation of

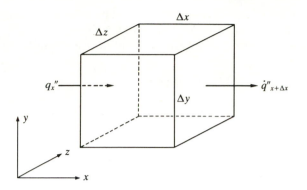

F<small>IG</small>. 4.1. Elemental volume through which heat flows in the x direction.

thermal energy requires that

<table>
<tr><td>Rate of heat input
− Rate of heat output</td><td>+</td><td>Rate of heat
generation in
elemental volume</td><td>=</td><td>Rate of heat
accumulation in
elemental volume</td></tr>
<tr><td>(1)</td><td></td><td>(2)</td><td></td><td>(3)</td></tr>
</table>

(1) The rate of heat input less the rate of heat output in the x direction will be equal to the x component heat flux at x, $\dot{q}''_x$, multiplied by the cross-sectional area $\Delta y \Delta z$, less the x component heat flux at $x + \Delta x$, $\dot{q}''_{x+\Delta x}$, multiplied by the same cross-sectional area, $\Delta y \Delta z$. Hence for term (1):

$$(\dot{q}''_x - \dot{q}''_{x+\Delta x})\Delta y \Delta z = \left\{ \dot{q}''_x - \left(\dot{q}''_x + \frac{\partial \dot{q}''}{\partial x} \cdot \Delta x \right) \right\} \Delta y \Delta z$$

$$= -\frac{\partial \dot{q}''_x}{\partial x} \cdot \Delta x \, \Delta y \Delta z. \qquad (4.1)$$

Thus, summing the x, y and z components of the heat flux, yields

$$-\left(\frac{\partial \dot{q}''_x}{\partial x} + \frac{\partial \dot{q}''_y}{\partial y} + \frac{\partial \dot{q}''_z}{\partial x} \right) \Delta V = -(\nabla \dot{q}'')\Delta V. \qquad (4.2)$$

(2) The rate of heat generation (if any) in the elemental volume ΔV can be expressed as $\dot{q}''' \, \Delta V$ where $\dot{q}'''$ is the rate of heat generation per unit volume.

(3) Finally, the rate of heat accumulation in the elemental volume ΔV will be equal to the element volume multiplied by its heat capacity per unit volume and its rise in temperature per unit time. However, since heat capacities are usually expressed in terms of the amount of heat required to raise the temperature of unit mass (rather than unit volume) by one temperature unit, it

follows that $C_V = \rho C$ (ρ = density, C = heat capacity per unit mass) and that the rate of heat accumulation is equal to $\rho C \Delta V (\partial \theta / \partial t)$.

The equation of energy conservation may thus be formulated:

$$-\nabla \cdot \dot{q}'' + \dot{q}''' = \rho C \frac{\partial \theta}{\partial t} \qquad (4.3)$$

The *full* form of the energy equation contains convective terms, as described in Chapters 5 and 6.

(b)   Mass conservation (or continuity) equation

Mass units

We will now use a similar accounting procedure to develop the equation for mass conservation, commonly referred to as the *continuity equation*. Consider the transport of a solute A through solvent B. A dynamic mass balance for A as shown in Fig. 4.2 requires that

Mass of A entering per unit time − Mass of A leaving per unit time	+	Rate of mass generation of A in elemental volume	=	Rate of mass accumulation of A in elemental volume
(1)		(2)		(3)

(1) The transport term (1): is given by

$$(\dot{n}''_{A,x} - \dot{n}''_{A,x+\Delta x}) \Delta y \, \Delta z = -\frac{\partial \dot{n}''_{A,x}}{\partial x} \Delta x \, \Delta y \, \Delta z = -\frac{\partial \dot{n}''_{A,x}}{\partial x} \Delta V. \qquad (4.4a)$$

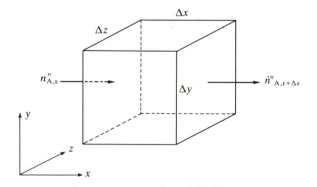

FIG. 4.2.   Elemental volume through which mass of a substance A flows in the x direction.

Including the y and z components of the mass flux (if any) yields

$$-\left(\frac{\partial \dot{n}''_{A,x}}{\partial x} + \frac{\partial \dot{n}''_{A,y}}{\partial y} + \frac{\partial \dot{n}''_{A,z}}{\partial z}\right)\Delta V \equiv -(\nabla \cdot \dot{n}''_A)\Delta V \qquad (4.4b)$$

The generation term (2): the rate of mass generation of A (if any) in the elemental volume ΔV can be expressed as $\dot{r}'''_A \Delta V$, where $\dot{r}'''_A$ is the rate of mass of A generated in unit volume.

The transient term (3): finally, the rate of mass accumulation in the elemental volume ΔV will be equal to the element volume $\times$ the change in the specific density of A with respect to time: $\Delta V(\partial \rho_A/\partial t)$.

The equation of mass conservation may thus be formulated:

$$-\nabla \cdot \dot{n}''_A + \dot{r}'''_A = \frac{\partial \rho_A}{\partial t}. \qquad (4.5)$$

In many circumstances, it is more convenient to express the continuity equation in terms of molar units. For example, when we are looking at an interfacial reaction, say between two diffusing solutes which combine to form a new species in the other phase, then they will react together at the two phase interface in simple molecular ratios. Thus, for the case of dissolved carbon and oxygen in steel reacting to form carbon monoxide gas, we have the reaction

$$(C)_{Fe} + (O)_{Fe} \rightarrow CO$$

occurring at the interface, such that the molecular (atomic) fluxes of carbon, oxygen and carbon monoxide will all be equal, according to stoichiometry. Thus $\dot{N}''_C = \dot{N}''_O = \dot{N}''_{CO}$, whereas the mass fluxes are in the ratios

$$\dot{n}''_{(C)} = (12/16)\dot{n}''_{(O)} = (12/28)\dot{n}''_{CO}$$

Molar units

Following identical procedures to those used for the mass units, we can express the mass conservation equation for A in molar or atomic notation. In molar units for solvent A:

$$-\nabla \cdot \dot{N}''_A + \dot{R}'''_A = \frac{\partial C_A}{\partial t} \qquad (4.6)$$

Similarly for solvent B:

$$-\nabla \cdot \dot{N}''_B + \dot{R}'''_B = \frac{\partial C_B}{\partial t}. \qquad (4.7)$$

(c)   Cylindrical and spherical coordinate systems

The inherent symmetry of engineering systems results in heat and mass transfer problems in which the relevant differential equations can often be expressed in terms of rectangular, cylindrical or spherical geometry. It is useful, therefore, briefly to include equations equivalent to those so far presented, in cylindrical and spherical coordinates. Referring to Fig. 4.3, we have the following:

(i)   Transfer equations: cylindrical and spherical coordinates

Conductive heat transfer:

$$\dot{q}_r'' = -k\frac{\partial\theta}{\partial r} \tag{4.8}$$

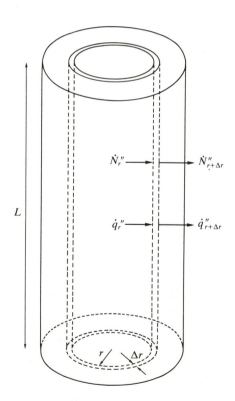

FIG. 4.3.   A segment of length L, through a cylinder of infinite length, showing radial transfer of heat by conduction and of mass by diffusion.

Diffusive mass transfer:

$$\dot{N}''_{A,r} = -D_{AB}\frac{\partial C_A}{\partial r} \tag{4.9}$$

(ii) Conservation equations: Cylindrical coordinates

For heat transfer, term (1) of the heat conservation equation is given by

$$\{\dot{q}''_r 2\pi r L - \dot{q}''_{r+\Delta r}2\pi(r+\Delta r)L\}$$

$$= \left\{-\frac{\partial \dot{q}''_r}{\partial r}\Delta r\, 2\pi r L - \dot{q}''_r 2\pi\,\Delta r L - \frac{\partial \dot{q}''_r}{\partial r}\Delta r^2 2\pi L\right\}.$$

Taking the Δr^2 term as being negligible,
The transport term (1)

$$= -2\pi r L\Delta r\left(\frac{\partial \dot{q}''_r}{\partial r}+\frac{\dot{q}''_r}{r}\right)= -2\pi r L\Delta r\left(\frac{1}{r}\frac{\partial}{\partial r}(r\dot{q}''_r)\right),$$

The generation term (2)

$$= \dot{q}'''(2\pi r L\Delta r)$$

and the transient term (3)

$$= \rho C\frac{\partial \theta}{\partial t}(2\pi r L\Delta r)$$

so that

$$-\frac{1}{r}\frac{\partial}{\partial r}(r\dot{q}''_r)+\dot{q}''' = \rho C\frac{\partial \theta}{\partial t}. \tag{4.10}$$

Similarly, for mass transfer:

$$-\frac{1}{r}\frac{\partial}{\partial r}(r\dot{N}''_{A,r})+\dot{R}'''_A = \frac{\partial C_A}{\partial t}. \tag{4.11}$$

(iii) Conservation equations: spherical coordinates

It is left as an exercise to derive the following conservation equations for spherical geometry.
For heat transfer:

$$-\frac{1}{r^2}\frac{\partial}{\partial r}(r^2\dot{q}''_r)+\dot{q}''' = \rho C\frac{\partial \theta}{\partial t}. \tag{4.12}$$

For mass transfer:

$$-\frac{1}{r^2}\frac{\partial}{\partial r}(r^2\dot{N}''_{Ar})+\dot{R}'''_A = \frac{\partial C_A}{\partial t}. \tag{4.13}$$

Finally, it should be noted that these equations are not completely generalized forms, since we have implicitly assumed that temperatures and concentrations are, at a given radius, everywhere equal and independent of angle. Thus a rod heated only on one side would require an extra term to describe the angular conductive heat transfer component.

4.2.   BOUNDARY CONDITIONS

As mentioned in Section 4.1, it is necessary that appropriate boundary conditions be specified so that the differential equation describing the particular problem at hand can be solved. These boundary conditions obviously depend on the physical and chemical characteristics of the system and typically take the following forms:

Heat transfer:
(1) temperature of boundaries known;
(2) heat fluxes through boundaries known;
(3) heat flux as a function of surface temperature known;
(4) interfacial temperature between two adjacent phases identical;
(5) interfacial instantaneous heat fluxes identical.

Mass transfer
(1) solute concentrations at boundaries known;
(2) mass/molar flux of solute through boundary known;
(3) interfacial activities of solute in the two adjacent phases identical;
(4) interfacial instantaneous molar fluxes stoichiometrically defined.

4.3.   EXAMPLES OF STEADY-STATE TRANSFER THROUGH MOTIONLESS MEDIA

We shall now go on to consider some simple steady-state heat and mass transfer problems to illustrate the general procedures and short-cuts to be adopted in tackling similar examples listed for the reader's use.

(a)   Steady-state heat conduction: single phase

A 200-tonne steelworks' ladle (Fig. 4.4) containing liquid pig iron is lined with a 0.25 m-thick layer of refractory bricks. The inside surface temperature of the bricks will be that of the liquid pig iron (1365°C) while the outer surface of the

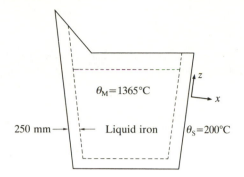

Fɪɢ. 4.4. Steelworks' transfer ladle containing molten pig iron (hot metal) at 1365 °C.

refractory remains at 200°C. Taking a typical refractory thermal conductivity of 2 W m^{-1} K^{-1}, calculate the steady-state rate of heat loss per unit area of ladle surface.

Solution

In its simplest form, this problem can be regarded as a one-dimensional heat conduction phenomenon in rectangular coordinates through a plane wall. Therefore, writing down the transfer and conservation equations and discarding those terms not relevant:

$$\dot{q}'' = -k\,\nabla\theta \rightarrow \dot{q}''_x = -k\frac{\partial\theta}{\partial x}, \tag{4.14}$$

$$-\nabla\cdot\dot{q}'' + \dot{q}''' = \rho C\frac{\partial\theta}{\partial t} \rightarrow -\frac{\partial\dot{q}''_x}{\partial x} = 0. \tag{4.15}$$

Thus, heat conduction in the y and z direction is taken to be zero, there being no significant temperature gradients in those directions. Similarly heat generation is zero and, finally, under steady-state conditions there can be no accumulation of heat in the brickwork, since at steady state, heat in equals heat out). Thus $\rho C(\partial\theta/\partial t) = 0$. Combining eqns (4.14) and (4.15):

$$-\frac{\partial}{\partial x}\left(-k\frac{\partial\theta}{\partial x}\right) = 0, \tag{4.16}$$

which is to be solved given the boundary conditions:

B.C.1:

$$x = 0, \ \forall t, \ \theta = \theta_M;$$

B.C.2:

$$x = L, \forall t, \theta = \theta_S.$$

Assuming k, the thermal conductivity of the refractory to be independent of temperature, we have

$$\frac{\partial^2 \theta}{\partial x^2} = 0, \tag{4.17}$$

which may be integrated to yield the general solution

$$\theta = Ax + B. \tag{4.18}$$

The particular solution is then obtained by application of the boundary conditions in this general equation:

B.C.1:

$$\theta_M = A(0) + B, \qquad B = \theta_M,$$

B.C.2:

$$\theta_S = AL + B, \qquad A = \frac{\theta_S - B}{L} = \frac{\theta_S - \theta_M}{L},$$

and the particular solution required becomes

$$\theta_x = \left(\frac{\theta_S - \theta_M}{L} \right) x + \theta_M. \tag{4.19}$$

Having obtained the appropriate temperature profile, which is linear in this case, we then use Fourier's law:

$$\dot{q}''_x = -k \left(\frac{\partial \theta}{\partial x} \right)_x$$

to obtain the corresponding heat flux. Thus, differentiating the temperature profile with respect to x, and multiplying by $-k$, we have

$$\dot{q}''_x = -k \left(\frac{\partial \theta}{\partial x} \right)_x = k \left(\frac{\theta_M - \theta_S}{L} \right)$$

$$= 2.0 \frac{(1365 - 200)}{0.25} \equiv 9.3 \text{ kW m}^{-2} \quad \text{or } 2.2 \text{ kcal m}^{-2} \text{s}^{-1}.$$

Although this approach is pedantic for such a simple problem, it is important to take note of the steps carried out in obtaining the solution, as this approach is basic to more complex problems.

Short cut

Since we have steady-state conditions, $\dot{q}''_x$ is a constant, so that we may immediately integrate Fourier's law:

$$\dot{q}''_x = -k\frac{\partial \theta}{\partial x} \tag{4.20}$$

$$\dot{q}''_x \int_0^L dx = -k \int_{\theta_M}^{\theta_S} d\theta$$

to give

$$\dot{q}''_x = +k\left(\frac{\theta_M - \theta_S}{L}\right).$$

N.B. In the case of cylindrical or spherical coordinates, one should take $\dot{q}_r$ rather than $\dot{q}''_r$, since $\dot{q}''_r$ is a function of r whereas $\dot{q}_r = (2\pi r L \dot{q}''_r)$ is a constant and independent of radial distance, for steady state heat flow.

(b)   Concept of resistance and driving forces in heat and mass transfer

The reader will no doubt recognize that there is an analogy between this result and the transfer of electrical energy through a resistor. In the latter case, we have (Ohm's law)

$$I = \frac{\Delta E}{R},$$

where I = current (amps); E = potential (volts); R = resistance (ohms). For this reason (L/k) may be called the resistive component to conductive heat transfer. Similarly, for the analogous case of diffusive mass transfer we would have

$$\dot{N}''_A = D\left(\frac{C_1 - C_2}{L}\right) \quad \text{or} \quad \left(\frac{C_1 - C_2}{L/D}\right), \tag{4.21}$$

so that L/D would be the resistive term in the equation and $(C_1 - C_2)$ would represent the "concentration driving force" in the same way that ΔE represents the "electric potential driving force."

(c)   Steady-state diffusive mass transport through composite media

Consider the case of steady-state diffusion of hydrogen (in atomic form) through a thin bimetallic composite of palladium and nickel as shown in Fig. 4.5. Pure hydrogen at 1 atm is in contact with the palladium surface while the other side (the nickel surface) is in contact with hydrogen at 0.1 atm. It is required to develop an expression for the rate of hydrogen leakage through the system based on a diffusion-controlled process. Setting up the relevant

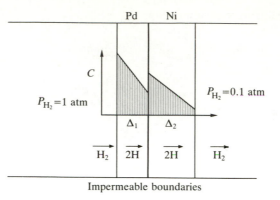

FIG. 4.5.   Transfer of hydrogen through a bimetallic composite of palladium and nickel.

differential equations and boundary conditions describing hydrogen transport through the two phases, we have by analogy with Example 4.3 (a).

Phase 1	Palladium $0 \leqslant x \leqslant \Delta_1$
Fick's law	$\dot{N}''_{H,x} = -D_{H-Pd}\dfrac{dC_H}{dx}$
Continuity	$\dfrac{\partial}{\partial x}(\dot{N}''_{H,x}) = 0$
B.C.1	$x = 0, \forall t, C_H = C_1^0 = q_{Pd}(P_{H_2}^{0,Pd})^{1/2}$
B.C.2	$x = \Delta_1, \forall t, C_H = C_1^*$

Phase 2	Nickel $\Delta_1 \leqslant x \leqslant \Delta_1 + \Delta_2$
Fick's law	$\dot{N}''_{H,x} = -D_{H-Ni}\dfrac{dC_H}{dx}$
Continuity	$\dfrac{\partial}{\partial x}(\dot{N}''_{H,x}) = 0$
B.C.1	$x = \Delta_1, \forall t, C_H = C_2^*$
B.C.2	$x = \Delta_1 + \Delta_2, \forall t, C_H = C_2^0 = q_{Ni}(P_{H_2}^{0,Ni})^{1/2}$

Following identical procedures to those outlined in Example 4.3(a), we obtain the following expressions for the hydrogen concentration profiles in the two phases:

$$C_{H,x}^{Pd} = C_1^0 + (C_1^* - C_1^0)\frac{x}{\Delta_1}, \text{ where } C_1^0 = q_{Pd}(P_{H_2}^{0,Pd})^{1/2} \qquad (4.22)$$

$$C_{H,x}^{Ni} = C_2^0 + (C_2^* - C_2^0)\frac{(\Delta_1 + \Delta_2 - x)}{\Delta_2}, \text{ where } C_2^0 = q_{Ni}(P_{H_2}^{0,Ni})^{1/2} \quad (4.23)$$

Examination of eqns (4.22) and (4.23) in terms of the problem as stated shows that the interfacial concentrations at the gas–metal interfaces are known through Sievert's relationship ($C = q(P_{H_2})^{1/2}$, $q =$ Sievert's constant), but that the interfacial concentrations of hydrogen in palladium and nickel at Δ_1 remain unspecified. However, it is important to note that two other boundary conditions at the metal junction remain to be satisfied, and these fix the values of the hydrogen concentrations at the bimetal interface. They are:

(1) that there are equal fluxes of hydrogen into and out of the interface (this follows from the concept that the interface has zero thickness so no accumulation is possible);
(2) that the activity of hydrogen in the respective phases immediately adjacent to the interface must be equal.

The second condition is equivalent to saying that at $x = \Delta_1$,

$$a_{H/Pd} = a_{H/Ni} \qquad (4.24)$$

or

$$\gamma_{H/Pd} X_{H/Pd}^* = \gamma_{H/Ni} X_{H/Ni}^*, \qquad (4.25)$$

where a is activity, X is mole (atom) fraction, and γ is activity coefficient.

Provided the activity coefficients remain constant over the range of hydrogen concentrations covered (which is the case), we can then say that our final boundary condition is that at $x = \Delta_1$, $\forall t^\dagger$,

$$\frac{C_2^*}{C_1^*} = \frac{\gamma_1}{\gamma_2}\frac{\text{mol.wt.Pd.}}{\text{mol.wt.Ni}}\frac{\rho_{Ni}}{\rho_{Pd}} = m; \qquad \begin{pmatrix} \text{a partition coefficient of} \\ \text{1.53 at } 620\,^\circ\text{C for Ni/Pd} \end{pmatrix}$$

$$m = \frac{\gamma_1}{\gamma_2}\frac{M_1}{M_2}\frac{\rho_2}{\rho_1}.$$

The problem is now completely defined, and fluxes and concentration profiles

---

† $C_1^* = \{X_1/(\text{mol.wt. Pd}/\rho_{Pd})\}$ g atoms of H per cm^3 of Pd.

can be determined. Thus, using the concept of equal fluxes:

$$\dot{N}''_H = -D_{H/Pd}\frac{dC_H}{dx}\bigg|_{x=\Delta_1} = -D_{H/Ni}\frac{dC_H}{dx}\bigg|_{x=\Delta_1}$$

$$D_{H/Pd}\frac{(C_1^0 - C_1^*)}{\Delta_1} = D_{H/Ni}\frac{(C_2^* - C_2^0)}{\Delta_2} \tag{4.26}$$

$$= D_{H/Ni}\frac{(mC_1^* - C_2^0)}{\Delta_2},$$

so that

$$C_1^* = \left(\frac{D_{H/Pd}}{\Delta_1}C_1^0 + \frac{D_{H/Ni}}{\Delta_2}C_2^0\right)\bigg/\left(\frac{D_{H/Pd}}{\Delta_1} + \frac{mD_{H/Ni}}{\Delta_2}\right),$$

which can be substituted in to eqn (4.22) to give the concentration profile of hydrogen in palladium:

$$C^{Pd}_{H,\,x} = C_1^0\left(1 - \frac{x}{\Delta_1}\right) + \frac{x}{\Delta_1}\left(\frac{D_{H/Pd}C_1^0}{\Delta_1} + \frac{D_{H/Ni}C_2^0}{\Delta_2}\right)\bigg/\left(\frac{D_{H/Pd}}{\Delta_1} + \frac{mD_{H/Ni}}{\Delta_2}\right).$$

If we differentiate eqn (4.26) with respect to x, we then obtain an expression for the flux of hydrogen through the bimetal diffusion couple:

$$\dot{N}''_{H,\,x} = -D_{H/Pd}\left(\frac{dC_H}{dx}\right)_x$$

$$= \left(C_1^0 - \frac{C_2^0}{m}\right)\bigg/\left(\frac{\Delta_1}{D_{H/Pd}} + \frac{\Delta_2}{mD_{H/Ni}}\right), \tag{4.27}$$

therefore

$$\dot{N}''_{H,\,x} = 2\dot{N}''_{H_2} = \left\{q_1\left(P^{o,\,Pd}_{H_2}\right)^{1/2} - q_2\left(P^{o,\,Ni}_{H_2}\right)^{1/2}\right\}\bigg/\left(\frac{\Delta_1}{D_{H/Pd}} + \frac{\Delta_2}{mD_{H/Ni}}\right) \tag{4.28}$$

N.B. $H_2 \rightarrow 2H$ at interface, so that $\dot{N}''_{H_2}/1 = \dot{N}''_H/2$ or the molecular flux of hydrogen $= 0.5\dot{N}''_H$.

Short cut

Since we have steady-state diffusion, the fluxes in the two phases are equal and independent of time, so that

$$\dot{N}''_H = D_{H/Pd}\frac{(C_1^0 - C_1^*)}{\Delta_1} = D_{H/Ni}\frac{(C_2^* - C_2^0)}{\Delta_2},$$

or, rearranging

$$\dot{N}_H'' \frac{\Delta_1}{D_{H/Pd}} - C_1^0 = -C_1^*$$

$$\dot{N}_H'' \frac{\Delta_2}{D_{H/Ni}} + C_2^0 = +C_2^* \text{ (or } mC_1^*\text{)}$$

Dividing by m and adding,

$$\dot{N}_H'' \left(\frac{\Delta_1}{D_{H/Pd}} + \frac{\Delta_2}{mD_{H/Ni}} \right) = \left(C_1^0 - \frac{C_2^0}{m} \right),$$

which is same expression as before (eqn 4.27).

Discussion of Example 4.3 (c)

Example 4.3(c) was chosen to demonstrate the important differences between heat and mass transfer phenomena, as well as to pave the way towards a treatment of slag/liquid metal and gas/liquid metal transport phenomena in Chapter 5.

Reference to Fig. 4.5 shows that there is a step change in the concentration of hydrogen at the palladium/nickel interface. Furthermore, there is a step up in the concentration level from left to right, even though both curves point downwards going from left to right. Although one may find this difficult to accept , particularly if one draws analogy with heat transfer, where temperatures θ_1^* and θ_2^* would be equal, it is important to remember that it is the activities, and not the concentrations, of hydrogen, $a_{H,1}^*$ and $a_{H,2}^*$, that are equal at the interface. The reader might wonder whether it would be more convenient to relate the flux of dissolved hydrogen to an activity gradient rather than a concentration gradient, so as to better 'marry' the disciplines of metallurgical thermodynamics and mass transport theory. In general, this procedure is not normally adopted, although expressions are available (e.g. Darken's equations†) and activity changes would be continuous at interfaces in the same way that temperature changes are continuous. It is important to realize, in this regard, that diffusion data on a given system are normally obtained experimentally. Thus, in general, a mass or molar flux, or chemical concentration gradient of the diffusing species is measured and related to time from the start of a diffusion experiment through appropriate transport equations describing the physical set-up. By this means, a chemical diffusion coefficient can be evaluated. For gases, good predictions of chemical diffusion

---

† For a good discussion on this subject, the reader is referred to pp. 431–449 of *Transport phenomena in metallurgy*, by Geiger and Poirier (see Further Reading).

coefficients are possible based on the kinetic theory of gases. In order to convert such data to a thermodynamic level, extra information would be required on the magnitude and way in which the activity coefficient of the diffusing species varied with composition of the system. Since this procedure is one stage removed from practical applications of mass transport theory (i.e., predictions of fluxes and concentration gradients), the approach is normally redundant.

4.4.   UPHILL DIFFUSION

Fick's laws contain the implicit assumption that the driving force for diffusion is the concentration gradient. However, should the activity gradient be of opposite sign, then it is possible for diffusion to occur against, or up, a concentration gradient. For example, should the initial hydrogen composition C_1 in the palladium be somewhat lower than C_2 in the nickel, then at $t=0$, on the basis of mass transport control, the interfacial concentrations C_1^* and C_2^* will immediately readjust to satisfy the condition $C_2^*/C_1^* = m$ at $t=0$. This will necessitate C_1^* dropping below C_1 and C_2^* rising above C_2, together with time-dependent concentration and activity gradients similar to those shown in Fig. 4.6. being set up until steady-state mass flux conditions are achieved. In this way, we see that there will be a nett flux of hydrogen from left to right even though the initial bulk concentration in the nickel was higher than in palladium. It is important to note that our equations are quite capable of handling similar, so-called 'uphill diffusion' problems in ternary or higher order systems provided the activity coefficients of the diffusing species in the respective phases remain constant.

For example, take the best-known case of uphill diffusion, given by Darken (1948), in which two pieces of carbon steel (see Fig. 4.7) were welded together and carbon allowed to diffuse between one bar containing 3.80 wt% Si and 0.48 wt% C and the other containing 0 wt% Si and 0.44 wt% C. Since silicon greatly increases the activity of carbon in iron, the carbon in the high-silicon bar was at a much greater 'potential' than it would appear to be in concentration terms, resulting in analogous concentration profiles to those in Fig. 4.6. Thus after 100 hours or so for carbon to diffuse, we see that the carbon activity is everywhere equal but that most of the carbon has migrated to the low-silicon iron bar.

Note that we could treat the problem quantitatively following the procedures outlined in the previous example. This is only valid provided the silicon concentration levels remain constant everywhere. Since $D_{C/Fe} \gg D_{Si/Fe}$, this is virtually true for the first hundred hours, after which the diffusion of silicon starts to become appreciable. The situation then becomes much more complex and necessitates series or numerical methods of solution. Since the

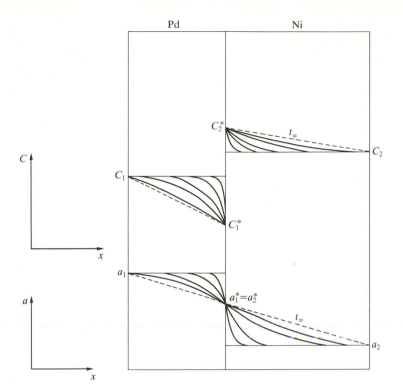

FIG. 4.6. A step change in the concentration of hydrogen at the palladium–nickel interface.

bars are finite in extent, concentrations of silicon and carbon will finally even out as shown in Fig. 4.7.

Uphill diffusion is a possibility even in a two-component system. Thus, in a metastable solid solution of a binary system, spinodal decomposition (Christian 1975; Cahn 1968) will lead to 'uphill' diffusion. In this, activity gradients become negative with respect to chemical gradients, leading to a spontaneous unmixing of the first phase by a diffusional clustering phenomenon. These diffusional processes precede the final precipitation of a fine-scale homogeneous mixture of the two equilibrium phases.

In summary, these complications do not normally arise and concentration gradients are generally in the same direction as activity gradients. Also, many multicomponents systems can be treated as pseudo-binary systems provided the solutes are present in small concentrations. This would apply to many metallurgical refining processes, such as copper- and steel-making, in which small amounts of dissolved gases and other impurities are removed to yield a

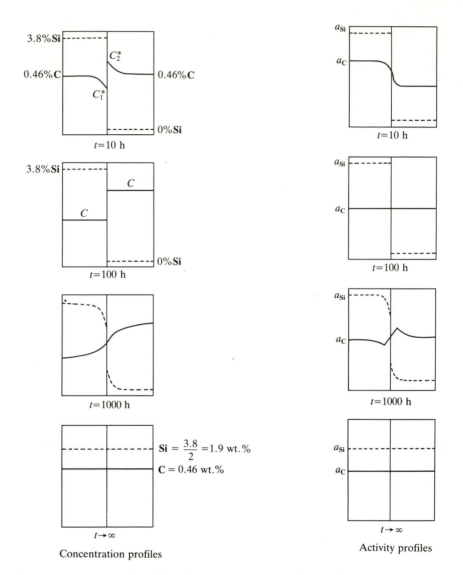

Concentration profiles

Activity profiles

FIG. 4.7.   Example of 'uphill diffusion'. Concentration and activity curves during transient interdiffusion of carbon and silicon through solid steel.

refined product. It is at the other extreme, however (alloy and alloy development work), where such phenomena become important.

4.5. EXERCISES ON STEADY-STATE CONDUCTION AND DIFFUSION

(1) Diffusion of hydrogen through nickel

Hydrogen dissolves in atomic form in nickel; the equation representing the solubility is

$$\log_{10}(S/P^{\frac{1}{2}}) = A/T + B$$

where S = solubility (cm^3 H$_2$ at 0°C, 1 atm H$_2$)/100 g nickel;
P = gas pressure (mm Hg);
T = temperature (K);
$A = 64.5$, $B = 0.082$.
A hydrogen partial pressure of 1 atm is maintained on one side of a thin nickel diaphragm, 5×10^{-4} m thick and 2 cm in diameter, while a partial pressure of zero is maintained at the other side. Assuming steady-state conditions, and that the mass transfer of hydrogen is diffusion-controlled in the metal phase, calculate (at 1223 K):

(a) the molar flux of hydrogen;
(b) the mass flux of hydrogen;
(c) the total mass permeation per hour.

Data:
$D_{H/Ni}$ (m^2 s^{-1}) = $6.7 \times 10^{-8} e^{-\Delta E/RT}$
$\Delta E = 2.2$ kcal gmole^{-1}
R = gas constant
    = 8.35 J mole^{-1} K^{-1} or 1.98 cal gmole^{-1} K^{-1}
ρ_{Ni} (density) = 8900 kg m^{-3}.

(2) Gas-phase diffusion through porous media

The molybdenum windings of a high-temperature furnace element are suspended inside a ceramic tube 0.4 m in length. In order to minimize oxidation of the windings, hydrogen at 1 atm pressure is passed into the tube via an inlet valve, and after circulating around the windings leaves at the same pressure via an outlet valve. Since the ceramic tube becomes slightly porous during operation (ε, the porosity is given by the volume of voids/total volume; $\varepsilon \simeq 9 \times 10^{-6}$), there is a tendency for oxygen in the air surrounding the hot furnace element to diffuse into the tube and oxidize the molybdenum windings. Calculate the hydrogen flow rate necessary to maintain the partial pressure of oxygen inside the tube continuously below 10^{-4} atm. What flow rate of hydrogen at NTP (273 K, 1 atm) does this represent?

Solution

In obtaining a solution, one can define an effective diffusion coefficient to take into account: (i) the reduced cross-sectional surface area available for diffusion (i.e., $A = \varepsilon A_{nominal}$), and (ii) the extra length of the diffusion path over and above ($r_{outside} - r_{inside}$) due to the tortuous nature of the interconnecting passageways between the outer and inner surfaces of the tube, i.e.

$$D_{effective} = \frac{\varepsilon D_{O_2/H_2}}{\tau},$$

where ε is the porosity given by pore volume/total volume, and $\tau =$ is the tortuosity given by

$$\tau = \frac{\text{actual effective thickness of tube}}{\text{nominal thickness of tube}}.$$

Hence, show that

$$\dot{N}_{O_2} = \frac{2\pi\varepsilon DL}{\tau \ln\left(\dfrac{r_0}{r_i}\right)} \left(\frac{P^0_{O_2}}{RT} - \frac{10^{-4}}{RT}\right) \text{mole s}^{-1}.$$

Data:

Operating temperature of alumina tube	1873 K
Inside radius of tube, r_i	20 mm
Outside radius of tube, r_0	25 mm
D_{O_2/H_2} diffusion coefficient of O_2/H_2, at 1600°C,	10^{-3} m^2 s^{-1}
Gas constant, R	8.35 J mole^{-1} K^{-1}
Tortuosity factor, τ	6
Length of tube, L	0.40 m

Note: It is permissible to treat cylindrical systems as planar, provided a logarithmic mean area is defined, i.e.

$$A = \frac{A_0 - A_i}{\ln(A_0/A_i)},$$

and provided A_0/A_i or $r_0/r_i < 2.0$. Assume outside surface of Al_2O_3 is at 1873 K.

(3) Diffusion of gas through two phases: effect of relative solubilities on fluxes

(a) If the gas on either side of a diaphragm, consisting of two metals sandwiched together, is at the same pressure, and the solubility of the gas in metal 1 at this pressure is 1×10^{-6} kg m^{-3}, what is the solubility of gas in metal 2 if the partition coefficient $m (= C_2/C_1)$ is 2? Sketch the concentration profile for equilibrium conditions (i.e., no diffusion).

(b) If now the gas pressure on side 1 is raised, while the gas pressure on side 2 is kept constant, a concentration gradient in 1 will be established and steady-state diffusion will eventually occur. Plot the concentration gradients in metals 1 and 2 for steady state diffusion for $C_1 = (1.1, 2.0$ and $10.0)$ $\times 10^{-6}$ k gm^{-3}, and assume zero interfacial resistance—i.e., chemical equilibrium (+ transport control in metal phase). For the purposes of the present calculation, take $D_1 = D_2$.

(4) Diffusion-controlled thermal decomposition of a hydride

A solid metal hydride MH_2 was contained in a bomb—a metal cylinder 12 mm in diameter—the flat ends of which were made of nickel 1.5 mm thick. The bomb was heated to 1200 K in an evacuated furnace, at which temperature the free energy of formation (ΔG_T^0) of the hydride is $+3000$ kcal kmole^{-1}.

Hydrogen solution in nickel follows Sievert's law, the diffusion coefficient of H atoms being 1.28×10^{-7} m^2 s^{-1}. It was found that 150 kmoles of hydrogen dissolved in 1 m^3 of nickel, under a partial pressure of hydrogen of 1.013×10^5 N m^{-2} (1 atm) although hydrogen did not dissolve in the metal of which the cylindrical walls were made.

Calculate the rate of decomposition of the hydride, assuming hydrogen diffusion in nickel is the rate-controlling step.

(5) Hot metal temperature losses in torpedo cars travelling between blast furnaces and steel-making plants

Since scrap can be a much cheaper source of steel than molten pig iron, it is necessary to maximize scrap usage in steel-making operations. To achieve the correct mass and heat balances for a BOF charge, it is necessary to predict the temperature of the hot metal charged into the BOF furnace. Since the hot metal is transferred from the blast furnaces to the melt shop in transfer cars, where it is poured into a transfer ladle for subsequent charging into the furnace, two major sources of heat loss are incurred: the first is conductive heat loss through the refractory walls of the transfer car; the second is radiative heat loss when the hot metal is poured from the transfer cars into the transfer ladle (see Fig. 4.8).

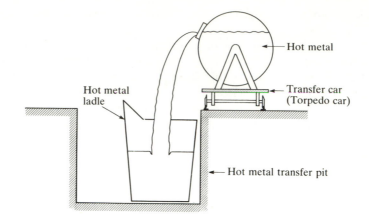

FIG. 4.8.   Hot metal temperature losses in torpedo cars travelling between blast furnace and steel-making plant.

Part I

A torpedo car containing 95 tonnes of hot metal at 1750 K is transferred from the blast furnaces to the melt shop by locomotive, the journey taking 1.5 hours. Calculate the final hot metal temperature when it arrives at the melt shop, knowing that the rate of heat loss from the hot metal depends on steady-state conduction of heat through a 0.23 m thick layer of firebrick whose mean thermal conductivity is $1.73 \ W \ m^{-1} \ K^{-1}$. The total internal surface area of brick equals 50 m^2 and the outer surface of the torpedo car is at 470 K. The heat capacity of the hot metal can be taken as $460 \ J \ kg^{-1} \ K^{-1}$.

Part II

Calculate the temperature drop (and hence final hot metal temperature) resulting from the radiative heat losses from the stream and continuously exposed hot metal surface in the filling ladle. The surface area of the stream is approximately 2.75 m^2 (3.6 m trajectory $\times$ 0.75 m perimeter), the surface area of the ladle is 11 m^2 (3.6 m mean diameter) and the total pouring time is 6 minutes. The emissivity of the hot metal can be taken as 0.4, the Stefan–Boltzmann constant is $5.146 \times 10^{-10} \ W \ m^{-2} \ K^{-4}$. You may ignore radiative heat losses after pouring has stopped since a protective slag/kish layer generally forms and insulates the metal from further radiative heat loss. Make any other assumptions you consider justified and state them explicitly.

  Hint: Part I. Since the thermal resistance to heat transfer lies in the brickwork rather than in the hot metal in the torpedo car, the hot metal temperature can, for all practical purposes, be assumed everywhere equal.

Thus:

Heat loss from metal/time = heat conducted through brickwork/time

or

$$\rho_{\text{H.M.}} C_{\text{H.M.}} \frac{\mathrm{d}\theta}{\mathrm{d}t} = k_{\text{B}} A_{\text{B}} \frac{(\theta - \theta_0)}{L}.$$

Part II may be treated in a similar fashion.

4.6. THE 'EXACT' FORM OF FICK'S LAW OF DIFFUSION

Since diffusion represents the physical motion of a species through a solvent, no binary system can ever be truly motionless unless any volume loss caused by the egress of a diffusion species from a control volume is made up by an equal and opposite volume increase by counterdiffusion of solvent atoms. To decide whether the bulk movement of a system is essentially zero, and pure diffusion applies, one can first define a bulk molar velocity $\bar{U}$ equal to

$$\bar{U} = \sum_{i=1}^{i=n} C_i U_i \bigg/ \sum_{i=1}^{i=n} C_i. \tag{4.29}$$

We see that, for a binary system,

$$\bar{U} = \left(\frac{C_A U_A + C_B U_B}{C_A + C_B} \right) = (X_A U_A + X_B U_B). \tag{4.30}$$

Clearly the terms $\bar{U}(C_A + C_B)$, $C_A U_A$ and $C_B U_B$ represent fluxes respectively for the total mixture, for component A, and for component B. If we now decide that, rather than knowing about the total flux $(C_A U_A)$, of A, we wish to know only about the *diffusive component* of that total flux, we would have to write

$$\dot{N}''_{A,\text{Diffusion}} = \dot{N}''_{A,\text{Total}} - \dot{N}''_{A,\text{Convection}}, \tag{4.31}$$

or

$$\dot{N}''_{A,\text{Diffusion}} = C_A U_A - C_A \bar{U}. \tag{4.32}$$

Replacing $\dot{N}''_{A,\text{Diffusion}}$ by a customary diffusion coefficient and concentration gradient, we can therefore write

$$-D_{AB} \frac{\mathrm{d}C_A}{\mathrm{d}x} = C_A U_A - C_A \bar{U}. \tag{4.33}$$

or, by definition of $\bar{U}$,

$$-D_{AB} \frac{\mathrm{d}C_A}{\mathrm{d}x} = \dot{N}''_A - C_A \left(\frac{C_A U_A + C_B U_B}{C_A + C_B} \right), \tag{4.34}$$

giving

$$-D_{AB}\frac{dC_A}{dx} = \dot{N}''_A - X_A(\dot{N}''_A + \dot{N}''_B),\qquad(4.35)$$

or

$$\dot{N}''_A = X_A(\dot{N}''_A + \dot{N}''_B) - D_{AB}\frac{dC_A}{dx}.\qquad(4.36)$$

This is known as the 'exact form' of Fick's first law, for the diffusion of A through a binary mixture of A and B.

(a) Simplifications

(i) Equimolar/equi-atom counterdiffusion

We see from eqn (4.36) that if $\dot{N}''_B = -\dot{N}''_A$ we revert to the approximate form of Fick's law we have used in the text. An example given later concerning the diffusion of the radioactive silver isotope ^{105}Ag through a solvent matrix of ^{108}Ag atoms will meet these conditions, since a process of substitutional diffusion takes place, every atom of ^{105}Ag being replaced with an ^{108}Ag, i.e. with no voids or density changes in the sample. Similarly, for equimolar counterdiffusion in binary gas systems, we would also have $\dot{N}''_B = -\dot{N}''_A$, so that

$$\dot{N}''_A = X_A(\dot{N}''_A - \dot{N}''_A) - D_{AB}\frac{dC_A}{dx},$$

or

$$\dot{N}''_A = -D_{AB}\frac{dC_A}{dx}.\qquad(4.37)$$

(ii) Solute present at low concentrations

Referring again to the exact form of Fick's first law of diffusion, if $X_A \ll 1$, the convective term is negligible provided $U_B = 0$, so that once more

$$\dot{N}''_A = -D_{AB}\frac{dC_A}{dx}.\qquad(4.38)$$

A practical example would be the diffusion of carbon through a steel bar. The carbon atoms, being at low concentration and smaller than their nearest-neighbour iron atoms, move interstitially through the matrix of iron atoms, for which $U_B = 0$.

(iii)   The diffusivity identity, $D_{AB} = D_{BA}$

For binary mixtures of gases at constant pressure everywhere, C_A and C_B are only space-dependent, rather than time, dependent. Consequently, since $C_T = C_A + C_B = \text{constant}$,

$$\frac{\partial C_A}{\partial x} + \frac{\partial C_B}{\partial x} = 0. \tag{4.39}$$

Writing Fick's law for species A and B and adding the result:

$$\dot{N}''_A = X_A(\dot{N}''_A + \dot{N}''_B) - D_{AB}\frac{dC_A}{dx},$$

$$\dot{N}''_B = X_B(\dot{N}''_B + \dot{N}''_A) - D_{BA}\frac{dC_B}{dx}$$

$$\overline{\dot{N}''_A + \dot{N}''_B = (\dot{N}''_A + \dot{N}''_B)1 - D_{AB}\frac{dC_A}{dx} - D_{BA}\frac{dC_B}{dx}}, \tag{4.40}$$

from which it is evident that

$$-D_{AB}\frac{dC_A}{dx} = D_{BA}\frac{dC_B}{dx},$$

Replacing $\dfrac{dC_B}{dx}$ with $-\dfrac{dC_A}{dx}$, on the basis of eqn (4.39), we arrive at:

$$D_{AB} = D_{BA}. \tag{4.41}$$

This is a curious result that one might not have anticipated. For example, $D_{H_2/He} \gg D_{H_2/Hg}$ since the mercury atom is much heavier and slower than a helium atom. However, the diffusive capabilities of mercury vapour in a large volume of hydrogen is as high as that of hydrogen atoms moving through mercury vapour. One can imagine the higher velocity hydrogen atoms colliding with the more slowly moving mercury atoms, hastening their diffusion.

(iv)   Dilute multicomponent systems

When all solutes are present at low concentration, their effect on each other can be ignored and each individually with the solvent can be treated as a simple binary system at low solute levels (case (ii)).

(b)   Worked example: evaporation of zinc

A sample of liquid zinc is held in a narrow alumina tube as shown in Fig. 4.9 and allowed to vaporize and diffuse. Argon at 1 atm pressure is flushed past

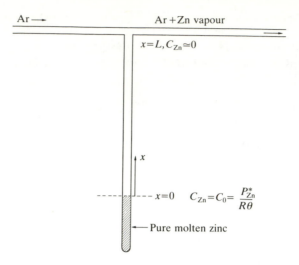

FIG. 4.9.   Evaporation and diffusion of zinc up a narrow vertical tube of alumina.

the neck of the tube so as to maintain the exhaust zinc vapour pressure at less than 0.2 per cent of its equilibrium value at temperature θ. Assuming no turbulence is generated at the 'T' section, show the importance of allowing for convection within the ascending column of argon gas and zinc vapour by estimating zinc fluxes with and without convection.

Solution

Writing the approximate form of Fick's law (i.e., neglecting convection) we have, for steady-state, one-dimensional diffusion, that

$$\dot{N}''_{Zn,\,approx} = -D_{Zn/Ar}\frac{dC_{Zn}}{dx}. \tag{4.42}$$

Given the boundary conditions, $x=0$, $C_{Zn}=P_{Zn}/R\theta$; $x=L$, $C_{Zn}=0$, we have the solution:

$$\dot{N}''_{Zn,\,approx} = \frac{-D_{Zn/A}}{L}C^*_{Zn}. \tag{4.43}$$

Taking the exact form for Fick's law, in which we recognize that the upward diffusion of zinc vapour can generate a finite bulk velocity, we have

$$\dot{N}''_{Zn} = X_{Zn}(\dot{N}''_{Zn} + \dot{N}''_{Ar}) - D_{Zn/Ar}\frac{dC_{Zn}}{dx}. \tag{4.44}$$

Since we know that argon has zero solubility in liquid zinc, it follows that the flux of argon vapour at $x=0$ must be zero, i.e. $\dot{N}''_{Ar}=0$. Since steady-state conditions apply throughout the tube, we have requirement that $\dot{N}''_{Ar}$ is everywhere zero between $x=0$ and $x=L$. Consequently, we have

$$\dot{N}''_{Zn}=X_{Zn}(\dot{N}''_{Zn}+0)-D_{Zn/Ar}\frac{dC_{Zn}}{dx},$$

or

$$\dot{N}''_{Zn}=\frac{D_{Zn/Ar}}{\left(1-\dfrac{C_{Zn}}{C_T}\right)}\frac{dC_{Zn}}{dx}. \qquad (4.45)$$

Taking the 'short-cut' route to solve this problem:

$$\dot{N}''_{Zn}\int_0^L dx=D_{Zn/Ar}\int_{C^*_{Zn}}^0\frac{dC_{Zn}}{(1-C_{Zn}/C_T)},$$

or

$$\dot{N}''_{Zn}L=-D_{Zn/Ar}C_T\left\{\ln\left(1-C_{Zn}/C_T\right)\right\}\Big|_{C^*_{Zn}}^0$$

or

$$\dot{N}''_{Zn}=\frac{D_{Zn/Ar}C_T}{L}\left\{+\ln\left(1-\frac{C^*_{Zn}}{C_T}\right)\right\}. \qquad (4.46)$$

Defining the error between $\dot{N}''_{Zn,\,approx}$ and $\dot{N}''_{Zn}$ as

$$E=\left(\frac{\dot{N}''_{exact}-\dot{N}''_{approx}}{\dot{N}''_{exact}}\right),$$

we have, for the per cent error %E,

$$\%\,E=\left(1-\frac{X^*}{\ln(1-X^*)}\right)\times 100. \qquad (4.47)$$

Figure 4.10 shows a plot of the errors involved versus the mole fraction of zinc vapour diffusing from the interface. As seen, errors are reasonable up to a zinc vapour mole fraction of about 0.1 (i.e., 5% error). Beyond this, errors become serious, and by the time the mole fraction of zinc vapour becomes unity, the liquid zinc is boiling, and zinc vapour moves away solely by convection. Our estimated error then becomes 100%, since the diffusive component is negligible.

Before leaving this problem, the more probing reader may wonder how a concentration gradient of argon can exist between $X_A=1$ at $x=L$ and $X_A=1-X^*_{Zn}$ at $x=0$, without any diffusion of argon actually occurring. The answer is that the upward convective flux of argon and zinc vapour exactly, and everywhere, compensates for the downward diffusive flux of argon, so

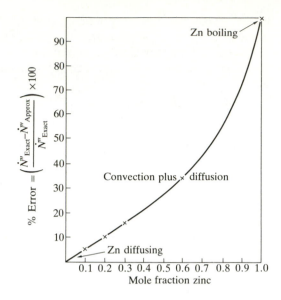

Fɪɢ. 4.10.   Errors caused through neglecting the convective flux of zinc vapour versus its mole fraction at the molten zinc surface.

that $U_{Ar}=0$ everywhere. It is left as an exercise to show the validity of this statement and to obtain a solution for the zinc concentration profile shown in Fig. 4.11, where the vapour pressure of zinc was taken as 0.5 atm, viz,

$$C_{Zn}=C_T\left\{1-\left(1-X_{Zn}^*\right)^{(L-x)/L}\right\}.\qquad(4.48)$$

Note that the requirement of constant pressure implies that $X_{Ar,x}+X_{Zn,x}=1$ everywhere.

4.7.   UNSTEADY-STATE TRANSFER THROUGH MOTIONLESS MEDIA: GENERAL

The term 'unsteady-state' is normally ascribed to systems in which temperature and/or concentrations vary with time at any particular location. Take for example, the case of a small element of solid or stagnant fluid through which solute A is transferring in the x direction, such as that shown in Fig. 4.2.

Then, using eqn (4.7) to write down the general form of the continuity equation, we may delete non-relevant terms:

$$\frac{\partial C_A}{\partial t}=\dot{R}'''-\frac{\partial \dot{N}''_{A,x}}{\partial x}-\frac{\partial \dot{N}''_{A,y}}{\partial y}-\frac{\partial \dot{N}''_{A,z}}{\partial z}.$$

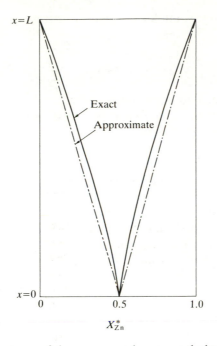

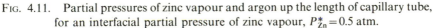

FIG. 4.11. Partial pressures of zinc vapour and argon up the length of capillary tube, for an interfacial partial pressure of zinc vapour, $P_{Zn}^* = 0.5$ atm.

This yields

$$\frac{\partial C_A}{\partial t} = -\frac{\partial}{\partial x}(\dot{N}_{A,x}''),$$

which, combined with Fick's first law of diffusion gives

$$\frac{\partial C_A}{\partial t} = -\frac{\partial}{\partial x}\left(-D\frac{\partial C_A}{\partial x}\right). \qquad (4.49)$$

This equation says that the difference between the output and input of solute A (r.h.s of equation) equals the amount of A which accumulates in the differential volume element.

If D (or D_{AB}) is a constant, independent of concentration, then:

$$\frac{\partial C}{\partial t} = D\frac{\partial^2 C}{\partial x^2}. \qquad (4.50)$$

This equation is known as *Fick's second law*.

The analogous expression for transient heat conduction, *Fourier's second law*, is:

$$\frac{\partial \theta}{\partial t} = \alpha \frac{\partial^2 \theta}{\partial x^2}, \tag{4.51}$$

where α is the thermal diffusivity ($m^2\,s^{-1}$). Note that α has the same units as molecular diffusivity ($m^2\,s^{-1}$). Again, the equation is a combination of the conservation of energy equation and Biot's or Fourier's first law and represents one-dimensional unsteady-state conduction of heat.

(a) Example of unsteady-state diffusion: plane source solution

Suppose that we wish to calculate the diffusion coefficient of radioactive silver in silver and intend to do this by measuring various concentration–time profiles developed when a thin sandwich of radioactive silver is inserted between two silver bars, as shown in Fig. 4.12. The types of concentration profiles we might intuitively expect are sketched in the diagram. Obviously, as

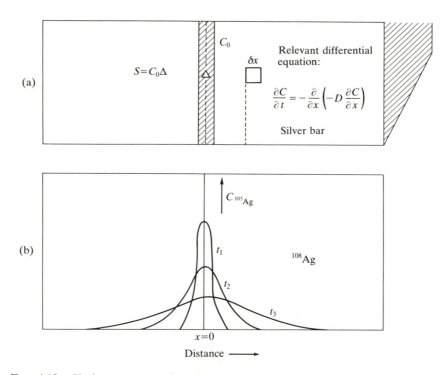

FIG. 4.12. Various concentration–time profiles (b) developed when a thin sandwich of radioactive silver, ^{105}Ag is inserted between two silver bars (a).

time proceeds, a greater dispersion of radioactive silver will take place so that concentration profiles will tend to flatten and widen. Similarly, at very great distances (e.g. 4 cm in typical diffusion work in solids), ^{105}Ag will not have penetrated so that $C_{105_{Ag}} \to 0$, while at the centre $x = 0$, of the diffusing source, the slope of the curves must be zero since there can be no net flux of ^{105}Ag from right to left or vice versa. This, of course, follows from Fick's law, since

$$\dot{N}''_{105_{Ag}}\bigg|_{\substack{x=0 \\ t=t}} = 0 = -D \frac{\partial C_{105_{Ag}}}{\partial x}\bigg|_{\substack{x=0 \\ t=t}}.$$

Finally, the area under each of the curves must be the same, since the total amount of ^{105}Ag remains constant (half-life $\gg$ time of experiment).

Let us suppose that a strip of ^{105}Ag, 0.01 mm thick, is plated at $x = 0$; then the number of grams deposited per unit cross-sectional area is equal to the (density of ^{105}Ag) $\times$ volume $= 9.5 \times 1 \times 1 \times 0.01 = 0.095$ g mm^{-2}. If we call the amount of silver plated per unit surface area, S (g mm^{-2}) and suppose that the thickness Δ of plated material is much less than diffusion penetration distances, we can regard it as being a plane source (i.e. of zero thickness).

Formulating the problem mathematically, we have

$$\frac{\partial C}{\partial t} + \frac{\partial \dot{N}''_x}{\partial x} = 0 \text{ and } \dot{N}''_x = -D \frac{\partial C}{\partial x},$$

which may be combined to give

$$\frac{\partial C}{\partial t} = D \frac{\partial^2 C}{\partial x^2},$$

assuming $D_{105_{Ag}/108_{Ag}} = $ constant (i.e. independent of concentration). Some boundary conditions that we can try using to obtain a solution to Fick's second law for the particular case in question are:

B.C.1 $\qquad\qquad\qquad C = 0; \; x = x, \; t \to \infty,$

B.C.2 $\qquad\qquad\qquad C = 0, \; x \to \infty, \; t = t,$

B.C.3 $\qquad\qquad\qquad C = \infty, \; x = 0, \; t = 0$

$\qquad\qquad\qquad$ (plane source, zero thickness)

B.C.4 $\qquad\qquad\qquad S = \displaystyle\int_{x=-\infty}^{x=+\infty} C \, dx \qquad$ (conservation of mass),

B.C.5 $\qquad\qquad\qquad \dfrac{\partial C}{\partial x}\bigg|_{\substack{x=0 \\ t=t}} = 0, \; x = 0, \; t = t.$

The general solution to this partial differential equation can be obtained mathematically by a method known as separation of variables, or by Laplace transform techniques. These show that

$$C_{x,t} = \frac{A}{\sqrt{t}} e^{-x^2/Bt} \tag{4.52}$$

is a general solution. It is readily shown that $B = 4D$ (see Exercise 4.8(1)). In order to evaluate the constant A in the above expression, we see that application of boundary conditions, 1, 2, 3 to eqn (4.52) yield indeterminate *particular* solutions, i.e.

B.C.1 $\quad 0 = \dfrac{A}{\infty} e^{-x^2/4D(\infty)};$ $\qquad$ B.C.2 $\quad 0 = \dfrac{A}{\sqrt{t}} e^{-\infty^2/4D(t)};$

B.C.3 $\quad \infty = \dfrac{A}{\sqrt{0}} e^{-0^2/4D(0)}.$

The fourth boundary condition, expressing conservation of mass, is

$$S = \int_{x=-\infty, t=t_1}^{x=+\infty, t=t_1} C_{x,t}\, dx.$$

Substituting for C_{x,t_1} we have

$$S = \int_{x=-\infty, t=t_1}^{x=\infty, t=t_1} \frac{A}{\sqrt{t_1}} e^{-x^2/4Dt_1}\, dx. \tag{4.53}$$

We can simplify this integral, by defining a new variable η, where:

$$\eta^2 = \frac{x^2}{4Dt}.$$

It follows that

$$\partial \eta = \frac{\partial x}{\sqrt{4Dt}}$$

and as

$$x \to \infty,\ t = t_1;\ \eta \to \infty,$$
$$x \to -\infty,\ t = t_1;\ \eta \to -\infty,$$

so that eqn (4.53) can be rewritten as

$$S = \frac{A}{\sqrt{t}} \sqrt{4Dt} \int_{\eta=-\infty}^{\eta=\infty} e^{-\eta^2}\, d\eta. \tag{4.54}$$

Figure 4.13 shows how the function $y = e^{-\eta^2}$ varies with η. The well known Gaussian error distribution curve is, incidentally, of this form. As seen, $y = 1$ at $\eta = 0$; $y = 0$ at $\eta = \pm\infty$; $y = 0.368$ at $\eta = \pm 1$. The area under the curve can be

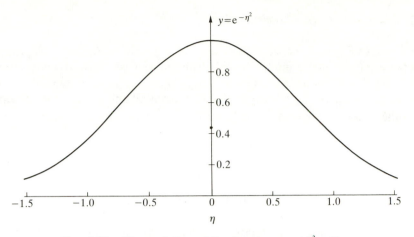

FIG. 4.13. The variation of the function $y = e^{-\eta^2}$ with η.

shown to equal $\sqrt{\pi}$. It follows therefore, that

$$S = \frac{A}{\sqrt{t}} \sqrt{4Dt} \sqrt{\pi},$$

so that

$$A = \frac{S}{\sqrt{4\pi D}},$$

and the particular solution is

$$C_{x,t} = \frac{S}{\sqrt{4\pi Dt}} e^{-x^2/4Dt}, \tag{4.55}$$

It can easily be verified that this solution also satisfies the other boundary conditions previously enumerated. Thus, by feeding in experimental data on $C_{x,t}$, a value for D, the diffusion coefficient, may be derived.

(b)   Plane source: initial finite bulk concentration of solute.

Referring to Fig. 4.12, suppose the initial bulk concentration of radioactive silver is C_B: then the same solution can be adjusted by defining an 'excess' plane source concentration, S_{xs};

$$C_{x,t} = C_B + \frac{S_{xs}}{\sqrt{4\pi Dt}} e^{-x^2/4Dt}. \tag{4.56}$$

(c)   Plane source: semi-infinite medium

Referring to Fig. 4.14, we see in this case that a layer of ^{105}Ag has been placed on the surface of a semi-infinite bar of silver and we wish to know, once again, how the concentrations of the diffusing solute vary with time at some particular location $C_{x,t}$. Clearly the same differential applies as before, except that the area under the curve in Fig. 4.13 equals $\sqrt{\pi}/2$, so that $A = 2S/\sqrt{(4\pi D)}$ and

$$C_{x,t} = \frac{S}{\sqrt{(\pi Dt)}} e^{-x^2/4Dt}. \tag{4.57}$$

Alternatively, one may invoke the principle of 'reflection and superposition' to arrive at the same result. Thus, one adds the contribution of the now non-existent left-hand source to the right-hand side, so that all concentrations on the r.h.s. are double those predicted for an inifinite bar of silver. This concept is depicted in Fig. 4.15.

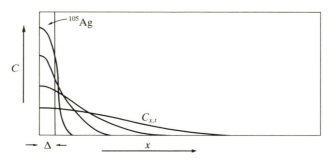

FIG. 4.14.   Variation with position and time of the concentrations of ^{105}Ag, the diffusing solute, in a semi-infinite body of silver.

4.8.   EXERCISES ON TRANSIENT CONDUCTION AND DIFFUSION WITH PLANE SOURCES

(1)   Verification of plane source analytical solution

Show that the solution $C_{x,t}(A/\sqrt{t})e^{-x^2/Bt}$ satisfies Fick's second law equation,

$$\frac{\partial C}{\partial t} = D \frac{\partial^2 C}{\partial x^2},$$

and find the necessary value of B.

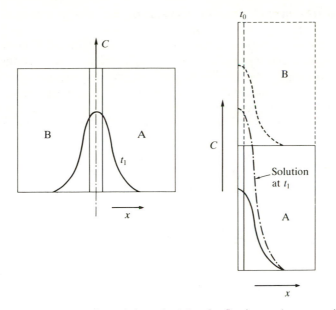

FIG. 4.15.   Demonstration of the principle of reflection and superposition.

(2)   Spot and butt welding problems

A butt weld is applied to two sheets of metal placed next to each other (see figure). Assume that an instantaneous heat supply of q'' J m^{-2} is supplied at time $t=0$ at $x=0$ (where the sheets are to be joined). Derive an expression for

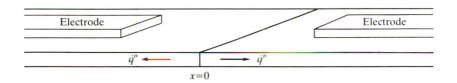

the temperature distribution at any time t, and any distance x away from the weld zone. You may assume:

(1) heat losses from the surfaces of the sheet negligible in comparison with unsteady state conduction of heat through the sheets;

(2) no end or edge effects, (i.e. one-dimensional flow);

(3) the sheets sufficiently wide in the x direction that they can be considered to be semi-infinite (i.e. $\theta=0$, $x\rightarrow\infty$, $t=t$).

(3)   Heat treatment of alloys for homogenization prior to rolling procedures

Solidification of industrial alloys invariably results in the phenomenon of microsegregation, in which unacceptably high solute concentrations occur in localized regions of the alloy matrix. This microsegregation can often be minimized by subsequent heat treatment at high temperature, whereby diffusional processes result in the transport of solute from regions of higher concentration to those of lower concentration. This diffusional process goes under the name 'homogenization'. Taking the simplest example of micro-segregation, involving the deposition of an enriched layer of solute B in the form of a very thin plate, of thickness ΔX, and concentration C_B^0, surrounded by solvent A (see figure), calculate the heat treatment times required to reduce

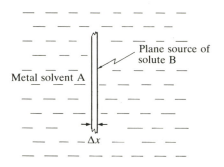

the concentration of solute B in the plate to half its original value, given:

(a) $\qquad\qquad \Delta x = 1\ \mu\text{m}, \qquad D_{B/A} = 10^{-16}\,\text{m}^2\,\text{s}^{-1};$

(b) $\qquad\qquad \Delta x = 4\ \mu\text{m}, \qquad D_{B/A} = 10^{-16}\,\text{m}^2\,\text{s}^{-1}.$

Comment on the results.

   Hint: Assume concentration profiles can be described by eqn (4.56).

(4)   Martensite formation resulting from rapid cooling.

Flash butt welding is commonly used for joining steel coils together to form continuous coil. In the process, a welding and clamping electrode is attached to either coil, the coils are brought together and a welded joint results. Since the welding is carried out very rapidly, the electrically generated heat may be regarded as an instantaneous heat source at $t=0$ and $x=0$ (see figure). Assume that the temperature varies with time and location as follows:

$$\theta_{x,t} = \frac{A}{\sqrt{t}}\,e^{-x^2/4\alpha t},$$

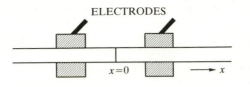

where $\theta_{x,t}$ is the temperature at x at time t, and α is the thermal diffusivity.

In the present case, a 0.8% carbon steel (eutectoid steel) was to be welded, and the possibility of martensite formation was considered. It is known that martensite will form if the instantaneous cooling rate exceeds 550 K s^{-1} as the temperature of the heated steel falls below 995 K (i.e. the A_1 temperature). Calculate the maximum width of the martensite zone that will be observed, assuming sufficient power is generated at the interface to momentarily melt a 10 μm thick layer of steel at the interface, $x = 0$.

Data:

Thermal diffusivity of steel, α	$10 \times 10^{-6} \text{ m}^2 \text{ s}^{-1}$
Heat capacity of steel	$586 \text{ J kg}^{-1} \text{ K}^{-1}$
Latent heat of fusion	274 kJ kg^{-1}
Mean melting temperature	1703 K

Convective and radiative heat losses from the surfaces can be ignored in comparison with conductive heat losses through the strip. Similarly, although the use of eqn (4.58) is not rigorous, in that it ignores the melting and solidification aspects, these times are short, so that it describes the process to a good first approximation.

4.9.   UNSTEADY-STATE DIFFUSION: EXTENDED SOURCES

An important class of problems dealing with unsteady-state transfer of heat or mass through motionless media involve systems in which an extended source of solute, rather than a plane source, exists. Take for example, the case of two semi-infinite bars of steel welded together at $x = 0$, one containing carbon at a concentration level of C_0, the other containing no carbon at all (Fig. 4.16(a)). Figure 4.16(b) shows the initial carbon concentration profiles before diffusion is allowed to occur. When the bars are heated to, say, 1273 K, carbon diffusion will become appreciable, and the type of concentration profiles that we can expect after various times are sketched in Fig. 4.16(c).

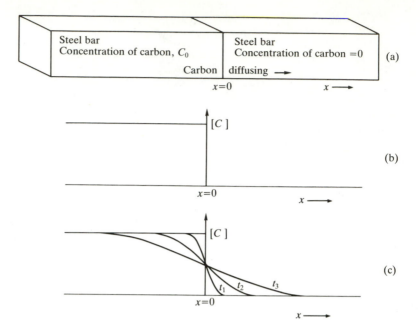

Fig. 4.16.   Example of unsteady-state diffusion: extended source solutions.

In order to calculate the rate of penetration of carbon into the initially carbon-free steel, we observe that the same partial differential equation as before (Fick's second law) has to be solved, using the boundary conditions:

B.C.1 $\qquad x \to +\infty,\ t = t,\ C = 0;$

B.C.2 $\qquad x \to -\infty,\ t = t,\ C = C_0.$

In order to solve

$$\frac{\partial C}{\partial t} = D \frac{\partial^2 C}{\partial x^2} \tag{4.58}$$

(D assumed constant) for the particular case at hand, we can make use of our previous solution for plane sources. Thus, by considering that our extended source (i.e., $C = C_0$, $-\infty < x < 0$, $t = 0$) is really equivalent to an infinite number of plane sources, we can superpose the corresponding infinite number of elementary solutions $\delta C_{x,t}$, for the carbon concentration at location x, and time t from the start of the diffusion processes, to arrive at the "summed" carbon concentration $C_{x,t}$. In this way we can obtain an expression describing how the carbon concentration varies with time and location in the two bars. Referring to Fig. 4.17, consider what effect the presence of a plane source located a distance x' away from a plane x_1 has on the carbon concentration at

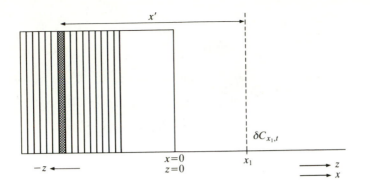

FIG. 4.17. The effect of the presence of a plane source located at distance x' on the carbon concentration at x_1.

x_1. According to eqn (4.55), we see that its contribution to the carbon content at x_1 will be

$$\delta C_{x_1,t} = \frac{S}{\sqrt{4\pi Dt}} e^{-x'^2/4Dt}. \qquad (4.59)$$

Suppose for example, that x_1 is located 2 cm from $x=0$, while the plane source is located at a distance $z=-3$ cm ($z=0$ at $x=0$); then $x'=5$ cm, or, mathematically,

$$x' = x_1 - z$$

so that

$$dx' = -dz \quad \text{or} \quad \delta x' = -\delta z.$$

We want to sum all contributions from imaginary plane sources located between $z=0$ and $z=-\infty$, i.e.

$$\delta C_{x_1,t} = \frac{C_0(-\delta z)e^{-(x_1-z)^2/4Dt}}{\sqrt{4\pi Dt}}, \qquad (4.60)$$

So that

$$\Sigma \delta C_{x_1,t} = C_{x,t} = \int_{z=0}^{z=-\infty} \frac{C_0(-\delta z)e^{-(x_1-z)^2/4Dt}}{\sqrt{4\pi Dt}},$$

or

$$C_{x_1,t} = + \int_{z=-\infty}^{z=0} \frac{C_0 e^{-(x_1-z)^2/4Dt}}{\sqrt{4\pi Dt}} \, dz. \qquad (4.61)$$

We can simplify the expression by defining η, where

$$\eta = \frac{x_1 - z}{\sqrt{4Dt}}.$$

Thus

$$d\eta = \frac{-dz}{\sqrt{4Dt}},$$

and also when $z = -\infty$, $\eta = \infty$, and when $z = 0$, $\eta_1 = x_1/\sqrt{4Dt}$.

The transformed integral therefore becomes

$$C_{x_1,t} = C_0 \int_{\eta = \infty}^{\eta = x_1/\sqrt{4Dt}} \frac{-e^{-\eta^2}\sqrt{4Dt}}{\sqrt{4\pi Dt}}\, d\eta$$

$$= \frac{C_0}{\sqrt{\pi}} \int_{\eta_1 = x_1/\sqrt{4D_1 t}}^{\eta = \infty} e^{-\eta^2}\, d\eta, \tag{4.62}$$

or, writing this somewhat differently, since we know the value of this integral between 0 and ∞ (Fig. 4.13) (i.e., $\sqrt{\pi}/2$):

$$C_{x_1,t} = \frac{C_0}{\sqrt{\pi}} \left(\int_{\eta = 0}^{\eta = \infty} e^{-\eta^2}\, d\eta - \int_{\eta = 0}^{\eta = x_1/\sqrt{4Dt}} e^{-\eta^2}\, d\eta \right)$$

$$= \frac{C_0}{\sqrt{\pi}} \left(\frac{\sqrt{\pi}}{2} - \int_{\eta = 0}^{\eta = x_1/\sqrt{4Dt}} e^{-\eta^2}\, d\eta \right)$$

$$= \frac{C_0}{2} \left(1 - \frac{2}{\sqrt{\pi}} \int_{\eta = 0}^{\eta = x_1/\sqrt{4Dt}} e^{-\eta^2}\, d\eta \right).$$

The integral

$$\frac{2}{\sqrt{\pi}} \int_{\eta = 0}^{\eta = \eta_1} e^{-\eta^2}\, d\eta$$

is known as Gauss's error function (or 'erf(η)' for short). Although it cannot be solved analytically, tables are available and are reproduced in Appendix 3, Table A3.1. Therefore, recasting our solution, we finally arrive at

$$C_{x_1,t} = \frac{C_0}{2}\{1 - \mathrm{erf}(\eta_1)\} = \frac{C_0}{2}\{1 - \mathrm{erf}(x_1/\sqrt{4Dt})\},$$

or

$$C_{x_1,t} = \frac{C_0}{2}\,\mathrm{erfc}\,(x_1/\sqrt{4Dt}). \tag{4.63}$$

Short cut

When tackling extended-source problems, similar to the carbon diffusion one just considered, it is more appropriate to assume that the general solution to

Fick's or Fourier's second laws for extended sources are of the form

$$C_{x,t} = A + B\,\mathrm{erf}(x_1/\sqrt{4Dt}) \tag{4.64}$$

and solve for A and B.

Thus, in the previous example, we can apply the two boundary conditions to obtain the required particular solution. Thus applying B.C.1 ($x = -\infty$, $t = t$, $C = C_0$) to eqn (4.64) gives

$$C_0 = A + B\,\mathrm{erf}(-\infty) = A + B(-1) = A - B.$$

Applying B.C.2 ($x = +\infty$, $t = t$, $C = 0$) gives

$$0 = A + B\,\mathrm{erf}(+\infty) = A + B(+1) = A + B.$$

Solving these two simultaneous equations, $A = C_0/2$, $B = -C_0/2$, and

$$C_{x,t} = \frac{C_0}{2}\{1 - \mathrm{erf}(x/\sqrt{4Dt})\}, \tag{4.65}$$

which is what we had before.

Finally, let us calculate the total amount of carbon that has diffused out per unit surface area during this time. Differentiating eqn (4.63) with respect to x, we have

$$\frac{\partial C_{x,t}}{\partial x} = -\frac{C_0}{2}\frac{\partial}{\partial x}\left(\frac{2}{\sqrt{\pi}}\int_{\eta=0}^{\eta=\eta} e^{-\eta^2}\,d\eta\right),$$

where $\eta = x/\sqrt{4Dt}$. Since $\partial/\partial x \equiv (\partial/\partial\eta)\cdot\partial\eta/\partial x$, we have

$$\frac{\partial}{\partial x} \equiv \frac{1}{\sqrt{4Dt}}\frac{\partial\eta}{\partial x}$$

Rewriting eqn (4.43), noting that $(\partial/\partial\eta)\int f(\eta)\,d\eta = f(\eta)$, we have

$$\frac{\partial C_{x,t}}{\partial x} = -\frac{C_0}{2}\frac{2}{\sqrt{\pi}}\frac{1}{\sqrt{4Dt}}e^{-\eta^2}.$$

We wish to evaluate $\partial C_{x,t}/\partial x$ at $x = 0$ (or $\eta = 0$), since we know from Fick's law that the instantaneous flux of carbon out of the surface is

$$\dot{N}''_{x=0,t=t} = -D\frac{\partial C}{\partial x}\bigg|_{x=0,t=t} = -D\frac{-C_0}{\sqrt{4\pi Dt}}e^0$$

$$= \frac{1}{2}\sqrt{\frac{D}{\pi t}}\,C_0. \tag{4.66}$$

The total flux of carbon to diffuse out over an exposure time period t_e is then

$$N_t'' = \int_0^{t_e} \dot{N}_t'' \, dt = \frac{C_0}{2} \sqrt{\left(\frac{D}{\pi}\right)} \int_0^{t_e} \frac{dt}{\sqrt{t}} = C_0 \sqrt{\left(\frac{Dt_e}{\pi}\right)}. \qquad (4.67)$$

(a) Worked example: decarburization of steel

Let us suppose a 1% carbon steel is exposed to an oxidizing atmosphere and that surface decarburization occurs as a result (Fig. 4.18). Our boundary conditions can be written:

B.C.1 $\qquad\qquad\qquad\qquad x = 0,\ C = 0;$

B.C.2 $\qquad\qquad\qquad\qquad x = -\infty,\ C = C_0.$

Inserting into the general solution for extended sources;

$$C_{x,t} = A + B \operatorname{erf}\left(\frac{x}{\sqrt{4Dt}}\right),$$

we have

$$0 = A + B \operatorname{erf}(0), \qquad A = 0,$$
$$C_0 = A - B, \qquad\qquad B = -C_0,$$

giving our particular solution:

$$C_{x,t} = -C_0 \operatorname{erf}\left(\frac{x}{\sqrt{4Dt}}\right).$$

Suppose we now want to know what the carbon content in the steel 1 mm away from the surface will be after 8 hours (we will ignore any problems with scale, i.e., iron oxide formation), we find that $D_{C/Fe} = 10^{-10} \text{ m}^2 \text{ s}^{-1}$ at 1728 K,

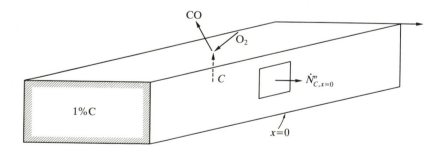

FIG. 4.18.   1% carbon steel billet exposed to an oxidizing atmosphere causing surface decarburization.

so that

$$\frac{C_{x,t}(\text{wt}\%)}{(1\,\text{wt}\%)} \simeq \text{erf}\left\{\frac{0.001}{(4\times10^{-10}\times8\times36000)^{1/2}}\right\} = \text{erf}(0.295).$$

From Table A3.1, $\text{erf}(0.295)=0.323$ so that the carbon concentration will have dropped to 0.323 wt.%, 1 mm from the surface, after 8 hours.

(b)   The 'diffusion distance'

It is particularly noteworthy that plots of dimensionless concentration, such as $C_{x,t}/C_0$ in the above example, are a function only of the single dimensionless variable $x/\sqrt{Dt}$. Thus, the diffusion penetration curve, or plot of dimensionless concentration against distance, is always of the same form. If we consider how long it will take for the concentration at a given depth in the steel sample to reach a value midway between the surface and the bulk, we see from the error function in Table A3.1, that for $C_{x,t}/C_0=0.5$, $x_{1/2}/\sqrt{Dt_{1/2}}=2 \times 0.48 = 0.96 \sim 1.00$. In other words, the diffusion distance $x_{1/2}$ is numerically equal to $\sqrt{Dt_{1/2}}$. Taking $D=10^{-10}\,\text{m}^2\,\text{s}^{-1}$, we see that $t=10^5\,\text{s}$, or about 28 h.

The concept of 'diffusion distance' is extremely important, as it enables one to perform rough order-of-magnitude calculations, and to decide whether the semi-infinite or infinite media approximations will suffice. The analogous relation for heat conduction would be $x_{1/2}/\sqrt{\alpha t_{1/2}} = 1$.

(c)   Diffusion and convection in solids

In our treatment of diffusion, we have so far side-stepped certain issues relating to diffusion and convection phenomena in solids. For interstitial diffusion for instance, any motion of the solvent atoms has been ignored, it being assumed that the diffusing atoms are small and cause little distortion of the lattices of solvent atoms. We have thereby ignored the possibility of enhanced diffusion along the grain boundaries of the crystals making up the solid, or the possible effect of a stressed lattice, or the effect of concentration-dependent diffusivities.

With substitutional diffusion processes, the picture becomes further clouded. For instance, if we extend our example of the silver isotope atoms exchanging positions with regular atoms within the silver lattice, by considering a more general substitutional alloy system, it is no longer possible to assume equivalent mobility between different atom species. Consider Fig. 4.19, showing a gold–nickel diffusion couple made by joining the two metals together. If we place an inert tungsten wire at their point of contact, and then

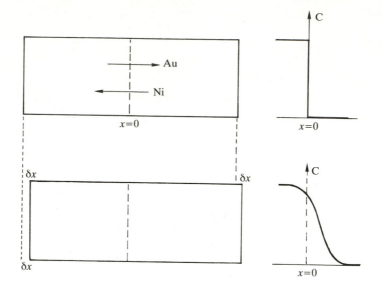

FIG. 4.19.   Diagram illustrating the 'Kirkendall effect', in which a greater molar flow of gold past an inert tracer compared to a counterflow of nickel, leads to an apparent displacement of the marker. Note nett volume of bar remains constant.

measure the subsequent interdiffusion of nickel and gold, an apparent shift in the marker's position will take place. This occurs because more gold atoms, being of significantly higher mobility than those (Ni) counterdiffusing, will have convected/diffused past the tungsten marker. The phenomenon is known as the *Kirkendall effect* and the apparent shift as the *Kirkendall shift* (Smigelskas and Kirkendall 1947).

   Providing the vacancy concentration in the alloy system remains constant (constant volume), as in this case, a similar analysis to that just presented for the argon–zinc vapour example, based on the total flux equations

$$\dot{N}''_{Au} = -D_{Au}\frac{\partial C_{Au}}{\partial x} + U_x C_{Au} \tag{4.68}$$

and

$$\dot{N}''_{Ni} = -D_{Ni}\frac{\partial C_{Ni}}{\partial x} + U_x C_{Ni}, \tag{4.69}$$

leads to

$$\dot{N}''_{Au,\,x} = -(X_{Ni}D_{Au} + X_{Au}D_{Ni})\frac{\partial C_{Au}}{\partial x}. \tag{4.70}$$

   Recalling Fick's first law of diffusion, we see that the chemical diffusivity defined in that relationship is related to the intrinsic diffusivities of gold and nickel, D_{Au}, D_{Ni}, according to the identity

$$D_{Au-Ni} = X_{Ni}D_{Au} + X_{Au}D_{Ni}. \tag{4.71}$$

Thus D_{Au-Ni} (or D_{Ni-Au}) the *chemical diffusion, interdiffusion,* or *mutual diffusion* coefficient of gold in nickel (or vice versa) is a function of the so-called *intrinsic diffusion* coefficients D_{Au} and D_{Ni}, of gold and nickel atoms respectively.

Another type of diffusion, described in Chapter 1, is *self-diffusion*. Equivalent to the motion of atoms or molecules in a gas, it relates to speed at which atoms of the material meander around the atomic lattice. The mechanisms whereby these movements take place are generally attributed to holes, or vacancies (defects) within the lattice. As mentioned, these vacancies increase dramatically with temperature, thereby leading to increased rates of atomic interchanges, and higher self-diffusivities. Other mechanisms whereby self-diffusion can occur, can be by interstitial displacements, or jumps.

Radioactive isotopes of the metal of interest can be used to measure self-diffusivities. While the isotope atoms will have the same mobility as their non-radioactive equivalents, measured *tracer diffusion* coefficients under-estimate *self-diffusion* coefficients somewhat, because the 'wandering vacancy' will afford more opportunities for the more numerous, regular atoms, to jump around. Thus, for f.c.c. metals, $D^t \simeq 0.72D^*$, and for b.c.c. metals, $D^t \simeq 0.78D^*$, where the superscript t refers to the tracer diffusion coefficient of the metal, and * represents the metal's self-diffusion coefficient (Compaan and Haven 1956).

(d)   Variable diffusivity

In the analyses so far, it has also been assumed that the diffusion coefficient is independent of concentration, etc. If not, Fick's second law must be written as

$$\frac{\partial C}{\partial t} = \frac{\partial}{\partial x}\left(D\frac{\partial C}{\partial x}\right), \tag{4.72a}$$

or

$$\frac{\partial C}{\partial t} = D\frac{\partial^2 C}{\partial x^2} + \frac{\partial D}{\partial x}\frac{\partial C}{\partial x}. \tag{4.72b}$$

Using the 'Boltzmann substitution', in which variable λ is used to replace $x/t^{1/2}$, we have

$$\frac{\partial C}{\partial t} = \frac{dC}{d\lambda}\frac{\partial \lambda}{\partial t} = -\frac{\lambda}{2t}\frac{dC}{d\lambda}.$$

Similarly,

$$\frac{\partial C}{\partial x} = \frac{dC}{d\lambda}\frac{\partial \lambda}{\partial x} = \frac{1}{t^{1/2}}\frac{dC}{d\lambda}$$

and

$$\frac{\partial D}{\partial x} = \frac{1}{t^{1/2}} \frac{dD}{d\lambda},$$

giving

$$-\frac{1}{2}\lambda \frac{dC}{d\lambda} = D\frac{d^2C}{d\lambda^2} + \frac{dD}{d\lambda}\frac{dC}{d\lambda} = \frac{d}{d\lambda}\left(D\frac{dC}{d\lambda}\right), \tag{4.73}$$

showing that the substitution of λ to replace $(x/t^{1/2})$ is correct. Multiplying through by $d\lambda$, and transposing,

$$d\left(D\frac{dC}{d\lambda}\right) = -\frac{1}{2}\lambda\,dC.$$

Integrating from $C=C_0$ (where $dC/d\lambda=0$) to C,

$$D\frac{dC}{d\lambda} = -\frac{1}{2}\int_{C_0}^{C} \lambda\,dC,$$

or

$$D = -\frac{1}{2}\int_{C_0}^{C} \lambda\,dC \Big/ \frac{dC}{d\lambda}. \tag{4.74}$$

This relationship can be used to evaluate D, the chemical diffusion coefficient, from a concentration–distance (diffusion penetration) curve, such as that illustrated in Fig. 4.20. In choosing a reference plane for diffusion, and in the light of the preceding discussion on the Kirkendall effect, it is customary to locate the reference frame with respect to the *Matano interface*. This is obtained by setting the areas of the C–x curve on either side of the 'Matano interface' equal. The variation in the chemical diffusivity with concentration can then be obtained by evaluating the integral and slope of the C–x curve, as illustrated in Fig. 4.20. Jost (1960) points out that this approach can be extended to two-phase interdiffusional penetration phenomena.

4.10.   EXERCISES ON TRANSIENT CONDUCTION AND DIFFUSION WITH EXTENDED SOURCES

(1)   Carbon diffusion between two bars

Two iron ingots containing different proportions of dissolved carbon were welded together at plane $x=0$, in a similar fashion to that shown in Fig. 4.16(a). Assuming a general solution for the carbon concentration of the form

$$C_{A,x,t} = A + B\,\text{erf}\left(\frac{x}{\sqrt{4Dt}}\right),$$

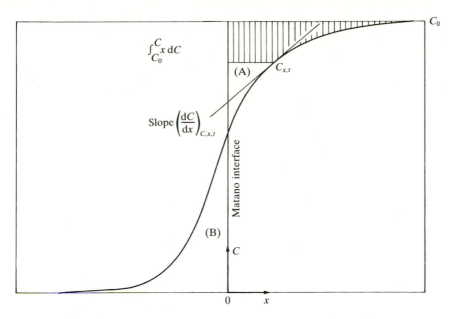

Fig. 4.20. A diffusion penetration curve following isothermal diffusional processes in a copper–aluminium couple. The Matano interface ($x=0$, $t=t$) is located so that the areas A and B on either side are equal. Chemical diffusion coefficients are then calculated via eqn (4.74).

write the appropriate boundary conditions to obtain an exact solution for the above example.

If one cylinder contains 0.2% C and the other 0.05% C, what is the concentration of carbon at $x=0$ at time t? Write down an expression for the mass flux of carbon at $x=0$, at time t.

(2)   Conductive heat losses during extrusion: water-cooled plunger

During the extrusion of an aluminium billet (see figure), a water-cooled plunger at a uniform temperature θ_1 was contacted with the hot aluminium billet initially at a uniform temperature of θ_2. Assuming the plunger temperature remains at θ_1 and heat losses from the sides of the billet may be neglected,

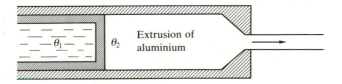

calculate how soon the temperature in the billet 76 mm away from the plunger drops by 20 K, or 95 per cent of its original temperature (in °C).

Data:

Initial temperature of billet	673 K
Temperature of plunger	293 K
Thermal diffusivity of aluminium α	$1 \times 10^{-4}\,\text{m}^2\,\text{s}^{-1}$

Assume a solution of form $\theta_{x,t} = A + B\,\text{erf}(x/\sqrt{4\alpha t})$.

(3) Conductive heat losses during extrusion of billets: steel die

A steel plunger initially at a uniform temperature θ_1 throughout, was contacted with an aluminium billet, initially at a uniform temperature θ_2 throughout (see figure), during an extrusion process. Neglecting any heat losses from the sides of the billet and plunger, calculate the interfacial temperature θ^*.

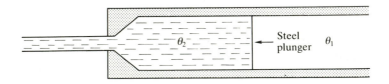

You may assume general solutions of the form

At $x<0$ $\qquad \theta = A_1 + B_1\,\text{erf}\left(\dfrac{x}{\sqrt{4\alpha_1 t}}\right)$ (steel plunger)

At $x>0$ $\qquad \theta = A_2 + B_2\,\text{erf}\left(\dfrac{x}{\sqrt{4\alpha_2 t}}\right)$ (aluminium billet)

with the boundary conditions

$$t=t,\ x \to -\infty,\ \theta = \theta_1;$$

$$x \to +\infty,\ \theta = \theta_2; \qquad x = 0,\ \theta = \theta^*; \qquad x = 0,\ \dot{q}_1'' = \dot{q}_2''$$
$$\text{(equal heat fluxes)}$$

Data:

α_1 (thermal diffusivity of steel)	$9.6 \times 10^{-6} \text{ m}^2 \text{ s}^{-1}$
k_1 thermal conductivity of steel	$4.18 \times 10^{-3} \text{ Wm}^{-1} \text{ K}^{-1}$
θ_1	298 K
α_2 thermal diffusivity of aluminium	$95 \times 10^{-6} \text{ m}^2 \text{ s}^{-1}$
k_2 thermal conductivity of aluminium	$23 \times 10^{-3} \text{ Wm}^{-1} \text{ K}^{-1}$
θ_2	673 K

$$\text{erf}(\eta_1) = \frac{2}{\sqrt{\pi}} \int_0^{\eta_1} e^{-\eta^2} \, d\eta$$

(4) Adsorption of surface active agents at interfaces: flotation of sulphides

Froth flotation procedures have been widely used in the mineral processing industry for the separation of lead, zinc, copper and nickel sulphides from gangue material (silicates, etc.). The principle of froth flotation is to condition gas–liquid interfaces by (i) adsorbing collecting agents to promote the attraction and attachment of sulphide particles on bubble interfaces, and (ii) adsorbing frothing agents on bubble surfaces so as to promote the formation of a stable froth on top of the flotation cell for rabbling off the particles.

One particular froth flotation procedure (see figure) incorporates the use of a frothing agent, decyl alcohol, and a collecting agent, lauric acid at bulk concentration levels of 7.5×10^{-5} mole m^{-3} and 1.4×10^{-6} mole m^{-3} respectively. It is desirable to estimate what 'conditioning times' will be required for a freshly introduced gas–liquid interface (i.e., a bubble) to achieve 60 per cent equilibrium adsorption of decyl alcohol and lauric acid respectively. The equilibrium adsorption densities Γ^0 of decyl alcohol and lauric acid may both be taken as 2×10^{-11} moles m^{-2} of surface, while the rate-controlling step in the overall adsorption process can be regarded as unsteady-state diffusion of frother, or collector, through a stagnant solvent (water) to the freshly created gas–liquid interface (i.e., instantaneous adsorption of material into interface; therefore $x=0$, $t=t$, $C=0$).

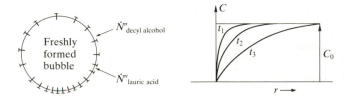

Data: diffusion coefficient of decyl alcohol through water $= 5 \times 10^{-10} \, \mathrm{m^2 s^{-1}}$; diffusion coefficient of lauric acid through water $= 5 \times 10^{-10} \, \mathrm{m^2 \, s^{-1}}$.

Assume a planar interface and unidirectional unsteady-state diffusive mass transfer.

(5)   Cooling water requirements for hot rolling mill operations

About 60 per cent of the total water requirements in a modern integrated steelworks plant is used in cooling operations in the hot rolling mills. A slab 1.5 m wide by 0.14 m thick at a uniform temperature of 1573 K enters the second roughing stand which is rotating at 0.17 rev s^{-1}; the roll dimensions are 1.3 m diameter and 2.5 m length. The separation between the upper and lower rolls is 0.13 m (see figure).

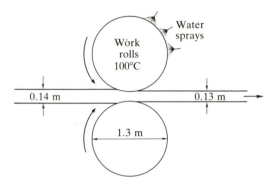

Calculate the total heat transferred to the two rolls (in watts), knowing that the inlet temperature of the water-cooled rolls is maintained at 373 K by water sprays. Calculate the water requirements for the stand based on the criterion that the bulk temperature of the exit cooling water from the stand should not increase by more than 10 K.

Hint: It is legitimate to assume unsteady-state conduction of heat between two semi-infinite bodies, the roll and the slab, and to locate $x = 0$ at the interface between the two. Ignore distortion effects on the solution, and also increases in heat energy resulting from mechanical work of deformation.

Data:

Thermal conductivity of slab and roll, k	$42 \, \mathrm{Wm^{-1} K^{-1}}$
Thermal diffusivity of slab and roll α	$71 \times 10^{-5} \, \mathrm{m^2 \, s^{-1}}$
Inlet slab temperature	1573 K
Inlet roll temperature	373 K
Natural cosine (6°)	0.99452

(6) Case-hardening of steel gear shafts

Six thousand cylindrical steel gear shafts 0.3 m long and 25 mm in diameter are to be case-hardened in a reducing atmosphere containing not less than 98 per cent carbon monoxide at 1200 K. The case-hardened depth required is 1.5 mm (i.e. the carbon concentration at 1.5 mm below the surfaces should be 0.07 wt.% higher than the core carbon content).

Calculate the time required to achieve such a case-hardened depth, the total consumption of carbon monoxide during the period due to depletion by the reaction $2CO \rightarrow C + CO_2$, and the mean CO flow rate (litres h^{-1} at NTP., 273 K, 1 atm) that must be provided to the furnace to maintain the CO_2 content in the gases at no more than 2 per cent.

Data:

Core carbon content	0.093 wt.%.
Diffusion coefficient of carbon in austenite	1.6×10^{-11} $m^2 s^{-1}$
Activity of carbon at the gas–steel interfaces	1.0 (unity)
Temperature	1200 K

Hint: Ignore radial and edge effects and treat as a one-dimensional unsteady-state diffusion system with $x=0$ located at the cylinders' surfaces. The Fe–C equilibrium phase diagram is needed, together with error function tables.

4.11. FINITE SOURCE: INFINITE MEDIUM

Finally let us consider the case of a finite (rather than plane or extended) source located between $-h$ and $+h$ as shown in Fig. 4.21. Adopting the same

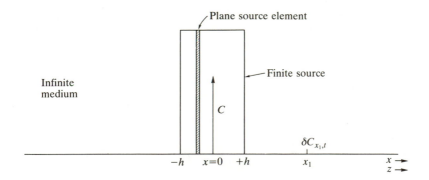

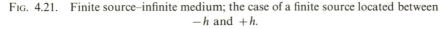

FIG. 4.21. Finite source–infinite medium; the case of a finite source located between $-h$ and $+h$.

set of coordinate systems as used in Fig. 4.17 in the derivation for extended sources in an infinite medium, it is evident that we can regard the finite source as being composed of a number of plane sources located between $z = +h$ and $z = -h$ and sum the corresponding number of elementary solutions to arrive at the concentration $C_{x_1,t}$ at position x, and time t. Thus, as before,

$$\delta C_{x_1,t} = \frac{C_0 \, \delta x'}{\sqrt{4\pi Dt}} e^{-x'^2/4Dt} = \frac{C_0(-\delta z)}{\sqrt{4\pi Dt}} e^{-(x_1-z)^2/4Dt}, \tag{4.75}$$

where $x' = x_1 - z$, and

$$C_{x_1,t} = \sum \delta C_{x_1,t} = \int_{z=h}^{z=-h} \frac{C_0(-\delta z) e^{-(x_1-z)^2/4Dt}}{\sqrt{4\pi Dt}}. \tag{4.76}$$

Putting

$$\eta = \frac{x_1 - z}{\sqrt{4Dt}}, \text{ then } d\eta = \frac{-dz}{\sqrt{4D}}.$$

When

$$z = h, \; \eta = \frac{x_1 - h}{\sqrt{4Dt}}; \text{ when } z = -h, \; \eta = \frac{x_1 + h}{\sqrt{4Dt}}.$$

Transforming the integral as before,

$$C_{x_1,t} = \int_{(x_1-h)/\sqrt{4Dt}}^{(x_1+h)/\sqrt{4Dt}} \frac{C_0}{\sqrt{\pi}} e^{-\eta^2} \, d\eta$$

$$= \frac{C_0}{\sqrt{\pi}} \int_{(x_1-h)/\sqrt{4Dt}}^{0} e^{-\eta^2} \, d\eta + \frac{C_0}{\sqrt{\pi}} \int_{0}^{(x_1+h)/\sqrt{4Dt}} e^{-\eta^2} \, d\eta$$

$$= \frac{C_0}{2} \left(-\frac{2}{\sqrt{\pi}} \int_{0}^{(x_1+h)/\sqrt{4Dt}} e^{-\eta^2} \, d\eta + \frac{2}{\sqrt{\pi}} \int_{0}^{(x_1+h)/\sqrt{4Dt}} e^{-\eta^2} \, d\eta \right)$$

$$= \frac{C_0}{2} \left\{ -\mathrm{erf}\left(\frac{x_1 - h}{\sqrt{4Dt}} \right) + \mathrm{erf}\left(\frac{x_1 + h}{\sqrt{4Dt}} \right) \right\}$$

$$C_{x_1,t} = \frac{C_0}{2} \left\{ \mathrm{erf}\left(\frac{h - x_1}{\sqrt{4Dt}} \right) + \mathrm{erf}\left(\frac{h + x_1}{\sqrt{4Dt}} \right) \right\} \tag{4.77}$$

Example

Recalculate the heat treatment times required for homogenizing as set out in Exercise 4.8.3.

Solution. In order that the concentration at $x=0$ shall drop to half the value required Exercise 4.8 (3), we now have that

$$C_{x=0,\,t=t_e}=\frac{C_0}{2}=\frac{C_0}{2}\,2\,\mathrm{erf}\!\left(\frac{h}{\sqrt{4Dt}}\right)\Bigg\},$$

i.e. $\frac{1}{2}=\mathrm{erf}(h/\sqrt{4Dt})$. From tables, we have:

$$\frac{h}{\sqrt{4Dt_e}}=0.48 \quad\text{so that}\quad \frac{\Delta x/2}{2\sqrt{D}\times0.48}=\sqrt{t_e},$$

i.e.

$$\frac{\Delta x^2}{16D\times(0.48)^2}=t_e,$$

whence $\Delta x^2/3.7D=t_e$, versus $\Delta x^2/\pi D$ from the plane source solution.

4.12.   EXAMPLES OF UNSTEADY-STATE HEAT (MASS) TRANSPORT IN FINITE SYSTEMS

(a)   Introduction

Our discussions and examples of unsteady-state diffusion and conduction have so far been limited to systems which may be regarded as being of infinite or of semi-infinite extent, in that the presence of a boundary located at distance $x=L$ (say) does not disturb the temperature or concentration profiles generated in the region of $x=0$, at least, over 'short' contact times. For these reasons, it is quite legitimate to use the simple error function solutions previously derived to calculate unsteady-state concentration profiles and mass fluxes around $x=0$, the boundary. This approach is particularly valid for cases involving diffusion into (or out of) a body of even small dimensions (e.g., 25 mm or so) since depths of mass penetration are small even for prolonged diffusion times (e.g., 10 h) on account of low solute diffusivities in solids (10^{-9}–10^{-16} m^2 s^{-1}).

Equally, however, there are many other practical cases, particularly in heat transfer, in which the size of the body plays an important part in determining unsteady-state temperature–time profiles. In such circumstances, the simple boundary conditions for semi-infinite and infinite media (e.g. $\theta=0$, $x=0$; $\theta=\theta^B$, $x=L$ (or ∞)) will only hold for relatively short contact times, after which temperature–time profiles begin to be affected by the size of the body.

Taking as an example the classic case of an ingot cooling in air as depicted in Fig. 1.3(a), the differential equation and boundary conditions are as follows.

$$\frac{\partial\theta_{x,t}}{\partial t}=\frac{k_s}{\rho C}\frac{\partial^2\theta_{x,t}}{\partial x^2}. \tag{4.78}$$

Initial condition: $\theta = \theta_0$, $-L \leqslant x \leqslant +L$.
Boundary conditions:

$$x = 0, \; \forall t, \; \frac{\partial \theta}{\partial x} = 0$$

(i.e., no net heat transfer across plane located at $x = 0$).

$$x = \pm L, \; \forall t, \; \dot{q}''_{x=L} = h(\theta_s - \theta_\infty).$$

Inspection of the parameters involved in the problem shows that

$$\theta_{x,t} = f(t, x, L, \rho, C, k_s, h, \theta_0, \theta_\infty). \tag{4.79}$$

Using methods of dimensional analysis presented in Chapter 3, one can readily express the dependence of the temperature at x and t (the dependent variable) on the remaining (independent) variables, as a series of *dimensionless* π groupings. Thus

$$\frac{\theta_{x,t} - \theta_\infty}{\theta_0 - \theta_\infty} = f\left(\frac{\alpha t}{L^2}, \frac{k_s}{hL}, \frac{x}{L}\right), \tag{4.80}$$

where $\alpha t/L^2 =$ the Fourier number, $hL/k_s =$ the Biot number (not to be confused with the Nusselt number, hL/k_L in which k_L represents the thermal conductivity of the fluid).

In fact, analytical solutions to these type of problems are available and take the form of series solutions. In the present case, the particular solution is

$$\frac{\theta_{x,t} - \theta_\infty}{\theta_0 - \theta_\infty} = 2 \sum_{n=1}^{\infty} \frac{\sin(\lambda_n L)}{\lambda_n L + \sin(\lambda_n L)\cos(\lambda_n L)} \times \exp(-\lambda_n^2 \alpha t \cos(\lambda_n x)), \tag{4.81}$$

where the λ_n, are the infinite number of roots to the transcendental equation

$$\cot(\lambda_n L) = \frac{1}{(hL/k_s)}(\lambda_n L). \tag{4.82}$$

Since the series converges quite rapidly, only 3 or 4 terms of the series need be taken. Details of the mathematical procedures in obtaining these types of solutions are to be found in most heat transfer texts including that by Carslaw and Jaeger (1947). and, for diffusive mass transfer, in Crank (1967).

It is appropriate to note here that numerical integration procedures used in conjunction with high-speed digital computers now enable solutions to some of the most complex heat and mass transfer situations to be readily obtained with minimal programming difficulty. In many instances, series solutions of the type shown above are inappropriate. Thus, if the convective heat transfer coefficient or thermal and physical properties of the body are functions of temperature one would normally proceed straight to numerical techniques of the type illustrated in Chapter 6.

Figures 4.22(a), (b) and (c) show graphical solutions to the case of unsteady-state conduction of heat through slabs, cylinders and spheres, and these are plotted in terms of the dimensionless groupings appearing in eqn (4.80). The 'Heisler' charts are restricted to values of the Fourier modulus greater than 0.2. For smaller values, the reader is referred to his original paper. To illustrate their use, let us consider an example similar to that described in Chapter 1.

(b) Temperature profiles inside a stainless steel slab cooling in air

A 0.12 m-thick slab of 18-8 stainless steel, initially at 1273 K is allowed to cool in air by radiative and convective heat transfer to the surroundings. Using charts provided, estimate the temperature distribution in the slab after 10 min of cooling.

Data:

Thermal diffusivity, α	5×10^{-6} m^2 s^{-1}
Density of steel, ρ	7600 kg m^{-3}
Thermal capacity, C	670 J kg^{-1} K^{-1}
Temperature of surrounding T_∞	273 K
Initial slab temperature, T_0	1273 K
Overall surface heat transfer coefficient, h (i.e. radiation + convection, assumed constant)	75 W m^{-2} K^{-1}

Solution

Using the data provided.

$$\text{Fourier number} = \frac{\alpha t}{L^2} = \frac{5 \times 10^{-6} \times 10 \times 60}{0.06^2} = 0.833;$$

$$\text{Biot number} = \frac{hL}{k} = \frac{75 \times 0.06}{5 \times 10^{-6} \times 7600 \times 670} = 0.177;$$

$$(\text{Biot number})^{-1} = \frac{k}{hL} = 5.66.$$

Referring to Fig. 4.22(a) (i) the dimensionless centre temperature would be about 0.87 of the initial value, i.e. $(\theta_{0,10} - 273)/(1273 - 273) = 0.87$, or $\theta_{0,10} = 870\,°C$. We may proceed to Fig. 4.22 (a) (ii) to deduce the temperature profile through the slab. The following table lists temperatures across the slab from the centreline ($x/L = 0$) to the surface ($x/L = 1$).

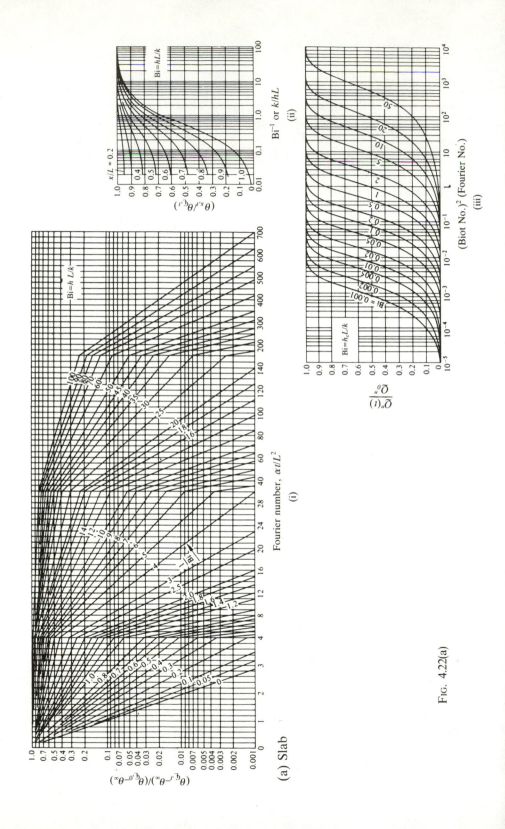

Fig. 4.22(a)

(a) Slab

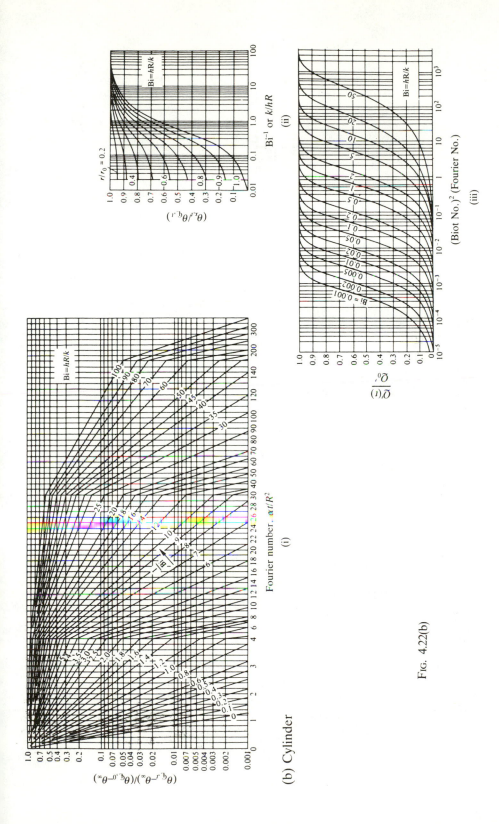

(b) Cylinder

Fig. 4.22(b)

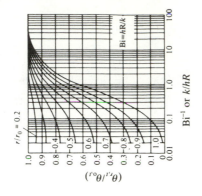

Bi^{-1} or k/hR

(ii)

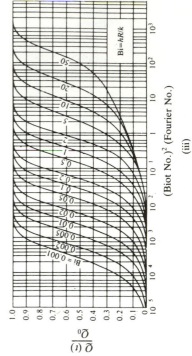

(Biot No.)^2 (Fourier No.)

(iii)

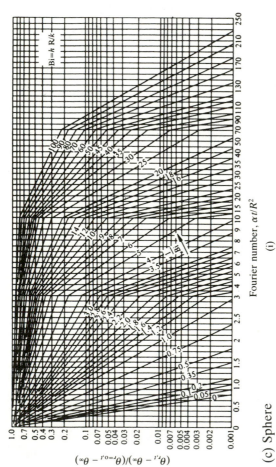

Fourier number, $\alpha t/R^2$

(i)

(c) Sphere

Fig. 4.22(c)

FIG. 4.22. (a) (i) Dimensionless midplane temperature difference with respect to ambient conditions for an infinite plate of thickness $2L$, plotted versus Fourier number. (ii) Dimensionless local temperature difference for an infinite plate of thickness $2L$, referenced to midplane temperature difference, versus inverse Biot number. (iii) Cumulative fraction heat transferred by time t for a half-plate (0 to L) versus $(Biot)^2$ (Fourier). b) (i) Dimensionless axial temperature difference with respect to ambient conditions for an infinite cylinder of radius R plotted versus Fourier number (ii) Dimensionless local temperature difference for an infinite cylinder of radius R, referenced to initial axial temperature difference, versus inverse Biot number. (iii) Cumulative fraction of heat transferred to ambient surroundings by time t versus $(Biot)^2$ (Fourier). c) (i) Dimensionless centre temperature for a sphere of radius R, plotted versus Fourier number. (ii) Dimensionless local temperature difference for a sphere, of radius R, referenced to the initial axial temperature difference, versus inverse Biot number. (iii) Cumulative fraction of heat transferred to ambient surroundings by time t versus $(Biot)^2$ (Fourier) (From M. P. Heisler (1947). *Trans. A.S.M.E.* **69**, 227–36).

x/L	$\dfrac{\theta_{x,\,10\ min} - \theta_\infty}{\theta_0 - \theta_\infty}$	x (mm)	$\theta_{x,\,10\ min}$ (°C)	$\theta_{x,\,10\ min}$ computed
0.0	0.87	0	870	892.77
0.2	0.87	12	870	889.9
0.4	0.86	24	860	880.8
0.6	0.85	36	850	865.0
0.8	0.82	48	820	843.7
1.0	0.80	60	800	818.04

The temperature profile is shown in Fig. 4.23 for comparison with that obtained using numerical methods discussed in Chapter 6. Similarly, Figure 4.22(a) (iii) can be used to deduce the fractional amount of heat initially contained within the slab, that has been lost to the surrounding (i.e. about 0.2).

Figures 4.22(b) and 4.22(c) respectively provide equivalent graphical solutions for infinite cylinders, and for spheres. One may note the difficulty of reading off the dimensionless temperatures to any great accuracy using these graphs, partly, in the last case, on account of the low Biot number and correspondingly shallow temperature gradients in the solid. Nevertheless they are useful in providing quick, rough and ready answers to problems of the type considered.

(c)   Limiting cases: high and low Biot numbers

The Biot number, hL/k_s, characterizes the relative ability of the solid and the fluid in which it is immersed to transport heat. A low Biot number (i.e. <0.1) indicates minimal resistance to heat in the solid compared with the fluid surrounding it. Similarly, a high Biot number indicates poor transfer of heat through the solid as compared with the fluid surrounding it. Since process metallurgists deal with systems containing metals, which all have high thermal conductivities, it frequently occurs that our systems approach one of these limiting conditions of the Biot number.

It is important to realize that under such circumstances great simplifications in the mathematical solutions are possible with very little loss in accuracy, as will be demonstrated in Chapter 6. In general, a low Biot number allows temperature gradients in the solid to be ignored. At the opposite extreme, a high Biot number would result in extremely steep temperature gradients in the region of the interface with little penetration of heat into the body over short contact times. Thus, error function solutions might, for instance, be appropriate depending on the circumstances involved.

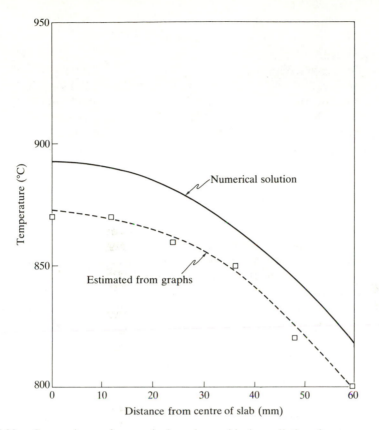

Fig. 4.23. Comparison of numerical and graphical prediction for temperature profiles within a stainless steel ingot following 10 min of air cooling.

(d)   Worked example of low Biot number: quench-hardening of metals

A thin metal sheet, 1.14 mm thick, was solution-treated at 1055 K and then quenched in water during an age hardening process (e.g. copper–beryllium and certain aluminium alloys). In order to prevent undesirable segregation, the temperature inside the sheet must everywhere fall below 755 K within 5 s. Calculate the time required for the maximum temperature within the sheet to fall below 755 K given the following typical data.

Data:

Surface heat transfer coefficient to water, h	$1.7 \text{ kW m}^{-2} \text{ K}^{-1}$
Density of metal (copper), ρ	8900 kg m^{-3}
Heat capacity, C	$385 \text{ J kg}^{-1} \text{ K}^{-1}$
Thermal conductivity, k	$380 \text{ W m}^{-1} \text{ s}^{-1}$
Thickness of sheet, Δ	1.14 mm
Temperature of body at location x, $t=0$, θ_x	1055 K
Surface temperature, $t=0$, $\theta*$	1055 K
Ambient temperature, θ_w	289 K

Solution

Since metal has a much greater ability in general to conduct heat in comparison to transfer through a non-metallic liquid, it is likely that temperature gradients in the sheet are negligible. We may confirm this by calculating the Biot number, noting that L represents the volume/surface area ratio of the body. For a wide sheet of thickness Δ, $L=(\Delta \times A)/2A=\Delta/2$,

$$\text{Bi}=\frac{hL}{k_s}=\frac{1.7 \times 10^3 \times (1.14/2) \times 10^{-3}}{380}=0.0026.$$

Since $\text{Bi} \ll 0.1$, temperature gradients in the strip are negligible. Proceeding now to the relevant statement of the macroscopic energy equation:

Rate of heat loss from surfaces $= -$ rate of heat accumulation in solid;

$$2hA_s(\theta_s - \theta_\infty) = -\rho C(A_s \Delta)\frac{d\theta}{dt};$$

$$\frac{2h}{\rho C \Delta}\int_0^t dt = -\int_{\theta_0}^\theta \frac{d\theta}{(\theta - \theta_\infty)};$$

$$\frac{-2ht}{\rho C \Delta} = \ln\left(\frac{\theta - \theta_\infty}{\theta_0 - \theta_\infty}\right);$$

$$t = -\frac{\rho C \Delta}{2h}\ln\left(\frac{\theta - \theta_\infty}{\theta_0 - \theta_\infty}\right)$$

$$= -\frac{8900 \times 385 \times 1.14 \times 10^{-3}}{2 \times 1.7 \times 10^3}\ln\left(\frac{755 - 289}{1055 \quad 289}\right)$$

$$= 0.57 \text{ s}.$$

Conclusion

The maximum time required for temperatures in the sheet to fall below 755 K is 0.57 s, which is well within the required time of 5 s. It is useful to note that the final equation obtained may be written in a slightly different form, so as to 'expose' the presence of the important dimensionless groupings given in eqn (4.53):

$$\ln\left(\frac{\theta-\theta_\infty}{\theta_0-\theta_\infty}\right) = \frac{-2ht}{\rho C \Delta} \equiv \frac{-ht}{\rho C L} \equiv -\left(\frac{k}{\rho C}\frac{t}{L^2}\right) \times \left(\frac{Lh}{k}\right) \equiv \frac{-\alpha t}{L^2}\frac{Lh}{k} \equiv -(Fo)(Bi).$$

We thus have the limiting relationship at low Biot numbers that

$$\frac{\theta_t-\theta_\infty}{\theta_0-\theta_\infty} = e^{-(Fo)(Bi)}.$$

4.13. RADIATIVE HEAT TRANSFER

(a) Introduction

Radiative heat transfer phenomena are commonplace in pyrometallurgical processing operations. Examples include heat losses from hot ingots, slabs, rod, or strip, as well as combustion systems, such as reheating furnaces, soaking pits, annealing furnaces, foundry furnaces, and their associated burners.

While some of the fundamental characteristics of radiation have been addressed in Chapter 1, additional information is needed with respect to the interchange of radiant energy between surfaces, before considering typical engineering examples.

(b) Reflection, transmission, and absorption

When radiant energy strikes a surface, part of this energy can be reflected, part absorbed, and the remainder transmitted through the material, as shown schematically in Fig. 4.24(a).

The reflectivity ρ can be defined as the fraction reflected, the absorptivity α as the fraction absorbed, and the transmissivity τ as the fraction transmitted. One therefore has the relationship

$$\rho+\alpha+\tau=1. \tag{4.83}$$

Ordinarily, one is concerned with situations in which τ, the transmissivity, is zero (i.e. opaque materials), and for which the reflective components of the incident radiation are more often *diffuse* than *specular*. An example of specular radiation would be that of a highly polished parabolic surface of

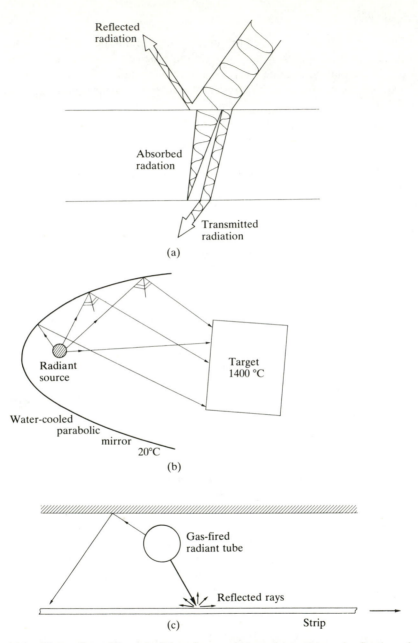

FIG. 4.24.   Illustration of the principles of transmission, absorption and reflection of radiation: (a) the three components; (b) an example of specular reflection—reflector furnaces; (c) an example of diffuse reflection—continuous strip annealing furnace.

aluminium or gold plate, for reflecting and focusing thermal energy from a furnace element (Fig. 4.24(b)). By contrast, thermal radiation from overhead heating elements striking adjacent refractory walls and underlying steel strip in the continuous annealing furnace, shown in Fig. 4.24(c), would be reflected from the refractory and the slightly oxidized steel surfaces in a more diffuse manner.

While incoming radiant energy is being reflected, transmitted or absorbed, the body itself will simultaneously emit radiation at an energy level character-istic of its own temperature and emissive properties. Supposing now that such a body is totally enclosed within a black-body enclosure, with which it is in thermal equilibrium, then the energy input from the enclosure, q_i, must equal the energy emitted by the body:

$$EA = q_i A \alpha, \tag{4.84}$$

where E is the energy emitted from body per unit area per unit time, and A is the surface area of the body. If an equivalent black body of the same shape and area replaces this, at the same enclosure temperature, we can write

$$E_b A = q_i A(1). \tag{4.85}$$

Here $\alpha = 1$, since by definition, a black body absorbs all incident radiation, making τ and ρ zero and α unity. For these two equations to be simulta-neously satisfied, we see that

$$\frac{E}{E_b} = \alpha. \tag{4.86}$$

However, the emissive power of an object to that of an equivalent black body at the same temperature, is known as the emissivity ratio, ε. Thus, from eqn 4.86, the emissivity of a body ($\varepsilon = E/E_b$) must equal its absorptivity (α). This equality, $\varepsilon = \alpha$, is known as Kirchoff's identity, and proves useful in estimating the absorptivity characteristics of a body, from more readily available data on emissivities.

(c)   Radiation shape factors

In our preliminary discussions of radiation in Chapter 1, an example was chosen in which the ingot emitting radiant heat to the melt shop was relatively small and was totally enclosed by the building. In such circumstances, all radiation emitted is intercepted by the enclosure and the object's view factor, or radiation shape factor, F_{12}, is therefore unity. By convention, F_{12} represents the fraction of energy leaving surface 1 which reaches surface 2. Similarly, F_{21} represents the fraction of energy leaving surface 2 and reaching surface 1.

Considering two black-body surfaces, both at the same temperature and interchanging thermal radiation, one may write

$$\dot{q}_{12} = E_{b1} A_1 F_{12} \tag{4.87}$$

and

$$\dot{q}_{21} = E_{b2} A_2 F_{21}. \tag{4.88}$$

Realizing that the nett interchange of energy $\dot{q}_{1-2} (= \dot{q}_{12} - \dot{q}_{21})$ must be zero under isothermal conditions, and that E_{b1} and E_{b2} are identically equal to $\sigma \theta^4$, we see that

$$A_1 F_{12} = A_2 F_{21}. \tag{4.89}$$

This identity, known as the *reciprocity theorem*, also applies to diffuse, grey-body systems. *Grey bodies* are defined as those for which the absorptivity coefficient α_λ for a specific wavelength λ, is independent of λ (and by implication of temperature), and equal to the body's overall absorptivity α. Using the reciprocity theorem, we can write that the nett interchange of radiant energy between two black bodies is

$$\dot{q}_{1-2} = A_1 F_{12}(E_{b1} - E_{b2}) = A_2 F_{21}(E_{b2} - E_{b1}). \tag{4.90}$$

Figure 4.25 provides information on geometric view factors, F_{12}, for systems of general engineering interest. One sees that enclosed bodies (a) or infinitely long parallel surfaces (b), enjoy view factors of 1.0, since all radiation emitted by them is intercepted by the second surface. All other view factors are less than unity. Figures 4.25(c) and (d) give F_{12} values for parallel discs and rectangles, with more general information on these geometries given graphically in Fig. 4.25(e). Finally, Fig. 4.25(f) gives relevant data on view factors for furnace environments. These data could be used, for instance, in estimating heat input to a charge from a single, or double line, of radiant tube furnace heating elements within an enclosing refractory environment.

(d)   Heat exchange between grey bodies

With the above shape factors, radiant energy interchange between black-body surfaces can readily be calculated with the aid of eqn (4.90).

However, most surfaces are not black bodies, so that part of the incident radiation will be reflected rather than totally absorbed. The problem of radiant energy interchange then becomes more complex, since multiple reflections between surfaces can take place. Defining the irradiation G as the total radiation incident upon a surface/unit time/unit area, and J as the radiosity, equal to the total radiation leaving a surface/unit time/unit area, we can sum the two components of radiation leaving a real (grey) surface,

Geometry	View Factor	Comments

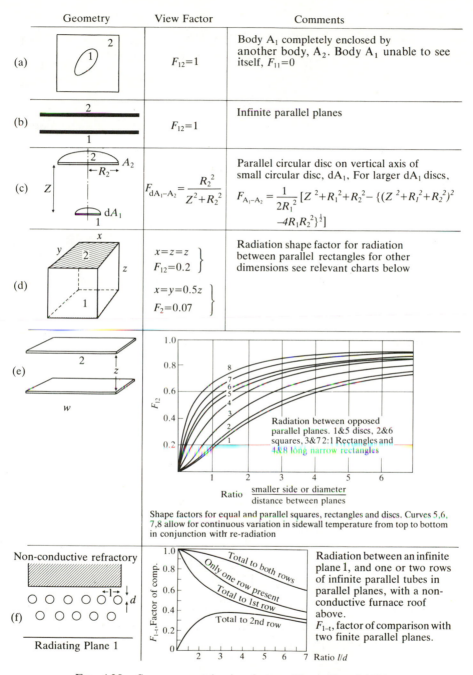

(a) — $F_{12}=1$ — Body A_1 completely enclosed by another body, A_2. Body A_1 unable to see itself, $F_{11}=0$

(b) — $F_{12}=1$ — Infinite parallel planes

(c) — $F_{dA_1-A_2}=\dfrac{R_2^2}{Z^2+R_2^2}$ — Parallel circular disc on vertical axis of small circular disc, dA_1, For larger dA_1 discs,

$$F_{A_1-A_2}=\frac{1}{2R_1^2}[Z^2+R_1^2+R_2^2-\{(Z^2+R_1^2+R_2^2)^2-4R_1R_2^2\}^{\frac{1}{2}}]$$

(d) — $\left.\begin{array}{l}x=z=z\\F_{12}=0.2\end{array}\right\}$ $\left.\begin{array}{l}x=y=0.5z\\F_2=0.07\end{array}\right\}$ — Radiation shape factor for radiation between parallel rectangles for other dimensions see relevant charts below

(e)

Shape factors for equal and parallel squares, rectangles and discs. Curves 5,6, 7,8 allow for continuous variation in sidewall temperature from top to bottom in conjunction with re-radiation

Radiation between opposed parallel planes. 1&5 discs, 2&6 squares, 3&7 2:1 Rectangles and 4&8 long narrow rectangles

Ratio $\dfrac{\text{smaller side or diameter}}{\text{distance between planes}}$

F_{12}

(f) Non-conductive refractory

Radiating Plane 1

F_{1-t}, Factor of comp.

Total to both rows
Only one row present
Total to 1st row
Total to 2nd row

Ratio l/d

Radiation between an infinite plane 1, and one or two rows of infinite parallel tubes in parallel planes, with a non-conductive furnace roof above.
F_{1-t}, factor of comparison with two finite parallel planes.

FIG. 4.25. Some geometric view factors. (From Hottel 1930.)

according to

$$\begin{pmatrix} \text{total efflux} \\ \text{of radiation} \end{pmatrix} = \begin{pmatrix} \text{emitted radiative heat} \\ \text{flux from body's surface} \end{pmatrix} + \begin{pmatrix} \text{reflected radiative} \\ \text{heat flux} \end{pmatrix}$$

(radiosity)

or

$$J = \varepsilon E_b + \rho G. \tag{4.91}$$

Here, the possible transmission of radiation through the irradiated body is to be zero, so that $\tau = 0$. Consequently, making use of the relationship $\rho + \tau + \alpha = 1$, we have

$$\rho = (1 - \alpha) = (1 - \varepsilon),$$

giving

$$J = \varepsilon E_b + (1 - \varepsilon) G. \tag{4.92}$$

Knowing that the nett energy flux leaving the surface must equal $J - G$, we have

$$\dot{q}_1'' = J_1 - G_1 = \varepsilon_1 E_{b,1} + (1 - \varepsilon_1) G_1 - G_1 = \varepsilon_1 (E_{b,1} - G_1). \tag{4.93}$$

Substitution for G from eqn (4.92) in the above yields

$$\dot{q}_1'' = \frac{\varepsilon_1}{1 - \varepsilon_1} (E_{b,1} - J_1), \tag{4.94}$$

or

$$\dot{q}_1 = \frac{E_{b1} - J}{\{(1 - \varepsilon_1)/\varepsilon_1 A_1\}} \equiv \frac{\text{potential difference}}{\text{'surface resistance'}}. \tag{4.95}$$

Let us suppose that part of this radiant heat flux strikes a second surface which, in exchange, emits its own heat flux $\varepsilon_2 E_{b,2}$ while reflecting back some of the incident flux from surface 1. We can write

$$\dot{q}_{1-2} = q_{12} - q_{21} = A_1 F_{12} J_1 - A_2 F_{21} J_2, \tag{4.96}$$

or

$$\dot{q}_{1-2} = \frac{J_1 - J_2}{(1/A_1 F_{12})} \equiv \frac{\text{potential difference}}{\text{'space resistance'}}, \tag{4.97}$$

since $A_1 F_{12} = A_2 F_{21}$. Consequently, for the interchange of radiant energy between two grey surfaces, we can draw the radiation network shown in Fig. 4.26(a) and write

$$\text{Current} = \frac{\text{potential difference}}{\Sigma \text{ (resistances)}},$$

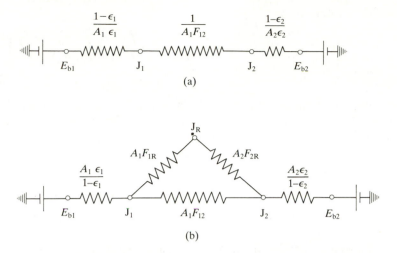

FIG. 4.26.   Equivalent network for radiation exchange: (a) between two grey surfaces and (b), between two grey surfaces connected by an adiabatic, re-radiating surface. (Note that E_R is a floating potential while E_1, and E_2, are fixed.)

or

$$\dot{q}_{1-2} = (E_{b1} - E_{b2}) \Big/ \left\{ \left(\frac{1-\varepsilon_1}{\varepsilon_1 A_1}\right) + \frac{1}{A_1 F_{12}} + \left(\frac{1-\varepsilon_2}{\varepsilon_2 A_2}\right) \right\} \qquad (4.98a)$$

$$\equiv A_1 \bar{F}_{12}(E_{b1} - E_{b2})$$

corresponding to eqn (4.90), where $\bar{F}_{12}$, the total exchange factor is given by

$$\bar{F}_{12} = \left\{ \frac{1-\varepsilon_1}{\varepsilon_1} + \frac{1}{F_{12}} + \frac{A_1}{A_2}\left(\frac{1-\varepsilon_2}{\varepsilon_2}\right) \right\}^{-1}. \qquad (4.98b)$$

Note that eqn (4.98a) reduces to eqn (4.90) for black bodies when ε_1 and ε_2 are set equal to unity in eqn (4.98b). It can be applied to long parallel plates or concentric cylinders, as well as concentric spheres, through substitution of appropriate view factors for F_{12}.

For three-body, and multibody problems, the same approach can be used, using the same radiation network scheme in conjunction with Kirchoff's current law, which states that the *net current* (heat flow) *entering* a node is zero under steady-state conditions (inflow = outflow). For example, for two flat grey surfaces, the first a source, the second a sink, enclosed or connected by a third re-radiating surface (e.g., the refractory walls of furnace), heat is exchange according to

$$\dot{q}_{1-2} = \sigma A_1 (\theta_1^4 - \theta_2^4) \Big/ \left\{ \frac{A_1 + A_2 - 2A_1 F_{12}}{A_2 - A_1 (F_{12})^2} + \left(\frac{1 - \varepsilon_1}{\varepsilon_1} \right) + \frac{A_1}{A_2} \left(\frac{1 - \varepsilon_2}{\varepsilon_2} \right) \right\}.$$

$$(4.99a)$$

The associated radiation network is shown in Fig. 4.26(b). As for the last equation, the reciprocal of the denominator can be defined as $\bar{F}$, the total exchange factor, and eqn (4.92a) can be written in the form

$$\dot{q}_{1-2} = \sigma A_1 \bar{F}_{12} (\theta_1^4 - \theta_2^4).$$

$$(4.99b)$$

4.14.   EXAMPLES OF RADIANT ENERGY TRANSPORT

(a)   Cooling of slabs on the table rolls of a hot strip mill

A steel ingot is rolled down to a 3.0 cm thick slab of steel before entering a multistand hot strip finishing mill. The maximum delay time before the tail end of the slab enters the mill is 5 min, during which time it loses heat by radiation to its surroundings. Given an exit temperature of 690 °C from the roughing mill, estimate whether the trailing edge of the slab can meet hot strip rolling specifications; these demand a minimum entry temperature of 650 °C into the first stand. Take the emmissivity of strip to be 0.4, the slab's density as 7600 kg m^{-3}, its heat emissivity as 0.15 kcal kg^{-1} °C^{-1}, and isothermal conductivity as 54 W m^{-1} K^{-1}.

Solution

Referring to Fig. 4.27, we can assume a view factor of unity for heat losses to the building, while back-radiation from the building will be negligible and can

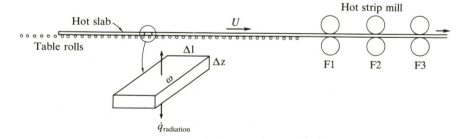

FIG. 4.27.   Schematic of slab cooling on table rolls during reduction to strip in a hot strip mill.

be ignored. Consequently, eqn (4.98) reduces to ($A_2 \rightarrow \infty$):

$$\dot{q}_{1-2} = \frac{E_{b1}}{\left[\dfrac{1-\varepsilon_1}{\varepsilon_1 A_1}\right] + \dfrac{1}{A_1}} = \varepsilon_1 \sigma \theta_1^4.$$

Since the slab is relatively thin, we can ignore temperature gradients

$$\left(\text{Bi} = \frac{hL}{k} = \frac{[\varepsilon_1 \sigma \theta_1^3]L}{k} \cong \frac{0.4 \times 5.7 \times 10^{-8} \times (963)^3 \times 0.03}{54} = 0.011\right)$$

so that a heat balance on a differential element Δl of the strip yields

$$0 - (\varepsilon_1 \sigma \theta_1^4) 2\Delta l \omega = \rho C \Delta l \omega \Delta z \frac{d\theta}{dt}.$$

Separating variables, and integrating between specified limits,

$$\int_{\theta_0}^{\theta_F} \frac{d\theta_1}{\theta_1^4} = -\frac{2\varepsilon_1 \sigma_2}{\rho C \Delta z} \int_{t_0}^{t_f} dt,$$

yields

$$\left[-\frac{1}{3\theta^3}\right]_{\theta_0}^{\theta_F} = -\frac{2\varepsilon\sigma\Delta t}{\rho C \Delta z}.$$

Rearrangement gives

$$\frac{\theta_F}{\theta_0} = 1 \left/ \left[1 + \frac{6\varepsilon_1 \sigma \theta_0^3 \Delta t}{\rho C \Delta z}\right]^{0.33}\right. .$$

Inserting physical properties and constants:

$$\frac{\theta_F}{\theta_0} = 1 \left/ \left[1 + \frac{6 \times 0.4 \times 5.7 \times 10^{-8} \times (963)^3 (5 \times 60)}{7600 \times (0.15 \times 10^3 \times 4.18)0.03}\right]^{0.33}\right. = 0.92674,$$

so that

$$\theta_F = 0.92674 \times 963 = 892 \text{ K or } 619\,°\text{C}.$$

Conclusion

Under such operating conditions, the tail end of the steel cannot meet rolling specifications, as it will enter the hot strip mill, 31 °C below the minimum entry temperature.

(b)   Radiative heat losses from the mouth of a BOF vessel

Estimate heat losses by radiation through the mouth of a BOF vessel shown in Fig. 4.28. The steel/slag mixture at turndown can be taken to be at 1630 °C:

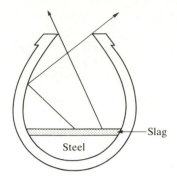

FIG. 4.28.   Radiative heat losses from a BOF vessel.

the diameter at the slag surface is 4 m and at the mouth of the vessel 2 m. The separation distance between these two planes is 7 m. The emissivity of molten slag is 0.95.

Solution

Referring to Figure 4.25(c), the view factor for a 'slag disc' of area A_1 in relation to a disc A_2 representing the furnace aperture, distance z apart, is

$$F_{1-2} = \frac{1}{2R_1^2}[z^2 + R_1^2 + R_2^2 - \{(z^2 + R_1^2 + R_2^2)^2 - 4R_1^2 R_2^2\}^{1/2}]$$

$$= \frac{1}{2 \times 2^2}[7^2 + 2^2 + 1^2 - \{(7^2 + 2^2 + 1^2) - 4(2^2 1^2)\}^{1/2}]$$

$$= 0.019.$$

Applying eqn (4.99), taking surface 1 as the slag and surface 2 as the furnace mouth, and discounting back-radiation from this imaginary (black) surface plane:

$$\dot{q}_{1-2} = \sigma A_1(\theta_1^2) \bigg/ \left[\frac{A_1 + A_2 - 2A_1 F_{12}}{A_2 - A_1(F_{12})^2} + \left\{ \frac{1}{\varepsilon_1} - 1 \right\} + \frac{A_1}{A_2}\left\{ \frac{1}{\varepsilon_2} - 1 \right\} \right]$$

$$= \sigma(12.57)(1903)^4 \bigg/ \left[\frac{12.57 + 3.14 - 0.24}{3.14 - 0.004} + \left\{ \frac{1}{0.95} - 1 \right\} \right]$$

$$= 1887 \text{ kW}$$

A simple alternative would be to approximate the furnace mouth as a black body plane emitting radiation to the melt shop with a view factor of unity. The heat loss is then

$$\dot{q} = \sigma \theta^4 A_2 = 2350 \text{ kW},$$

in reasonable accord with the previous calculation.

4.15. EXERCISES ON RADIANT HEAT TRANSFER

(1) Radiation heat shields

Two very large parallel plates of metal with emissivities of 0.4 are maintained at temperatures θ_1 and θ_2 respectively. Calculate the percentage reduction in heat flow between the two plates, when a thin radiation shield of polished aluminium ($\varepsilon = 0.04$) is placed between them, and construct the corresponding radiation network. Finally, show that for n radiation shields the reduction in heat flow is given by

$$(\dot{q}'')_{\text{with shields}} = \{1/(1+n)\} \dot{q}''_{\text{without shields}}$$

if all surfaces have the same emissivities.

(2) Transmission and absorption of radiation by glass plate

A glass plate is used to view the inside of a furnace. Assuming the latter to be equivalent to a black body at 1700 °C, plot the monochromatic emissive power E_λ in kW m^{-2} μm^{-1} versus wavelength λ in μm for this temperature. Use the graph to estimate the amount of energy absorbed by the glass (kW m^{-2}) and the amount of energy transmitted, given that transmissivity $\tau = 0.5$ for $0.2 \leqslant \lambda \leqslant 3.5$ μm and that $\varepsilon = 0.3$ for $0 \leqslant \lambda < 3.5$ μm and $\varepsilon = 0.9$ for $\lambda > 3.5$ μm.

Data:

$$E_{b,\lambda} = C_1 \lambda^{-5}/(e^{C_2/\lambda\theta} - 1), \quad E = \int_0^\infty E_{b,\lambda} \, d\lambda$$

Monochromatic emissive power, $E_{b,\lambda}$	W m^{-2} μm^{-1}
Wavelength, λ	μm
Temperature, θ	K
Constant C_1	3.743×10^8 μm^4 m^{-2}
Constant C_2	1.4387×10^4 μm K
Emissive power, E	W m^{-2} μm^{-1}

(3)   Liquid oxygen storage tanks

Liquid oxygen for a pyrometallurgical refining operation is stored in a 2.5 m O.D. sphere of aluminium ($\varepsilon = 0.09$), enclosed by a concentric 3.0 m O.D. sphere partially filled, in turn, with liquid nitrogen. The latter sphere is encased by a protective, third, concentric sphere of stainless steel, of 3.2 m I.D. and emissivity 0.3. The 10 cm space between the two outer spheres is evacuated. Assuming that the temperature of the outer shell is ambient (293 K), and that the two inner spheres are held at the boiling point of nitrogen, 77 K, estimate the heat (W) passing into the second storage tank, and the mass of sacrificial nitrogen which must evaporate from this middle tank (in $kg\,h^{-1}$) in order to maintain the oxygen below its boiling point of 90 K.

Data:

ε_1 (middle sphere)	=	0.09
Temperature (inner spheres)	=	77 K
ε_2 (outer sphere)	=	0.03
Surface temperature (outer sphere)	=	293 K
L_{e,N_2}	=	1.33 kcal/mole at 77 K
σ	=	$5.7 \times 10^{-8}\ W\,m^{-2}\,K^{-4}$

(4)   Continuous annealing lines: heating of strip

In a continuous annealing operation, steel strip 1 m wide, and 3 mm gauge, is fed through a long rectangular furnace filled with reducing HNX gas. The strip is heated by overhead electrical heaters, operating at 1400 °C. The furnace is 50 m long and well insulated by refractory.

Assuming the strip must reach 1100 °C before leaving the heating and annealing zones and entering the cooling section, write a short computer program illustrating the variation in exit strip temperature as a function of strip velocity, and determine the maximum throughput of material. Show how gauge (thickness) will affect metal throughput (t.p.h.).

You should assume a view factor of 1 between the elements and the strip and suppose that 40% of the refactory roof area facing the strip is covered by furnace elements of emissivity 0.8. Use appropriate thermal and physical properties for the strip.

4.16. CONCLUSIONS

This chapter has dealt with conduction and diffusion phenomena in solids, and associated boundary phenomena, including radiation. Gross convective flows, and examples in which fluid motion play a major role in distributing heat, or mass, within a phase, have been reserved for the next chapter.

The reader will find an extensive range of more advanced analytical solutions and examples than those considered in the present text in the classic works of Crank, Carslaw and Jaeger, and Jakob, listed in Further Reading.

FURTHER READING

1. Crank, J. O. (1967). *Mathematics of diffusion*. Oxford University Press.
2. Carslaw, M. S. and Jaeger, J. C. (1947). *Conduction of heat in solids*. Oxford University Press, London.
3. Jakob, M. (1949). *Heat transfer*. Wiley, New York.
4. Darken, L. S. and Gurry R. W. (1953) *Physical chemistry of metals*. McGraw-Hill, New York, Toronto, London.
5. Geiger and Poirier (1980). *Transport phenomena in metallurgy*, pp. 431–49. Addison-Wesley, Reading, MA.

5

HEAT AND MASS TRANSFER IN
CONVECTIVE FLOW SYSTEMS

5.0. INTRODUCTION

In the preceding chapter we were exclusively concerned with metallurgical examples of systems in which there was no bulk motion of the substance through which or across which heat or mass was being transferred. There are, however, a wide variety of systems in which heat or mass transfer occurs as a result of both convection and diffusion processes, as briefly explained in Chapter 1.

5.1. CONVECTIVE COMPONENTS IN HEAT AND
MASS TRANSFER OPERATIONS

(a) Bulk flow of fluids

In any system involving convection, convective components of heat or mass flow generally dominate molecular transport components, except in those regions, such as boundary layers, where the speed of flow is much reduced. Consider, for example, the transfer of heat or mass through a stream tube in which fluid is travelling at a velocity U ($m\,s^{-1}$). Figure 5.1 shows the two components involved (i.e. diffusion and convection). Taking heat transfer, the conductive component (*the conductive heat flux*) across a plane located at x is given by $\dot{q}''_x = -k\dfrac{\partial \theta}{\partial x}$ (Fourier's first law), while the convective component is defined as that heat carried across plane x resulting from the motion of the fluid itself. Supposing therefore, that the temperature of the fluid passing plane x is θ_x, the heat content of the fluid is equal to $\rho C\theta$ ($kcal\,m^{-3}$ of fluid) and the *rate* of convective heat transfer is equal to $U(\rho C\theta)$. Its units are therefore $kcal\,m^{-2}\,s^{-1}$ (or equivalent) and correspond to a *convective heat flux*.

For convective mass transfer, if U is the bulk velocity of the fluid, and D_A, is the number of moles of solute A per cubic metre of liquid, then the number of moles of A transferred across plane x as a result of convection is UC_A (moles of A per m^2). We see that if $UC_A \gg -D\dfrac{\partial C_A}{\partial x}\Big|_x$, the convective mass transfer component will dominate. Similarly, if $\rho C\theta U \gg -k\dfrac{\partial \theta}{\partial x}\Big|_x$, convective heat transfer will dominate.

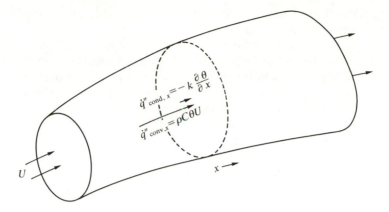

FIG. 5.1. Flow of liquid through a stream tube showing the conductive and convective components of heat fluxes across plane at x.

If we now turn our attention to heat and mass transfer at right angles to the direction of flow, the opposite set of circumstances exists. Since a stream tube (by definition) has zero transverse velocity across its boundary walls, the convective mass and heat transfer components must be zero and only equimolar interdiffusion and/or conduction can account for interchanges of mass and heat with fluid surrounding the stream tube.

(b) Flow of fluids over solid surfaces: boundary layers

An analogous set of circumstances exists when we look at liquid flowing over a solid surface. Figure 5.2 shows the type of velocity, concentration, and temperature profiles that will generally exist over a flat surface. Provided the liquid is in turbulent rather than laminar flow, there will be a central core of

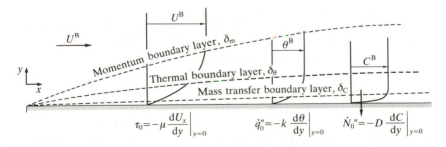

FIG. 5.2. Flow of fluid over a flat plate, showing the development of momentum, thermal, and mass transfer boundary layers adjacent to surface, together with appropriate expressions for fluxes of momentum, energy and mass flow normal to surface. Note that diffusional processes dominate immediately adjacent to surfaces.

fluid in which transverse and longitudinal velocity fluctuations in the bulk of the liquid ensure good mixing (Chapter 2).

Such mixing leads to relatively uniform bulk velocities, uniform bulk concentrations, and uniform bulk temperatures. Similarly, velocities will generally be high enough to make the convective components outweigh even axial *eddy* diffusion or *eddy* conductive transfer components. Close to the surfaces of the solid, however, the liquid is slowed, eddies are damped, and at the interface itself, all flow of liquid ceases (the no-slip condition).

Since fluid at a solid, impermeable, stationary interface is motionless, the vertical convective component will be negligible and molecular conduction /diffusion processes alone must limit any transfer rates into the solid. The velocity-affected region of flow is known as the momentum boundary layer δ_m, defined as that region where velocities are 99 per cent, or less, of the bulk flow velocity U^B. Equivalent regions close to the solid interface where the concentration and temperature profiles become affected can be represented by δ_C and δ_θ, the mass and heat transfer boundary layers respectively.

Referring to Fig. 5.2, it is then a very straightforward matter to write down expressions for *normal* fluxes crossing the interface into a solid. The inside surface of a pipe, for instance, at $r = R$,

$$\dot{N}''_{A, r=R} = -D_A \frac{dC_A}{dr}\bigg|_{r=R} \tag{5.1}$$

and

$$\dot{q}''_{r=R} = -k \frac{d\theta}{dr}\bigg|_{r=R}. \tag{5.2}$$

or, in the case of flow across flat surfaces, the equivalent expressions for fluxes are shown in Fig. 5.2. Thus, as in the case of transfer through motionless media, we use Fick's and Fourier's first laws. There is, however, a fundamental difference, in that the *slopes* of the gradients at the interface are affected by the rate of fluid flow past the surface.

Qualitatively, the greater the flow-rate, the steeper will be the velocity gradient at the solid interface, which in turn will lead to steeper concentration and temperature gradients and higher fluxes. It now becomes apparent that quantitative predictions of transfer rates between a solid surface and fluid require an exact knowledge of flow, temperature, and concentrations fields.

In cases of laminar flow, it is possible to 'quantify' the flow field and obtain exact expressions for velocity, concentration, and temperature boundary layer profiles by simultaneous solution of the continuity, momentum (Navier–Stokes), and energy/mass conservation equations.

When the flow field is turbulent, which is normally the case for industrial processes, exact mathematical solutions are not yet possible, and experimental measurements of heat or mass transfer between the liquid and the solids have to be made. These measurements involve measuring the amount of heat or mass transferred in a given time, for a given temperature or concentration driving force. Thus, recalling the concept of a heat and mass transfer coefficient, h and k_L respectively, the results are expressed in terms of h or k_L:

$$h = \frac{\dot{q}_0''}{(\theta^B - \theta^*)} \text{ W m}^{-2}\text{ K}^{-1}; \quad k_L = \frac{\dot{N}_{A,0}''}{(C_A^B - C_A^*)} \text{ m s}^{-1}. \tag{5.3}$$

The way in which h or k varies with flow conditions, fluid properties, and system geometry are usually presented in terms of correlations among the various dimensionless π groupings listed in Table 3.2. Some of these will be presented later in this chapter in Section 5.2.

Concept of a stagnant boundary layer thickness

Referring to Fig. 5.3, showing transfer of heat from a flowing liquid to the cold walls of a pipe, we see that three equivalent expressions can be formulated to express heat fluxes into the solid. For heat, we have

$$\dot{q}_{r=R}'' = -k\frac{d\theta}{dr}\bigg|_{r=R} = h(\theta^B - \theta^*) = k\frac{(\theta^B - \theta^*)}{\delta_\theta'}, \tag{5.4a}$$

where k is the thermal conductivity. Similarly for mass flow into the pipe wall,

$$\dot{N}_{r=R}'' = -D\frac{dC}{dr}\bigg|_{r=R} = k_L(C^B - C^*) = D\frac{(C^B - C^*)}{\delta_C'} \tag{5.4b}$$

Here δ_θ' and δ_C' are the fictitious thermal and mass transfer *stagnant boundary layer thicknesses* (typically $\simeq 10^{-5}$–10^{-6} m in low-viscosity liquids such as water and liquid metals). Note that they are mathematical artefacts with little real physical meaning and merely replace the interfacial gradients $(d\theta/dr)_{r=R}$

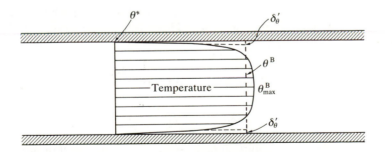

FIG. 5.3.   Transfer of heat from a hotter liquid, flowing through a pipe, to colder pipe walls, illustrating a local temperature profile and an equivalent stagnant boundary layer thickness, δ_θ.

or $(dC/dr)_{r=R}$ with the geometrical constructions $(\theta^B - \theta^*)/\delta'_\theta$ or $(C^B - C^*)/\delta'_C$. These terms are often found in older publications in the scientific literature. They are order-of-magnitude correct, and still useful conceptually, provided one realizes that they are not only a function of liquid velocity, but of a number of other variables (e.g., k, D, x, C and t). For laminar boundary layers, $Sc = v/D \simeq \delta_m/\delta_C$ while $Pr = v/\alpha \simeq \delta_m/\delta_\theta$ for $Pr \leq 1$. The momentum boundary layer thickness δ_m for flow parallel to a flat plate is given by:

$$\delta_m = 4.9 \left(\frac{\mu x}{\rho U^B} \right)^{1/2} \tag{5.5}$$

Finally, it is worth remarking that liquid metals have very low Prandtl numbers (~ 0.1–0.01), signifying thermal boundary layers, δ_θ, some ten times thicker than associated momentum boundary layers, δ_m. By contrast, mass transfer boundary layers in metals tend to be much thinner than momentum boundary layer thickness, since $D \ll v$ (or Sc high).

(c) Worked example: Convective flow of liquids past solid surfaces in a liquid-metal-cooled nuclear reactor

As a representative example of metallurgical convective mass transfer phenomena at solid/liquid boundaries, we will consider liquid metal flow in a nuclear reactor. The heat transfer loop for this type of reactor system is shown schematically in Fig. 5.4. Liquid lead–bismuth eutectic alloy is pumped around the closed circuit by a centrifugal pump and the nuclear heat produced within the reactor core is used to heat the liquid metal to 500°C. This heat is then removed in the cold leg of the loop using a heat exchanger system coupled to a steam turbine, and the temperature is everywhere maintained at 375 °C in the cold section of the loop. Most of the pipework for pumping the lead–bismuth alloy is constructed of steel, which has a small but significant solubility in the liquid alloy. This solubility increases with temperature, the relation being

$$\log_{10}(\text{p.p.m. Fe}) = 6.03 - \frac{4440}{T},$$

where p.p.m. = parts per million *atomic basis*, and T is in K.

Since the bulk concentration of iron leaving the hot zone is supersaturated with respect to the solubility limit of iron in the cold section, some mass transfer occurs resulting in 'corrosion' of piping in the hot leg section and deposition in the cold leg section. Obtain a relationship between the bulk concentration of iron entering the hot leg section, i.e. $C^B_{x=0}$, and the following

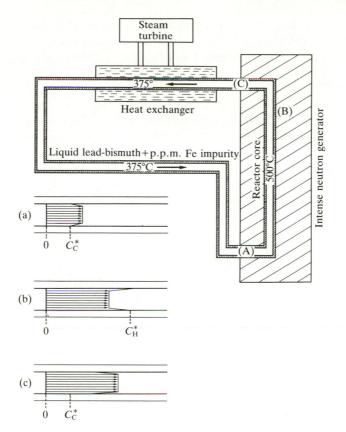

FIG. 5.4. A heat exchange system for a liquid metal cooled nuclear reactor. Concentration profiles for dissolved iron from the pipes are shown (a) just before molten alloy flows into reactor core, (b) midway through reactor core, and (c), immediately after exiting.

variables:

L	Length of hot zone	3 m
l	Length of cold zone	10 m
k_L	Liquid phase mass transfer coefficient of Fe in Pb–Bi alloy.	1×10^{-4} m s^{-1}
C_H^*, C_C^*	Equilibrium concentrations of Fe in Pb–Bi at the two temperatures, H = hot leg, C = cold leg	
U	Bulk velocity of liquid Pb–Bi coolant	0.3 m s^{-1}
d	Diameter of pipe	76 mm

In particular, calculate the maximum increase in internal diameter of the hot leg during a 2-year period of operation. You may assume that the solution process is mass-transfer-controlled and that the liquid Pb–Bi is in plug flow.

Solution

Figure 5.4 has been drawn so as to illustrate the type of concentration profiles that we can expect immediately before the lead alloy enters the hot leg (A), while it is in the hot leg itself (B), and just after leaving the hot leg (C). As we see, the direction of mass transfer reverses owing to the step change in the interfacial concentration of iron at $x=0$ and $x=L$. Also, note that in the bulk of the pipe, velocity and concentration profiles are taken to be flat (i.e. $\neq f(r)$) thanks to good turbulent mixing in the radial sense, and fully developed flow.

A. Hot leg

Writing down the steady-state macroscopic continuity equation for iron dissolving from the pipe into the volume element shown, we have:

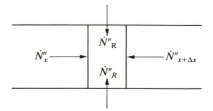

Axial input of A + radial input of A = axial output of A

or, in symbols,

$$\dot{N}''_x A_{pipe} = \dot{N}''_{x+\Delta x} A_{pipe} - \dot{N}''_{Fe,r=R} \pi d \, \Delta x,$$

or

$$(\dot{N}''_x - \dot{N}''_{x+\Delta x}) \frac{\pi d^2}{4} = -\pi d \, \Delta x \, k_L (C^*_{hot} - C^B),$$

$$\frac{\partial \dot{N}''_x}{\partial x} \Delta x = 4 \frac{\Delta x \, k_L}{d} (C^*_{hot} - C^B).$$

We can safely assume that the convective flux far outweighs the diffusive flux in the x direction, so that replacing $\dot{N}''_x$ with $U_x C^B_x$ and noting that U_x is independent of x (i.e. is constant) we obtain

$$U \frac{\partial C^B}{\partial x} = \frac{4 k_L}{d} (C^*_H - C^B),$$

which is readily integrated over the limits $0 \leqslant x \leqslant L$, $C_0^B \leqslant C \leqslant C_L^B$ to give

$$\frac{C_H^* - C_L^B}{C_H^* - C_0^B} = e^{-4k_L L/U d}.$$

B. Cold leg

Carrying out an equivalent procedure for a volume element of liquid metal in the cold leg of the reactor,

$$\dot{N}_x'' A = \dot{N}_{x+\Delta x}'' A + k_L \pi d \, \Delta x (C^B - C_C^*),$$

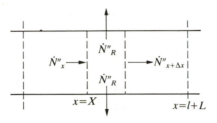

so that

$$-U \frac{dC^B}{dx} = \frac{4k_L}{d} (C^B - C_C^*),$$

which, on integrating gives:

$$\int_{C_L^B}^{C_0^B} \frac{dC^B}{(C^B - C_C^*)} = -\frac{4k_L}{Ud} \int_{x=L}^{x=L+l} dx.$$

Noting that $C_0^B = C_{L+l}^B$ gives

$$\frac{C_0^B - C_C^*}{C_L^B - C_C^*} = e^{-4k_L l/U d}.$$

Substituting for C_L^B in the above relationship, using the hot leg solution for C_L^B, we finally obtain an explicit expression for the concentration iron in the bulk of the liquid lead just as it enters the hot leg:

$$C_0^B = C_C^*(1 - e^{-4kl/Ud}) + C_H^*(1 - e^{-4kL/Ud})e^{-4kl/Ud}$$

Using the data provided, we then find $C_0^B = 0.56$ p.p.m.

Amount of corrosion

In order to calculate the rate of corrosion and thence the maximum increase in internal diameter of the hot leg during a 2-year time period, T, we can relate the flux of iron to the amount of iron lost from unit area of the pipe:

$$\dot{N}''_{r=R} T = \frac{\rho_{Fe}}{M_{Fe}} (1 \times 1 \times \Delta r),$$

so that

$$\Delta d = 2 \frac{M_{Fe}}{\rho_{Fe}} T k_L (C_H^* - C_0^B).$$

We therefore have to convert our concentrations C_H^* and C_0^B of dissolved iron expressed in atomic parts per million into units of kg-atoms of iron per cubic metre of lead–bismuth alloy.

Noting that $M_{Bi} = 209$, $M_{Pb} = 207$, $\rho_{Bi} = 10\,060\ \mathrm{kg\,m}^{-3}$, $\rho_{Pb} = 10\,653\ \mathrm{kg\,m}^{-3}$, $\rho_{Fe} = 7866\ \mathrm{kg\,m}^{-3}$, we need first to calculate the volume occupied by 1 kg mole of Pb–Bi alloy for a eutectic alloy of lead and bismuth ($X_{Pb} = 0.45$, $X_{Bi} = 0.55$):

$$V_M = 0.45 \times 207.2 \frac{\mathrm{kg}}{\mathrm{kg.mole}} \times \frac{1\ \mathrm{m}^3}{10653\ \mathrm{kg}} + 0.55 \times 209 \frac{\mathrm{kg}}{\mathrm{kg\text{-}mole}} \times \frac{1\ \mathrm{m}^3}{10060\ \mathrm{kg}}$$

$$= 0.00875 + 0.01143 = 0.02018\ \mathrm{m}^3\ \mathrm{kg\,mole}^{-1}$$

where we ignore the contribution made by contaminant iron. Therefore

$$1\ \text{p.p.m. Fe} \equiv \frac{1 \times 10^{-6}}{0.0202}\ \text{kg-atom Fe/m}^3 \text{ of alloy}.$$

Consequently,

$$\Delta d = 2 \times \frac{55.85}{7866} \times (2 \times 365 \times 24 \times 3600) \times 1 \times 10^{-4} \times (1.93 - 0.56) \times \frac{10^{-6}}{0.02}$$

$$= 0.006\ \mathrm{m}.$$

Conclusion

The maximum increase in diameter will be at the entry to the hot section, inside the nuclear core, and could amount to about 6 mm over a 2-year period!

5.2. CORRELATIONS FOR HEAT AND MASS TRANSPORT BETWEEN SOLIDS AND FLUIDS

In the last section, the corrosion of the steel pipe could only be calculated provided the mass transfer coefficient between the bounding walls of the tube and molten lead flowing through the circuit was known. Unfortunately analytical expressions for such transport coefficients are rarely available since flows are complex, or turbulent. None the less, convection heat and mass transfer data can be conveniently correlated in terms of dimensionless numbers such as the Nusselt (Nu), Sherwood (Sh), Reynolds (Re), Prandtl (Pr), Schmidt (Sc), Grashof (Gr), and Raleigh (Ra) numbers.

For many situations, heat transfer correlations can be applied to mass transfer if Sh and Sc substitute for Nu and Pr, and if the mass transfer forms of Gr and Ra are used in place of their thermal equivalents. Hence,

$$\text{Sh} = \frac{k_M L}{D_{AB}} \text{ substitutes for } \text{Nu} = \frac{hL}{k}, \tag{5.6}$$

$$\text{Sc} = \frac{v}{D_{AB}} \text{ substitutes for } \text{Pr} = \frac{v}{\alpha}, \tag{5.7}$$

$$\text{Gr}_M = \frac{\rho^2 \beta_M g \Delta X L^3}{\mu^2} \text{ substitutes for } \text{Gr}_\theta = \frac{\rho^2 \beta_\theta g \Delta \theta L^3}{\mu^2}, \tag{5.8}$$

$$\text{Ra}_M = \text{Gr}_M \text{ Sc substitutes for } \text{Ra}_\theta = \text{Gr}_\theta \text{ Pr}. \tag{5.9}$$

The characteristic length L is the diameter for circular geometries such as pipes, cylinders and spheres. For non-circular ducts, the characteristic length is taken as $(4 \times$ cross-sectional area)/(wetted perimeter). For flat surfaces, such as plates, the characteristic length is the length of the plate in contact with the fluid flow.

In the following correlations, the average heat or mass transfer coefficients for the surface as a whole are described, rather than the local values at distinct points on the surface.

It is worth noting that typical Prandtl (Pr), values for gases are $0.6 \sim 1.0$, for most liquids $1 \sim 10$. For liquid metals Pr values are usually very low, around 10^{-2}. Typical Schmidt (Sc), values are $1 \sim 2$ for gases, $10^2 \sim 10^3$ for most liquids, and around 10^3 for liquid metals.

(a) Forced convective heat and mass transfer correlations

(i) Flow over a plate

For laminar flow over isothermal plates, and $\text{Pr} > 0.6$, the average heat transfer coefficient is given by (Kreith and Black 1980):

$$\text{Nu} = 0.66 \, \text{Re}^{0.5} \, \text{Pr}^{0.33}. \tag{5.10}$$

For mass transfer, this correlation becomes:

$$\text{Sh} = 0.66 \, \text{Re}^{0.5} \, \text{Sc}^{0.33}. \tag{5.11}$$

The transition from laminar flow to one of turbulence for fluid passing over plates begins at $Re = 5 \times 10^5$. For turbulent flow over an isothermal plate, at $Pr = 0.7$ (Kreith and Black 1980).

$$Nu = 0.036 \, Re^{0.8} \, Pr^{0.33}. \tag{5.12}$$

For liquid metals, and laminar forced convective heat transfer for flow over a plate, one has the limiting condition for low Prandtl values (see worked solution), that

$$Nu = 1.12 \, (Re. \, Pr)^{0.5} \tag{5.13}$$

(ii)   Flow in pipes

For laminar flow in pipes (Sieder 1936):

$$Nu = 1.86 \, Re^{0.33} \, Pr^{0.33} \left(\frac{D}{L}\right)^{0.33} \left(\frac{\mu}{\mu_s}\right)^{0.14}, \tag{5.14}$$

where (D/L) is the ratio of the pipe diameter to its length, and μ and μ_s refer to the fluid viscosity at the average bulk temperature and wall temperature respectively. This correlation is valid for $Re \, Pr \, (D/L) > 10$.

The transition to turbulence for flow in pipes begins at $Re \cong 2000$. For turbulent forced convective heat transfer in pipes, for $Pr > 0.6$ (Colburn 1933):

$$Nu = 0.026 \, Re^{0.8} \, Pr^{0.33} \left(\frac{\mu}{\mu_s}\right)^{0.14}. \tag{5.15}$$

For developing flow, for pipe lengths $10 \sim 400$ times the pipe diameter, the entrance effects are accounted for as follows (Nusselt 1931),

$$Nu = 0.036 \, Re^{0.8} \, Pr^{0.33} \left(\frac{D}{L}\right)^{0.055}. \tag{5.16}$$

For turbulent forced convective heat transfer for liquid metals in pipes with constant wall temperature, and $Re \, Pr > 100$, $L/D > 60$ (Seban 1951).

$$Nu = 5.0 + 0.025(Re \, Pr)^{0.8}. \tag{5.17}$$

(iii)   Flow around cylinders

For the forced convective flow of gases and most liquids around a submerged cylinder set perpendicular to the flow direction, the liquid phase heat transfer coefficient, embodied in the Nusselt number, can be correlated in the form (Kreith 1980):

$$Nu = C \, Re^n Pr^{0.33}, \tag{5.18}$$

where the constant C and exponent n are dependent on Re as follows:

Re	C	n
0.4–4	0.989	0.330
4–40	0.911	0.385
40–4000	0.683	0.466
$4000–4 \times 10^4$	0.193	0.618
$4 \times 10^4–4 \times 10^5$	0.027	0.805

The same form of correlation applies to flow over a bank of cylinders, as in heat exchangers, but C and n are further dependent on tube spacing and alignment (Grimison 1937).

(iv) Flow around spheres

For forced convective flow around a rigid sphere (Whitaker 1972):

$$Nu = 2 + (0.4 \, Re^{0.5} + 0.06 \, Re^{0.67}) \, Pr^{0.4} \left(\frac{\mu}{\mu_s} \right)^{0.25}. \qquad (5.19)$$

For liquid metal flow over a sphere (Witte 1968):

$$Nu = 2 + 0.38(Re \, Pr)^{0.5} \qquad (5.20)$$

(b) Natural convective heat and mass transfer correlations

Natural convective heat transfer data is usually correlated in terms of the Rayleigh number (Ra) which is the product of Gr and Pr. In the following correlations, fluid properties such as density, conductivity, and viscosity, are evaluated at the film temperature, which is the arithmetic average of the bulk fluid temperature and the wall surface.

(i) Vertical plates and cylinders

Laminar natural convective heat transfer to vertical plates (and cylinders if $(D/L) > 35/Gr^{0.25}$) is given by (Gebhart 1970):

$$Nu = 0.56 \, Ra^{0.25}. \qquad (5.21)$$

Here, the characteristic length is taken as the length or height of the surface in contact with the fluid. Turbulence begins at around $Ra = 10^9$.

Heat transfer with turbulent natural convection to vertical planes and cylinders is (Gebhart 1970):

$$Nu = 0.02 \, Ra^{0.4}.$$ (5.22a)

or (Jakob, 1964)

$$Nu = 0.13 \, Ra^{0.33}$$ (5.22b)

(ii) Horizontal plates and cylinders

For laminar natural convection of fluid above the upper surface of warm horizontal plates or below the lower surfaces of cool plates, for Ra in the range $2 \times 10^4 \sim 8 \times 10^5$ (Fujii 1972).

$$Nu = 0.54 \, Ra^{0.25}.$$ (5.23)

This type of natural convection is known as Bénard convection.

For the same conditions, but higher Ra in the range $8 \times 10^6 \sim 1 \times 10^{11}$, in the turbulent regime:

$$Nu = 0.15 \, Ra^{0.33}.$$ (5.24)

The above equations can be used for horizontal cylinders if the diameter is taken as the characteristic length. For natural convective heat transfer from horizontal cylinders to liquid metals, (low Pr):

$$Nu = 0.53 \, Gr^{0.25} \, Pr^{0.5}.$$ (5.25)

For natural convection to the lower surface of warm plates or the upper surface of cold plates, in the Ra range 10^5–10^{11} (Singh 1969):

$$Nu = 0.58 \, Ra^{0.2}.$$ (5.26)

(iii) Natural convection around spheres

Heat transfer by free convection around spheres is given by (Yuge 1960):

$$Nu = 2 + 0.45 \, Ra^{0.25},$$ (5.27)

where the correlation is based on the sphere diameter. One can readily show that the steady-state conduction of heat between a sphere and a fluid, in the absence of any convection, corresponds to the limiting value of 2.0.

In metallurgy, one commonly encounters situations in which there is mass transfer between a fluid and a sphere. Examples include oxide pellets descending in countercurrent flow to reducing gases in a blast furnace, or metal drops falling through gases or liquid slags. Mass transfer in the continuous phase has been correlated as follows (Steinberger and Treybal 1960):

$$Sh = 2 + 0.35 \, Re^{0.6} \, Sc^{0.3} + 0.57(Gr'Sc)^{0.25},$$ (5.28)

where

$$Gr' = Gr_M + \left(\frac{Sc}{Pr}\right)^{0.5} Gr_\theta. \tag{5.29}$$

It is interesting that the effects on mass transfer rates of diffusion, forced convection and natural convection are taken as additive in the relationship. The transition to turbulent flow is around $Re = 10$.

(iv) Enclosed spaces

Convective heat transfer in an enclosed space between vertical parallel plates separated by a distance δ can be correlated in terms of Ra, where δ is the characteristic length, and the difference in plate temperatures is $\Delta\theta$ (Perry 1984).

For $Ra_c < 1700$, the fluid remains stagnant and $Nu = 1$. With Ra from 2×10^4 to 2×10^5, laminar natural convection in vertical spaces leads to:

$$Nu = 0.2 \, Ra^{0.25} \left(\frac{L}{\delta}\right)^{0.014}, \tag{5.30}$$

where L is the cell or cavity height.

For turbulent convection in vertical spaces (Ra from 2×10^5 to 1×10^7):

$$Nu = 0.07 \, Ra^{0.33} \left(\frac{L}{\delta}\right)^{0.014}. \tag{5.31}$$

Laminar convection in horizontal spaces, Ra from 1×10^4 to 3×10^5, is described by

$$Nu = 0.21 \, Ra^{0.25} \left(\frac{L}{\delta}\right)^{0.25}, \tag{5.32}$$

whereas turbulent convection in horizontal spaces, Ra from 3×10^5 to 1×10^7, correlated as

$$Nu = 0.075 \, Ra^{0.33}. \tag{5.33}$$

Note that the heat transfer coefficient is independent of plate separation under these conditions.

For liquid metals in vertical cylindrical containers, or cavities, turbulent Bénard-type convection in the bath is described by (Chiesa and Guthrie 1971):

$$Nu = 0.078 \, Ra^{0.33} (0.68)^{L/D}, \tag{5.34}$$

where the metal depth L is used as the characteristic length and D is the diameter of the container. This correlation applies for Ra from 2×10^4 to 2×10^7. Note similarity with eqn 5.33 for higher Prandtl number fluids. Again, the critical Rayleigh number for the onset of natural convection was found to be $Ra_c = 1700$.

(c)   Development of heat transfer correlation for flow of liquid metal over a flat surface

As explained, heat transfer phenomena in liquid metals are unusual in comparison to other systems, on account of their high thermal conductivities and attendant low Prandtl numbers. A low Prandtl number indicates that the thickness of the thermal boundary layer is much greater than an associated momentum boundary layer. As a result, the flow of liquid metal over a flat plate can be treated as being in plug flow, i.e., one can ignore the small portion of the thermal boundary layer where the flow of metal is slowed (inside the momentum boundary layer).

Consider liquid metal at a temperature θ_B, passing over a flat surface at temperature θ_s. Assuming plug flow, we need to solve the partial differential equation.

$$\frac{\partial \theta}{\partial t} = \alpha \frac{\partial^2 \theta}{\partial y^2}, \tag{5.35}$$

with the boundary conditions

$$\forall t, y = 0, \quad \theta = \theta_s; \quad \forall t, y = +\infty, \quad \theta = \theta_B.$$

The solution is

$$\theta_{y,t} = \theta_s + (\theta_\infty - \theta_s)\, \mathrm{erf}\left(\frac{y}{\sqrt{4\alpha t}}\right). \tag{5.36}$$

The heat flux at $y=0$ can be obtained from

$$\dot{q}'' = -k\frac{\partial \theta}{\partial y}\bigg|_{y=0} = -\frac{k(\theta_\infty - \theta_s)}{\sqrt{\pi \alpha t}}. \tag{5.37}$$

This analytical solution can be cast in the form of a Nusselt number as shown below:

$$\frac{\dot{q}''}{(\theta_s - \theta_\infty)} = h = \frac{k}{\sqrt{\pi \alpha t}}. \tag{5.38}$$

Replacing t by x/U_∞ and multiplying either side by x/k, one obtains

$$\frac{hx}{k} = \frac{kx}{k\sqrt{\pi \alpha x/U_\infty}} \tag{5.39}$$

$$= \sqrt{\left(\frac{1}{\pi}\frac{xU_\infty}{\nu}\frac{\nu}{\alpha}\right)},$$

or

$$Nu_x = \sqrt{\left(\frac{(Re_x)(Pr)}{\pi}\right)}. \qquad (5.40)$$

This shows that the local heat transfer coefficient Nu_x rises in proportion to the square root of velocity and inversely with x. Integration of h_x between 0 and L shows that the mean Nusselt number Nu_L over a plate of length L, is equal to $(2/\sqrt{\pi})[(Re_L)(Pr)]^{1/2}$, in agreement with eqn (5.13).

5.3. WORKED EXAMPLES EMPLOYING HEAT/MASS TRANSFER CORRELATIONS FOR TRANSPORT COEFFICIENTS

(a) The stability of thermal accretions in Q-BOP (OBM) steel-making technology

In order to protect adjacent brickwork from rapid erosion by submerged jets of oxygen, Savarde and and Lee (personal 1980) developed a co-axial tuyère shown schematically in Fig. 5.5. Natural gas, blown through an annular space, is heated and thermally cracked as it passes through pores within the accretion (commercially termed 'mushroom'). These endothermic effects can provide sufficient cooling capacity to compensate for convective heat input from the bath of molten steel. Calculate the efficiency of natural gas utilization in a specific commercial operation, given that the radial surface width of the accretion around the rim of the nozzle was 15 cm, that natural gas was blown at a rate of 5 per cent that of the central jet of oxygen, and the following data:

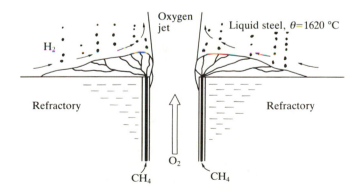

FIG. 5.5.   Schematic of thermal accretions formed in Q-BOP steel making operations using the Savarde–Lee shrouded tuyère.

Data:

Physical properties of steel (μ, ρ, C, k)	
Inward (average radial) velocity of steel adjacent to nozzle	50 cm s^{-1}
Bath temperature	1620°C
Melting point of steel	1490°C
Cooling capacity of natural gas	43.6 kcal mole^{-1}
Oxygen flow rate through central tuyère	32 Nm^3 min^{-1}
Inside diameter of accretion/tuyere	10 cm

Solution

Ignoring minimal heat flow by conduction from the accretion to the underlying refractory brick, a simple, steady-state heat balance requires that:

$$\text{Heat input from bath by convection to the exposed surface} \atop \text{of the accretion in contact with bath} \qquad (1)$$

$$=$$

$$\text{sensible and endothermic heat requirements for} \atop \text{methane gas cracking and heating during passage up} \atop \text{annular space and through accretion} \qquad (2)$$

i.e.

$$h A(\theta^{\text{bath}} - \theta^{\text{m.p.}}_{\text{Fe}}) = \dot{N} \Delta H.$$

To estimate the heat transfer coefficient h due to liquid steel washing over the accretion, we may approximate the system to flow over a flat plate and use the relationship

$$\text{Nu}_{\text{L}} = 1.12 \, \text{Re}^{1/2} \, \text{Pr}^{1/2}.$$

Taking $k = 41.8$ W m^{-1} K^{-1}, $\rho = 7000$ kg m^{-3}, $\mu = 6.7$ mPa s, $C = 750$ J kg^{-1}K^{-1}, we have

$$\text{Re}_{\text{L}} = \frac{7000 \times 0.5 \times 0.15}{6.7 \times 10^{-3}} = 78 \, 360,$$

$$\text{Pr} = \frac{\nu}{\alpha} = \frac{\mu C}{k} = \frac{6.7 \times 10^{-3} \times 750}{41.8} = 0.12.$$

Therefore

$$\text{Nu}_{\text{L}} = 1.12(78360)^{1/2}(0.12)^{1/2} = 108.6,$$

and

$$h = \text{Nu}_{\text{L}} \times \frac{k}{L} = 108.6 \times \frac{41.8}{0.15} \equiv 30.26 \text{ kW m}^{-2} \text{ K}^{-1}.$$

Consequently, the net heat flowing into the thermal accretion will equal

$$\dot{q} = hA(\theta^B - \theta_{Fe}^{m.p.}) = 30.26 \times 10^3 \times \frac{\pi}{4} \times (0.25^2 - 0.10^2)(1620 - 1490),$$

Therefore,

$$\dot{q}_{IN} \equiv 162.2 \text{ kW}.$$

The cooling component from the natural gas's cracking and sensible heat requirements is

$$\dot{N}\Delta H = \left(\frac{5}{100} \frac{32}{(22.4 \times 10^{-3})60}\right)(43.6 \times 4.18)$$

$$= 217 \text{ kW}.$$

Therefore the utilization efficiency of shroud gas cooling, $\eta = \dfrac{162.2}{217} \times 100 = 74$ per cent.

(b) Dissolution of steel scrap in molten iron

During studies into the operating characteristics of a small continuous-flow steel-making reactor (IRSID), it was proposed that the larger pieces of more bulky scrap could be dissolved in the hot metal rather than attempting to feed them directly into the reaction zone under the oxygen jet. Industrial tests were therefore carried out to see whether large pieces of scrap (hot mill crop ends) could first be dissolved in molten iron from the blast furnace, and the liquid mix then fed to the continuous reactor. The procedure is shown in Fig. 5.6. Estimate the time that should be allowed after pouring carbon-saturated hot metal over the 4-cm-thick pieces of scrap slab, so as to ensure that all has dissolved. Use correlations for vertical plates relating to natural convection phenomena, together with the following data, to arrive at an estimate.

Data:

Thickness of slab crop ends, Δ	5 cm
Amount of scrap charged, M_s	10 000 kg
Amount of hot metal charged, $M_{H.M.}$	104 550 kg
Initial hot metal temperature, $\theta_{H.M.}$	1480°C
Diffusion coefficient of carbon in iron, $D_{C/Fe}$	10×10^{-9} m^2 s^{-1}
Iron–carbon phase diagram	
(Appendix III, Fig A3-1)	
Heat capacity of scrap	627 J kg^{-1} K^{-1}
Heat capacity of hot metal	752 J kg^{-1} K^{-1}
Bulk carbon concentration	4.3 wt. %

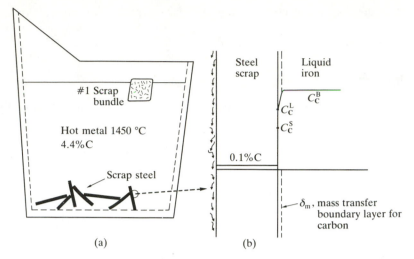

F<small>IG</small>. 5.6.   Schematic of scrap dissolution in molten iron in the transfer ladle prior to steel making operations: (a) Ladle and contents; (b) mechanism of scrap dissolution.

Solution

Scrap can dissolve below its thermal melting point temperature, since carbon diffusing through molten iron can raise the surface carbon content above carbon liquidus values. When this takes place, the steel will dissolve.

Clearly, once the carbon-saturated iron has been poured over the scrap, thermal equilibrium between the two will rapidly be established. The temperature of the heated scrap and cooled hot metal can therefore be deduced from the simple heat balance:

$$M_{\text{H.M.}} C_{\text{H.M.}} (\theta^0_{\text{H.M.}} - \theta^F_{\text{H.M.}}) = M_S C_S [\theta^F_S - \theta^0_S]$$

where θ^F_S, the final scrap temperature, is equal to $\theta^F_{\text{H.M.}}$, the final hot metal temperature. Sorting terms,

$$\theta^F = \frac{M_S C_S \theta^0_S + M_{\text{H.M.}} C_{\text{H.M.}} \theta^0_{\text{H.M.}}}{M_S C_S + M_{\text{H.M.}} C_{\text{H.M.}}},$$

giving 1372 °C. We can therefore assume that this temperature will prevail during the course of scrap dissolution and use the Fe–C diagram to show us that the solidus carbon composition (at low silicon) will remain at 0.86 wt.% (2.9 Atom %), and liquidus carbon composition C^L_C, at 1.93 wt.% (8.3 Atom %).

Consequently, and referring to Fig. 5.6(b), we can argue that

No. of moles of carbon diffusing through iron
to scrap interfaces/area/time

=

moles of iron dissolving/area/time × mole fraction of
carbon required to dissolve one mole of iron

or, mathematically,

$$k_C(C_C^B - C_C^L) \cong \dot{N}_{Fe}''[X_C^L - X_{C,s}^0].$$

In order to dissolve 1 mole of scrap, the carbon content of the scrap must be raised from C° ($\equiv 0.1$ wt. % C) to the liquidus composition C_C^L. A mass balance shows that

$$\dot{N}_{Fe}'' \equiv -\frac{\rho_s^0(dx/dt)}{M_{Fe}}.$$

Combining these two equations, the rate of retreat of the scrap surface $-dx/dt$, can be written

$$\frac{dx}{dt} = -k_C\frac{[C_C^B - C_C^L]}{[X_C^L - X_{C,s}^0]}\left(\frac{M_{Fe}}{\rho_s^0}\right).$$

Integrating between $\Delta/2$ and zero (the scrap is dissolved on two sides), and noting that k represents the mass transfer coefficient for carbon diffusion, we find

$$\Delta t = \frac{(X_C^L - X_{C,s}^0)}{k_C(C_C^B - C_C^L)}\left(\frac{\rho_s}{M_{Fe}}\right)\left(\frac{\Delta}{2}\right).$$

All that now remains to be done is to estimate the natural convective mass transfer coefficient for carbon diffusing to the (assumed-vertical) surfaces of plate-like scrap. Evidently, as the scrap dissolves, the density of liquid adjacent to the scrap surfaces will be higher than the density of hot metal. This causes downflow, with less-dense hot metal flowing into the boundary layer from above, generating turbulent natural convection phenomena. Referring to Section 5.3, the relevant correlation for mass transfer is eqn (5.22b):

$$Sh = 0.1\,Ra^{1/3}, \qquad 10^9 < Ra < 10^{13}.$$

Note, we choose the second correlation which renders our solution independent of characterstic length, L. Similarly, for simplicity, we ignore a thickening of the boundary layer by dissolving iron atoms (see Guthrie and Stubbs 1973):

$$\frac{k_C L}{D_{C/Fe}} = 0.1\left(\frac{\rho\beta_{Mg}\Delta X_C L^3}{\mu.D_{C/Fe}}\right)^{1/3},$$

$$\beta_M = \frac{1}{\rho}\left|\frac{\partial \rho}{\partial X_C}\right| \cong \frac{1}{6500}\left(\frac{7000-6000}{0-0.17}\right) \cong 0.9 \text{ per mole fraction carbon,}$$

since the density of molten steel (at 1540 °C) is 7000 $kg\,m^{-3}$, and the density of hot metal containing 4 per cent carbon at 1400 °C is 6000 $kg\,m^{-3}$.

Inserting appropriate values:

$$k_C = 0.1 D_{C/Fe}\left(\frac{6500 \times 0.9 \times 9.81 \times [0.183-0.083]}{6.7 \times 10^{-3} \times (10 \times 10^{-9})}\right)^{1/3}$$

$$= 4.4 \times 10^{-5}\,m\,s^{-1}.$$

Consequently,

$$\Delta t = \frac{(0.083-0.004)}{4.4 \times 10^{-5}(C_C^B - C_C^L)}\left(\frac{7000}{56}\right)\left(\frac{0.04}{2}\right).$$

Since,

$$C_C^B = \frac{\text{no. of moles } C}{\text{unit volume of bulk iron}} = \frac{\text{wt.}\% C/(M_C \times 100)}{1/\rho}$$

$$= \frac{4.3/(12 \times 100)}{1/6000},$$

$$C_C^B = 21.5 \text{ mole m}^{-3}.$$

Similarly, $C_C^L = [1.93/(12 \times 100)]/[1/6500] = 10.45 \text{ mole m}^{-3}$,

$$\Delta t = \left(\frac{0.083-0.004}{4.4 \times 10^{-5}(21.5-10.45)}\right)\left(\frac{7000}{56}\right)\left(\frac{0.025}{2}\right) = 254 \text{ s or } 4.2 \text{ min.}$$

Thickening of the boundary layer by iron dissolution would increase these estimates by about 50%. In actual plant tests, all the scrap was found to have dissolved on emptying the transfer ladle into a BOF after allowing 20 minutes for safety.

5.4.   INTERPHASE TRANSPORT IN FLUID–FLUID SYSTEMS

As shown in Chapter 1, a variety of the more important metallurgical processes involve the transfer of heat and mass between two or more fluid phases. For instance, most pyrometallurgical refining operations rely on oxidation of the less noble metalloids present in the impure metal product either to oxidize them into an upper slag phase or, if the products are gaseous, to carry them up out of the system and into the exhaust furnace gases. As we have seen, many of these processes tend to be batch rather than continuous.

There are many ways in which two fluids may come into contact and react with one another. For instance, the gas phase can contact the liquid as a submerged gas jet, an impinging jet, a bubble column, etc. Alternatively, liquid streams or droplets can contact a bulk gas phase such as flash smelting of copper sulphide concentrates. Similarly liquid-liquid contact can occur as droplets of one phase dispersed in a second phase, as, for instance, the slag–metal emulsion in steel-making.

It should be stated immediately that quantitative theoretical predictions of heat or mass transfer rates are normally impossible owing to the complicated hydrodynamic phenomena involved in all such commercial processes and also, more fundamentally, to a lack of knowledge of the basic kinetic mechanisms involved, particularly in mass transfer operations. However, since there are a number of theoretical models available for fluid–fluid interactions which lead to a good qualitative description of reaction rates, and sometimes also lead to good order of magnitude estimates of heat and mass transfer rates, these will now be considered in the following sections.

(a)   The Lewis–Whitman two-film theory

The Lewis–Whitman two-film theory represents the earliest model of solute transfer between two bulk fluid phases in contact with one another. In 1924, Lewis and Whitman proposed that, in general, the rate of solute transfer would depend upon the rate of molecular-type diffusion of solute through two laminar or stagnant layers of the two solvents adjacent to their common interface.

They postulated that convection currents (e.g., currents resulting from vibration, thermal gradients, turbulent flow of the fluids and bubble agitation) would ensure uniformity within a given phase except across those 'laminar films' adjacent to the fluid–fluid interface. If one accepts this notion, the flux of A through the interface from the metal phase into a slag or gas is simply given by:

$$\dot{N}''_A = D_{A,M} \frac{(C^B - C^*)}{\delta'_M}, \tag{5.41}$$

where δ'_M is the stagnant film thickness, * indicates the interface and B the bulk metal and $D_{A,M}$ is the molecular diffusivity of A through the metal, M. Since the concept of laminar films of discrete thickness separating the slag–metal phases in a bubble-stirred open hearth furnace (for example), is highly questionable, the equation above is better cast in the form:

$$\dot{N}''_A = k_L(C^B_A - C^*_A), \tag{5.42}$$

where k_L = mass transfer coefficient (m s^{-1}). This avoids having to ascribe any deep physical significance to δ'_m, as discussed in Section 5.1(a) for equivalent fluid–solid interphase transport (eqns (5.4) and (5.5)). The mass transfer coefficient is presently the commonest way of expressing the relationship between a solute flux and a 'concentration driving force' in systems involving convective mass transfer.

(b) Application of the two-film theory to slag–metal reactions

We will consider the refining period in open hearth steel-making, depicted in Fig. 5.7, using the two-film theory proposed by Lewis and Whitman to describe the removal of metalloids from molten iron into the upper slag phase.

Thus, the diagram illustrates three ways whereby oxygen is transferred to the metal both to oxidise manganese, silicon, phosphorus, etc, from the metal into a clearly delineated upper slag phase. The oxidation of carbon leads to a bubbling bath, CO being nucleated at heterogeneous sites along the furnace bottom. This bubbling ensures well mixed conditions within the bulk of the metal and slag phases, and helps flush out N_2 and H_2.

The model assumes chemical equilibrium at the interface and steady-state diffusion across two stagnant layers of slag and metal at the interface. Furthermore, in its application to slag metal systems, it is assumed that the non-volatile metalloids combine with dissolved oxygen at the slag–metal interface to form the various ionic oxide species present in the slag. Homogeneous nucleation of an oxide within the molten metal is, on the basis of

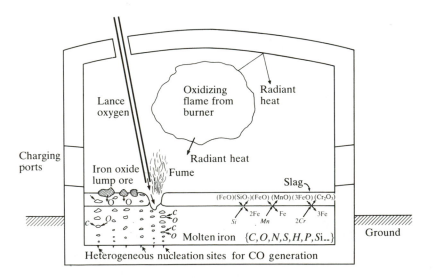

Fig. 5.7. A schematic illustrating the principal reactions taking place during refining in an open hearth furnace for steel-making (cross-sectional view of furnace).

nucleation theory, much less likely to occur than heterogeneous nucleation at the slag–metal interface.

The equations indicate that in the case of a simple transfer of a metalloid A from metal to slag, the rate of transfer is greatly affected by its mutual solubility in the two phases. Thus, in the case of a metalloid that is much more soluble in the metal phase, for example, the step limiting its transfer to the slag is likely to be diffusion of A through the slag diffusion layer (and vice versa). This follows from the equation

$$\dot{N}'' = \frac{C_{A,M} - mC_{A,Sl}}{(1/k_M + m/k_{Sl})}.$$ (5.43)

Since $m \to \infty$ in this case, $m = C_M^*/C_{Sl}^*$, the diffusional resistance term in the metal (k_M^{-1}) is negligibly small in comparison with the diffusional resistance in the slag (mk_{Sl}^{-1}), provided $k_M \simeq k_{Sl}$. Note that eqn (5.43) was first developed for steady-state diffusion through two adjacent solid phases in Chapter 4 (eqn 4.27).

We can illustrate the concept of mixed control, metal-phase control, and slag-phase control by considering two extreme cases: that of manganese transfer from the metal to the slag phase where its stability and solubility (as MnO) is appreciable; and that of nickel, for which the stability of NiO is very small.

Let us suppose that a ferromanganese addition and a pure nickel plate addition are made to the bath and that on melting, the solutes become thoroughly mixed, thanks to the 'carbon boil', resulting in bulk metal concentrations of $C_{(Mn)_{Fe}}^B$ and $C_{(Ni)_{Fe}}^B$. At the same time, let us suppose that dissolved manganese and nickel transfer into the slag by heterogeneous double-decomposition reactions at the slag–metal interface:

$$(Mn)_{Fe} + (FeO)_{slag} \to Fe + (MnO)_{slag},$$ (5.44)

$$(Ni)_{Fe} + (FeO)_{slag} \to Fe + (NiO)_{slag},$$ (5.45)

such that electroneutrality in the ionic slag phase is maintained. The basic transfer mechanisms envisaged are illustrated in Fig. 5.7, where we ignore any possible coupling between the set of simultaneous reactions. Assuming the processes to be transport-controlled, thermodynamic equilibrium between the species at the slag–metal interface will exist so that we can write

$$\Delta G_T^0 = -RT \ln\left(\frac{a_{MnO}^* \quad a_{Fe}^*}{a_{Mn}^* \quad a_{FeO}^*}\right),$$ (5.46)

$$\Delta G_T^0 = -RT \ln\left(\frac{a_{NiO}^* \quad a_{Fe}^*}{a_{Ni}^* \quad a_{FeO}^*}\right).$$ (5.47)

ΔG_T^0 is the standard Gibbs free energy of formation of the liquid oxides at 1600 °C:

$$Fe(l) + \tfrac{1}{2}O_2(g) \rightarrow FeO(l) \quad \Delta G^0 = -37.1 \text{ kcal mole}^{-1} \qquad (5.48)$$

$$Mn(l) + \tfrac{1}{2}O_2(g) \rightarrow MnO(l) \quad \Delta G^0 = -56.8 \text{ kcal mole}^{-1}, \qquad (5.49)$$

$$Ni(l) + \tfrac{1}{2}O_2(g) \rightarrow NiO(l) \quad \Delta G^0 \approx -16.0 \text{ kcal mole}^{-1}. \qquad (5.50)$$

Subtracting eqn (5.48) from (5.49) and eqn (5.48) from (5.50) respectively, we obtain

$$Mn + FeO \rightarrow MnO + Fe \quad \Delta G^0 = -19.7 \text{ kcal mole}^{-1}, \qquad (5.51)$$

and

$$Ni + FeO \rightarrow NiO + Fe \quad \Delta G^0 = +21.1 \text{ kcal mole}^{-1}. \qquad (5.52)$$

Assuming that we can apply the above thermodynamic data, obtained for equilibria between bulk phases, to interfacial equilibrium, we have *for manganese*:

$$-\frac{\Delta G^0}{RT} = \ln\left(\frac{a_{Fe}^* \; a_{MnO}^*}{a_{Mn}^* \; a_{FeO}^*}\right). \qquad (5.53)$$

Because liquid manganese and iron form a continuous solution, as do liquid iron oxide and manganese oxides, we can assume ideal behaviour (Raoultian) in the slag and metal phases. (Raoultian) activity coefficients for the solutes are then unity, and

$$-\frac{\Delta G^0}{RT} = \ln\left(\frac{X_{Fe}^* \; X_{MnO}^*}{X_{Mn}^* \; X_{FeO}^*}\right), \qquad (5.54)$$

where $X \equiv$ mole fraction.

Under typical open hearth operating conditions, the weight percentages of the constituents in the *bulk* of the two phases will be about 99% for Fe, 15% FeO, 0.4% Mn (say) and 10% MnO. If we consider this situation in terms of the type of concentration gradients that will be set up by the transfer of manganese up into the slag phase, we first note that continuity requires that at $x = 0$, the slag–metal interface:

$$\dot{N}_{Mn}'' = \dot{N}_{MnO}'' = -\dot{N}_{Fe}'' = -\dot{N}_{FeO}''. \qquad (5.55)$$

Here, subscripts to denote solute fluxes in the slag and iron phases have been dropped for convenience. Since the manganese in the steel is in short supply (0.4 wt.%) any concentration gradient across the stagnant metal film will cause an appreciable drop in the interfacial manganese concentration. We see that the concentration gradient of manganese in the stagnant concentration boundary layer, δ_M', certainly cannot exceed $(0.4 - 0)/\delta_M'$. By comparison, the concentration changes of the other three components will be practically negligible across their respective boundary layers and their interfacial concentrations will, to all intents, be equal to their bulk concentrations. We may readily see this by supposing that the concentration of Mn at the interface is at its minimum value.

The continuity equation then shows

$$D_{Mn/Fe}\frac{(0.4-0)}{\delta'_M} = D_{MnO/Sl}\frac{(C^*_{MnO}-10)}{\delta'_{Sl}} = -D_{FeO/Sl}\frac{(15-C^*_{FeO})}{\delta'_{Sl}} = -D_{Fe}\frac{(C^*_{Fe}-99)}{\delta'_M}$$

Taking $\delta'_M = \delta'_{Sl}$ and equal diffusivities, we then have, with suitable adjustments for molecular weights and slag/metal density differences,

$$C^*_{MnO} \approx 11.2 \,(\text{vs } 10), \quad C^*_{FeO} = 13.8 \,(\text{vs } 15), \quad C^*_{Fe} = 99.4 \,(\text{vs } 99).$$

These concentration profiles are illustrated in Fig. 5.8, where we see that it becomes quite legitimate to write

$$X^*_{FeO} \sim X^B_{FeO} \sim 0.15 \quad \text{and} \quad X^*_{Fe} \sim X^B_{Fe} \sim 1.$$

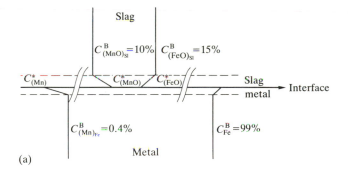

(a)

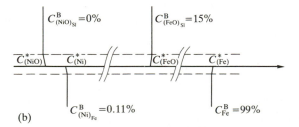

(b)

FIG. 5.8. Concentration profiles for solute transferring between metal and slag phases during refining operations in an open hearth steel-making furnace. (a) Manganese interchange; (b) nickel interchange.

Equation (5.54) can therefore be rewritten as

$$-\frac{\Delta G^0}{RT} = \frac{19\,700}{1.98 \times 1873} = \ln\left(\frac{1 \times X^*_{\text{MnO}}}{0.15 \times X^*_{\text{Mn}}}\right)$$

Evaluating, we obtain $X^*_{\text{Mn}}/X^*_{\text{MnO}} \simeq 0.033 \equiv m$. Here, m is equivalent to a partition coefficient.

The result shows that the interfacial concentration of manganese on the metal side will be only 3 per cent of the interfacial MnO concentration. In other words, the interfacial Mn concentration will always be pinned at zero when any serious imbalance between the bulk manganese contents in the slag and metal result in a net interchange of manganese. *Since the gradient is then at its maximum, so also is the flux, and the overall process therefore metal-phase controlled.* Looking at it another way, the equation describing the rate of interchange is (5.43) and we see that the net resistance r to mass transfer is

$$r = \frac{1}{k_{\text{overall}}} = \frac{1}{k_M} + \frac{m}{k_{\text{Sl}}}. \tag{5.56}$$

Taking $k_M \simeq k_{\text{Sl}}$, $r = 1.03/k_M$. For this reason we can ignore the slag-phase resistance term at three per cent of the overall resistance and again talk about 'liquid-metal-phase transport control'.

Nickel transfer

Going through similar arguments for nickel, let us suppose that initially the NiO content in the slag is zero, while in the metal it is running at 0.11 wt.% (typical of a nickel grade heat). Taking FeO $= 15\%$ and Fe $= 99\%$, we find

$$\frac{X^*_{\text{Ni}}}{X^*_{\text{(NiO)}}} \approx 1970.$$

Here we see that it is now the interfacial concentration of (NiO)* that is pinned down to a value approaching zero. Thus, the potential concentration gradient in the slag phase will always be limited before that in the metal phase, so that the kinetics of nickel transport are *slag-phase-controlled*. These concentration profiles are illustrated in Fig. 5.8(b).

These two examples illustrate the marked effect that relative solubility of the solute can have in determining where the bottle neck, or rate-controlling step, will lie. Sometimes such analyses are not so clear-cut, and both phases will present some resistance to mass transport (i.e. mixed control). Similar arguments or analyses can be applied to gas–liquid systems. A good discussion of these principles is to be found in the original paper by Lewis and Whitman (1924).

Finally, it should be noted that experimental results on interphase transport can only be expressed in terms of an overall transfer coefficient, since it is not possible to measure interfacial concentrations but only bulk solute concentrations. Very frequently authors then attempt to identify the individual transfer coefficients based on theoretical arguments similar to those developed above.

(c)   Worked example: refining silver by top jetting of oxygen

A 150 kg bath of molten silver is lanced with pure oxygen to oxidize out trace impurities of tin and bismuth which are then skimmed off the surface. Calculate how long it will take the silver to become semi-saturated with oxygen, given the following data:

Density of molten silver, ρ	$10\,200 \text{ kg m}^{-3}$
Overall surface mass transfer coefficient, k_L	$5 \times 10^{-4} \text{ m s}^{-1}$
Depth of bath, h	0.25 m

Assume that the bath is well stirred by the action of the oxygen jet and natural convection current and that the bath is a vertical cylinder.

Solution

Since the gas phase is *pure* oxygen, *gas-phase control is impossible.* Consequently we need only be concerned with the diffusion boundary layer at the liquid–silver–oxygen interface. Carrying out a macroscopic mass balance for oxygen:

$$\begin{pmatrix} \text{Input} \\ \text{of oxygen} \end{pmatrix} - \begin{pmatrix} \text{output} \\ \text{of oxygen} \end{pmatrix} = \begin{pmatrix} \text{Rate of accumulation} \\ \text{of oxygen in bath} \end{pmatrix}$$

i.e.,

$$\dot{N}''_{O,x=0}(A_{\text{bath}}) = \frac{\text{d}}{\text{d}t}(MC^{\text{B}}),$$

where M is the mass of silver, and C^{B} is the bulk concentration of oxygen in the bath.

Following a Lewis–Whitman type approach, we can replace the flux of oxygen entering the silver, according to:

$$\dot{N}'_{O} = D_{\text{O/Ag}} \frac{(C^{*} - C^{\text{B}})}{\delta'} = k_{\text{L}}(C^{*} - C^{\text{B}}),$$

where C^* is the interfacial concentration of oxygen dissolved in silver. (Silver oxide, incidentally, is only thermodynamically stable under standard conditions at temperatures below 200 °C, as can be seen from the oxide free energy diagram in the Appendix, i.e. from Fig. A3–2, $\Delta G^0 = 0$ at $T = 473$ K.) Substitution for $\dot{N}''$ then gives,

$$k_L A(C^* - C^B) = \frac{d}{dt}(VC^B).$$

Since V, the volume of the liquid bath remains constant, we can write $V = hA$, so that

$$k_L(C^* - C^B) = h\frac{dC^B}{dt},$$

$$\int_0^{T_{1/2}} \frac{k_L\, dt}{h} = \int_{C^B=0}^{C^B=C^*/2} \frac{dC^B}{(C^* - C^B)},$$

$$\frac{k_L T_{\frac{1}{2}}}{h} = [-\ln(C^* - C^B)]_0^{C^B=C^*/2}$$

$$= -\ln(C^*/2) + \ln(C^*)$$

$$= \ln 2.$$

Hence, $T_{\frac{1}{2}} = h/k_L \ln 2 = 0.25/(5 \times 10^{-4}) \times 0.693 = 346.5$ s or 5.8 min.

Conclusion

Under the conditions described, the silver would become half-saturated with oxygen within 6 min of jetting.

5.5.   EXERCISES ON INTERPHASE MASS AND HEAT TRANSFER OPERATIONS

(1)   Oxygen pick-up during casting of copper through a launder

A stream of pure molten copper passes through a long, narrow launder exposed to air in a wire bar casting operation. Assuming that the oxygen pick-up that will inevitably occur is controlled by natural convection of air above the melt surface, estimate the oxygen pick-up (wt.% dissolved in copper) in the final wire bar ingots.

Assume the flow to be plug flow horizontally along the launder but well mixed in the vertical direction (i.e. $D_{E/Uh} \gg D_{E/UL}$).

Data:

Length of launder, L	2 m
Width of launder, w	0.05 m
Depth of copper stream, h	0.05 m
Velocity of copper stream	$1\ \mathrm{m\,s^{-1}}$
Density of copper	$8\,330\ \mathrm{kg\,m^{-3}}$
Atomic weight of copper	162.5
Bulk gas temperature	366 K
Bulk copper temperature	1 422 K
Gas phase convective mass transfer coefficient	$0.22\ \mathrm{m\,s^{-1}}$

(2) Temperature losses during casting of copper through a launder

Estimate the temperature drop in the copper that will occur as a result of radiative and convective heat losses from its exposed copper surface during its flow down the launder. Take the effective gas phase convective heat transfer coefficient as

$$h_{\mathrm{eff}} = 57\ \mathrm{W\,m^{-2}\,K^{-1}}.$$

(3) Heterogeneous nucleation of bubbles in the open hearth process

In open hearth steel-making the 'carbon boil' is observed to originate mainly at the hearth of the furnace, where heterogeneous nucleation of CO bubbles takes place (see Figure 5.7).

(a) If the bath contains 2 wt.% carbon, what oxygen concentration at 1600 °C will the bath attain prior to nucleation at a 0.1 mm diameter crevice?
For the reaction

$$(C)_{Fe} + (O)_{Fe} \rightarrow CO_{(gas)}$$

$$K_P = 500 = \frac{P_{CO}(\mathrm{atm})}{(\mathrm{wt.\%\ C})(\mathrm{wt.\%\ O})}.$$

(b) Sampling of the open hearth bath shows that the $[C] \times [O]$ product is approximately twice that expected for a bath in equilibrium with 1 atm of CO ($1.01 \times 10^5\ \mathrm{N\,m^{-2}}$). It has been observed that a $[C] \times [O]$ product some 15 times the 1 atm equilibrium value could be obtained in a glazed silica crucible. What size crevices would serve as nucleation sites in these two cases?

Data:

$\Delta P_R = 2\sigma/R$; $N\,m^{-2}$, $P_{\text{bubble}} = P_{\text{static}} + 2\sigma/R$, ($R$ is the radius of nucleating bubble.)

Depth of steel in bath	0.25 m
Depth of slag	0.13 m
Density steel	7200 kg m^{-3}.
Density slag	3700 kg m^{-3}
Surface tension of molten steel	1.5 Nm^{-1}

---

(4) Ferromanganese additions in open hearth steel-making

During an open hearth steel-making operation for the production of a low-carbon steel, equilibrium was established between manganese dissolved in the metal and slag phases, resulting in a concentration of 0.1% by weight manganese in the molten metal phase. Subsequently, a manganese ore addition was made to the melt, causing the manganese content in the metal to rise to 4 wt.%. It has been shown that the rate of manganese transfer is governed by diffusion of manganese from the bulk of the metal phase through a thin 'boundary layer' of molten metal to the metal–slag interface (i.e., transport control in metal phase).

Calculate the final manganese content in the iron 20 min after the ore addition, and the total mass of manganese transferred into the slag during this time. You may assume that the interfacial concentration of manganese in the metal remains at 0.1 wt.% during this time period, and that the bath is rectangular in shape.

Data:

At $t=0$, C_{Mn}^B	0.4 wt.%
Depth of liquid iron	0.6 m
Mass transfer coefficient for manganese transport, k	1×10^{-3} m s^{-1}
Bath area	45 m^2
Density Mn $\approx$ Density Fe	7050 kg m^{-3} at 1867 K

---

(5) Development of a general equation to describe changes in solute concentrations in metal–slag phases during open-hearth refining

A solute A dissolved in the bulk of the metal phase diffuses into the bulk of the slag phase during a refining process. As time proceeds, the concentration of A in the metal becomes depleted, resulting in a decreased rate of mass transfer into the slag.

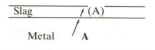

Given:

$$\dot{N}''_{A,M,t} = \dot{N}''_{A,Sl,t} = \frac{(C_{A,M,t} - MC_{A,Sl,t})}{(1/k_M) + (m/k_{Sl})},$$

where $m = C_M^*/C_{Sl}^*$: (equilibrium),

V_M = volume of metal phase; V_{Sl} = volume of slag phase; $C_{A,M,t}$ = bulk concentration of A in metal phase at time t (moles A/m^3 of metal); $C_{A,M,0}$ = initial bulk concentration of A in metal phase, $t = 0$, (moles A/m^3 of metal); Similar notation for slag phase; A = area of slag–metal interface.
Show:

(1) $V_M(C_{A,M,0} - C_{A,M,t}) = V_{Sl}(C_{A,Sl,t} - C_{A,Sl,0})$.
State any assumptions. Are they justifiable?

(2) $\dot{N}''_{A,M,t} = -\dfrac{V_M}{A}\dfrac{dC_{A,M,t}}{dt} = +\dfrac{V_{Sl}}{A}\dfrac{dC_{A,Sl,t}}{dt}$.

State assumptions made.

(3) $C_{A,M,t} = B + (C_{A,M,0} - B)\exp\left[\dfrac{-At}{(1/k_M) + (m/k_{Sl})}\left(\dfrac{1}{V_M} + \dfrac{m}{V_{Sl}}\right)\right]$

where

$$B = \frac{(V_M C_{A,M,0} + V_{Sl} C_{A,Sl,0})}{(V_{Sl,M} + V_M)}$$

Calculate limiting values of $C_{A,M,t}$ for: $t = 0$, for $k_{Sl} \to 0$, for $k_M \to 0$, for $V_M \gg V_{Sl}$ (note that the value of m will affect this result), and for $t \to \infty$, and qualitatively explain results.

Finally, write down an expression for the molar flux of A in terms of initial concentrations and other known quantities. Comment on the possibilities and shortcomings of using such an approach as a dynamic model of slag/metal refining processes.

5.6.   THE PENETRATION THEORY OF MASS TRANSFER BETWEEN FLUID PHASES

Although the Lewis–Whitman theory is useful from a conceptual point of view, it has nothing to say about the quantitative value of the resulting interphase mass transfer coefficients. There are a number of situations, however, which involve transport across fluid–fluid interfaces, in which some predictions can be made of appropriate values. The particular situation that Higbie (1935) was primarily interested in, involved the rate of gas absorption into freshly exposed surfaces of liquid, such as might occur during the flow of liquid over successive packing rings in a gas absorption tower or during the ascent of a gas bubble through a liquid. As Higbie pointed out, the rate of mass transfer would depend greatly upon the type of contact times that a freshly exposed liquid surface might have when in contact with the gas phase. Take, for example, the classic case of liquid flowing over a series of Rasching rings in a gas absorption tower contacted with an ascending gas that is soluble in the liquid phase.

If one envisages the situation shown in Fig. 5.9(a), the liquid is in laminar flow over a series of 'weirs', and becomes well mixed at the junction of each set of rings. Consider the case depicted in Fig. 5.9(b), involving the absorption of pure CO_2 by water that Higbie reports in the experimental section of his paper.

The system is subjected to the following conditions.

1. The depth of penetration of CO_2 into the water is so small in comparison with the thickness of the laminar film of water flowing over the piece of packing, length L, that U, the downwards velocity of the liquid, can be taken constant even though transverse velocity gradients will exist close to the solid bounding surfaces of the rings.

2. The rate of gas absorption into the water is liquid-phase diffusion-controlled and chemical equilibrium exists between the CO_2 in the gas phase and that on the liquid side of the interface.

3. Soluble gases such as CO_2 dissolving in water follow Raoult's law such that

$$C^*_{CO_2/H_2O} = kP^*_{CO_2} = kP_{CO_2}$$

taking $P^*_{CO_2} = P_{CO_2}$ for a liquid-phase controlled reaction.

A mass balance on the differential element $\Delta x \times \Delta z \times 1$ in Fig. 5.9(b) shows that

Vertical flux input + horizontal flux input =
$$\text{vertical flux output + horizontal flux output;}$$

$$\dot{N}''_z(\Delta xw) + \dot{N}''_x(\Delta zw) = \dot{N}''_{z+\Delta z}(\Delta xw) + \dot{N}_{x+\Delta x}(\Delta zw); \qquad (5.57)$$

or, mathematically,

$$\frac{\partial \dot{N}_{Ax}}{\partial x} + \frac{\partial \dot{N}''_{Az}}{\partial z} = 0. \qquad (5.58)$$

Note that elimination of $\dot{R}'''_A$, $\partial \dot{N}''_{Ay}/\partial y$, and $\partial C_{A,x,z}/\partial t$ in the conservation

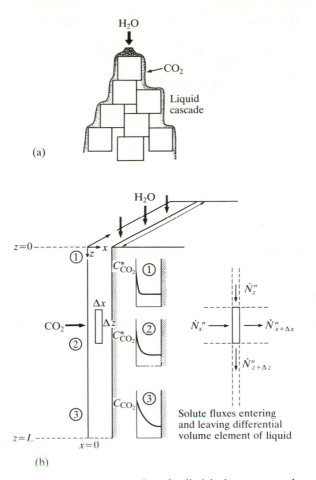

FIG. 5.9. (a) Flow of liquid over a series of cylindrical or rectangular objects. (b) Coordinate system for mathematical representation.

equation (4.6) immediately leads to the same result.

Replacing expressions for the vertical and horizontal fluxes of solute A (CO_2) through the solvent, we have:

$$\dot{N}''_{A,x} = U_x C_{A,x,z} - D_A \frac{\partial C_A}{\partial x} \quad \text{(zero convective component)} \quad (5.59)$$

and

$$\dot{N}''_{A,z} = U_z C_{A,x,z} - D_A \frac{\partial C_A}{\partial z} \quad \text{(convective component} \gg \quad (5.60)$$
$$\text{diffusive component)}$$

Hence, the appropriate differential equation requiring a solution is

$$\frac{\partial}{\partial x}\left(-D_A\frac{\partial C_A}{\partial x}\right)+\frac{\partial}{\partial z}(U_z C_A)=0, \tag{5.61}$$

subject to the boundary conditions

B.C.1   At $x=0$, $0\leqslant z\leqslant L$, $C_A=C_A^*$;

B.C.2   At $x\to\infty$, $0\leqslant z\leqslant L$, $C_A=C_A^B$.

Since U, the downwards velocity is constant and can be written as dz/dt, and D can also be taken constant, eqn (5.55) can be written under the form:

$$-D_A\frac{\partial^2 C_A}{\partial x^2}+\frac{dz}{dt}\frac{\partial C}{\partial z}=0 \quad \text{or} \quad \frac{\partial C}{\partial t}=D\frac{\partial^2 C}{\partial x^2}, \tag{5.62}$$

which is Fick's second law.

The general solution to this equation, with equivalent boundary conditions to those previously written, i.e.

B.C.1 $x=0$, $0\leqslant t\leqslant t_e$, $C=C_A^*$ ($t_e=$ exposure time
                                                                        of surface $=L/U_z$),

B.C.2 $x\to\infty$, $0\leqslant t\leqslant t_e$, $C=C_A^B$,

is

$$C_{A,x,t}=A+B\,\text{erf}\left(\frac{x}{\sqrt{4Dt}}\right). \tag{5.63}$$

Inserting B.C.1, $C_A^*=A$ and B.C.2, $C_A^B=A+B$, the particular solution is

$$C_{A,x,t}=C_A^*+(C_A^B-C_A^*)\text{erf}\left(\frac{x}{\sqrt{4Dt}}\right),$$

or, converting back to the fixed space coordinate system,

$$C_{A,x,z}=C_A^*+(C_A^B-C_A^*)\text{erf}\left(\frac{x}{\sqrt{(4Dz/U_z)}}\right). \tag{5.64}$$

Differentiating eqn (5.64) with respect to x, and evaluating at $x=0$, $z=z$, following equivalent procedures to those in Chapter 4, leads to

$$\left.\frac{\partial C_A}{\partial x}\right|_{\substack{x=0\\z=z}}=\frac{C_A^B-C_A^*}{\sqrt{\pi Dz/U}}, \tag{5.65}$$

so that the instantaneous flux of CO_2 into the surface at location $z=z$ is

$$\dot{N}''_{x=0,z=z}=\sqrt{\left(\frac{DU}{\pi z}\right)}(C_A^*-C_A^B), \tag{5.66}$$

and the mean flux into the surface of length L is given by

$$\bar{N}''_{A,x=0} = \int_{z=0}^{z=L} \dot{N}''_{A,x=0,z=z} dz \bigg/ \int_{z=0}^{z=L} dz = 2\left(\frac{DU}{\pi L}\right)^{1/2} (C_A^* - C_A^B). \quad (5.67)$$

We see that the term $2\sqrt{(DU/\pi L)}$ represents the overall mass transfer coefficient $k_L (LT^{-1})$.

As experience confirms, the shorter the exposure time of surface elements (L/U), the greater the mass transfer coefficient and rate of mass transfer.

(a) Worked example of typical penetration distances for gases dissolving in falling liquid films

Calculate the thickness of the mass transfer boundary layer δ (i.e., $C = 0.01C^*$), and the fictitious, or stagnant boundary layer, δ', corresponding to a distance L from initial contact between a freshly exposed planar liquid surface and a pure gas under the following conditions.

Data:

Length of liquid film exposed	0.1 m
Diffusion coefficient of gas in liquid	$5 \times 10^{-9} \, m^2 \, s^{-1}$
Surface velocity of film	$0.2 \, m \, s^{-1}$

Solution

Part I. Inserting appropriate values in the general solution of dissolved gas concentration as a function of depth and distance from initial contact between the two phases. i.e. eqn (5.64).

$$0.01C^* = C^* + (0 - C^*) \, \mathrm{erf}\left[\delta \bigg/ \left[\frac{4 \times 5 \times 10^{-9} \times 0.1}{0.2}\right]^{\frac{1}{2}}\right]$$

or

$$0.99 = \mathrm{erf}\left(\frac{\delta}{10^{-4}}\right)$$

or

$$\delta = 1.84 \times 10^{-4} \, m.$$

Part II. From eqn (5.65), the slope of the concentration–depth curve at the gas–liquid interface can be calculated, and this slope will be numerically equivalent to the concentration driving force divided by the fictitious boundary layer thickness; i.e.

$$-D\frac{dC}{dx}\bigg|_{x=0, z=0.1 \text{ m}} = \frac{D(C^*-0)}{\delta'}.$$

Consequently,

$$\frac{C^*}{(\pi Dz/U)^{\frac{1}{2}}} = \frac{C^*}{\delta},$$

giving

$$\delta' = \left(\frac{\pi Dz}{U}\right)^{\frac{1}{2}} = 0.09 \times 10^{-4} \text{ m}.$$

Conclusion: The real boundary layer thickness would equal 0.18 mm; the fictitious thickness would be half that value, 0.09 mm.

5.7.   BEHAVIOUR OF COMPOUND GASES IN LIQUID METALS

As noted in the preceding chapter, gases which dissolve in liquid metals do so in atomic form, contrary to the case for aqueous systems, in which for instance, a carbon dioxide *molecule* passes through the interface along the long axis of the molecule.In the following example, the desorption of carbon monoxide from a steel bath is considered in view of its technical and commercial importance to steel-making refining operations.

Consider the case of a bubble of carbon monoxide rising through a bath of steel containing dissolved oxygen and carbon in quantities such that continuous desorption of carbon monoxide occurs as the result of simultaneous diffusion of oxygen and carbon to the bubble interface. The situation is depicted in Fig. 5.10 and represents steel-making conditions in the open hearth or ladle vacuum degassing processes, in which small CO bubbles nucleated deep in the melt at crevices, etc., rapidly grow into large spherical cap bubbles during their ascent through the melt (e.g. see Fig. 5.7).

It is immediately evident that the situation is much more complex than the experimental situation investigated by Higbie, in that simultaneous diffusion of carbon and oxygen to the gas–liquid-metal interface is involved. Clearly, however, the transfer of carbon monoxide gas through the gas phase cannot be rate-limiting provided the bubble is composed entirely of carbon monoxide. Similarly, we can simplify the problem by approximating the true conditions of liquid flow around the front surface of a sphere to one of vertical planar flow of steel.

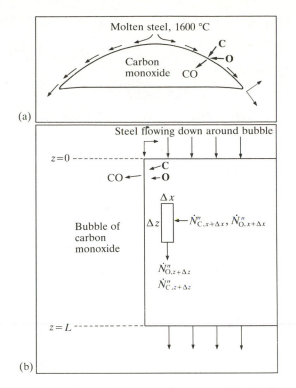

FIG. 5.10.   Removal of carbon and oxygen from a steel bath through growth of bubbles of carbon monoxide. (a) Environment (b) Coordinate system for mathematical representation.

As in the previous case, the relevant differential equations requiring solution are then:
For carbon,

$$U\frac{\partial C_C}{\partial z} = D_C \frac{\partial^2 C_C}{\partial x^2},$$
(5.68)

For oxygen,

$$U\frac{\partial C_O}{\partial z} = D_O \frac{\partial^2 C_O}{\partial x^2};$$
(5.69)

subject to the following boundary conditions:

B.C.1 $x = 0,\ 0 \leqslant z \leqslant L,\ C_C = C_C^*,\ C_O = C_O^*,$
B.C.2 $x = \infty,\ 0 \leqslant z \leqslant L,\ C_C = C_C^B,\ C_O = C_O^B,$

At $x=0$, we have equilibrium for the reaction $(C)_{Fe}+(O)_{Fe}=CO$, i.e.

B.C.3 $$C_C^* C_O^* = K P_{CO},$$

while, continuity at the interface requires:

B.C.4 $$x=0, \ \dot{N}_C'' = \dot{N}_O'' = \dot{N}_{CO}''.$$

The general solutions to this set of differential equations are, as before, error type functions, i.e.

$$C_{C,x,z} = A_C + B_C \ \text{erf} \left(\frac{x}{\sqrt{4 D_C z/U}} \right), \tag{5.70}$$

$$C_{O,x,z} = A_O + B_O \ \text{erf} \left(\frac{x}{\sqrt{4 D_O z/U}} \right). \tag{5.71}$$

The values of the four constants, A_C, B_C, A_O, B_O, can be obtained by successively applying boundary conditions 1 to 4 in the general solution. This then shows that

$$C_{C,x,z} = C_C^B + \tfrac{1}{2}[\sqrt{\{(\gamma C_O^B - C_C^B)^2 + 4\gamma KP\}} - (\gamma C_O^B + C_C^B)]\left(1 - \text{erf}\frac{x}{\sqrt{4 D_C z/U}} \right), \tag{5.72}$$

$$C_{O,x,z} = C_O^B + \frac{1}{2\gamma}[\sqrt{\{(\gamma C_O^B - C_C^B)^2 + 4\gamma KP\}} - (\gamma C_O^B + C_C^B)]\left(1 - \text{erf}\frac{x}{\sqrt{4 D_O z/U}} \right), \tag{5.73}$$

where

$$\gamma = \sqrt{(D_O/D_C)}.$$

Differentiation of either of the last two equations with respect to x and evaluating $-D_O \partial C_O/\partial x|_{x=0, z=z}$, then leads to an expression for the instantaneous flux of oxygen and carbon into the interface and of carbon monoxide out of the interface and into the bubble, from which the mean flux can be obtained:

$$\bar{N}_{CO}'' = \bar{N}_C'' = \dot{N}_O'' = \frac{2(D_O D_C)^{\frac{1}{4}}}{\sqrt{\pi L/U}} \sqrt{KP} \left[\sqrt{\left(\frac{(\gamma C_O^B - C_C^B)^2}{4\gamma KP} + 1 \right)} - \sqrt{\left(\frac{(\gamma C_O^B + C_C^B)^2}{4\gamma KP} \right)} \right] \tag{5.74}$$

It will be appreciated that this 'exact' expression is rather cumbersome for flux calculations.

Short cut

In many instances, it is possible to simplify these types of flux equations without introducing any significant errors provided one of the solutes is present in the melt at a higher concentration (say, four times or greater) than the other. Thus, following similar arguments to those developed for solute transfer in slag–metal systems using the Lewis–Whitman Theory, we can suppose that the interfacial concentration of that solute present at the highest concentration in the melt must remain close to its value in the bulk. In the case of open hearth steel-making operations, the carbon is at a higher concentration in the melt compared to oxygen throughout the major part of refining processes, as shown in Fig. 5.11, and transfer of oxygen is therefore rate-limiting.

Consequently, we can immediately write down an analogous expression to

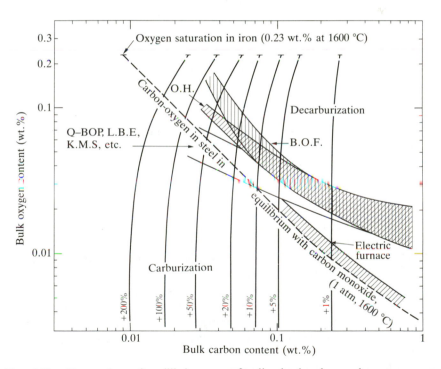

FIG. 5.11.   Comparison of equilibrium curve for dissolved carbon and oxygen versus operational C–O trajectories for various steel-making processes. Also shown are iso-percentage error lines to illustrate the relative importance of carbon to oxygen diffusion on CO kinetics.

eqn (5.67) for the flux of carbon monoxide into the rising bubble:

$$\bar{N}''_{CO,\,approx} = \bar{N}''_{O} = 2\sqrt{\left(\frac{D_O U_B}{\pi L}\right)}(C^B_O - C^*_O), \tag{5.75}$$

with

$$C^*_O = \frac{KP_{CO}}{C^*_C} \simeq \frac{KP_{CO}}{C^B_C}.$$

Figure 5.11 shows the various iso-percentage error lines, defined as $[\{(\bar{N}''_{CO,approx}/\bar{N}''_{CO,exact}) - 1\} \times 100]$, for the case of steel-making, taking $D_O = 5 \times 10^{-9}\ m^2\ s^{-1}$, $D_C = 40 \times 10^{-9}\ m^2\ s^{-1}$. Similar diagrams can be constructed for other important industrial examples involving compound gas formation (e.g. $(S)_{Cu} + 2(O)_{Cu} \rightarrow SO_2$ and $(S)_{Ni} + 2(O)_{Ni} \rightarrow SO_2$ in copper and nickel refining).

It has been shown that equations such as (5.67) and (5.75) fit experimentally obtained mass transfer coefficients quite well ($\pm 30\%$), where U_B is taken as the terminal rising velocity of the bubble and L is taken as the height of the bubble.

5.8.   MASS TRANSFER ACROSS BUBBLE-STIRRED INTERFACES

(a)   Introduction: vacuum ladle degassing

There are a number of metallurgical processes which involve the removal of metalloids into an upper slag or gas phase, in which the stirring action of large bubbles rising through the metal promotes the rate of impurity removal. Important examples of such processes include vacuum ladle degassing of steel, R.H. degassing of steel, lead softening (removal of Sb), and the removal of metalloids in the open hearth process.

In the case of vacuum ladle degassing of steel, dissolved hydrogen, nitrogen, and oxygen, typically running at 4–8, 40–60 and 200–600 p.p.m. respectively, are reduced to residual contents of (1–2, 20–30, and 20–50) p.p.m. respectively, after degassing at $\simeq 1$ Torr (0.13 kPa). Part of the gas removal is accomplished by surface evaporation of the gases at the gas–metal interface in the ladle, the remainder as a result of gas evaporation into rising argon bubbles, or, in the presence of carry-over slag containing FeO, by slag droplet entrainment into the steel, and reaction with dissolved carbon at the slag droplet/metal interfaces in all probability.

In order to make theoretical predictions about the rate of *surface evaporation*, a simplified model of the process may be postulated. Let us suppose that each rising bubble carries up fresh steel from the bulk, and deposits this material at the interface.

Since a vacuum is maintained over the melt surface, the gaseous solutes will diffuse to the liquid steel's surface and we may assume that the resistance to mass transfer will be located in the liquid phase on account of low solute concentrations and the vacuum in the upper gas phase. Finally, in our model, we may suppose that these *surface* elements are essentially stagnant during their residence time at the surface, since the depth of mass penetration will be much less than the scale of turbulence in the liquid. We discount the effects of any slag.

Thus, referring to Fig. 5.12 (a) and (b) the mass transfer mechanism for hydrogen removal, can, as an example, be considered. Thus, at the interface,

$$(H)_{Fe} + (H)_{Fe} \rightarrow H_2 \quad \text{where} \quad C_H^* = q\sqrt{P_{H_2}} \approx q\sqrt{0} = 0,$$

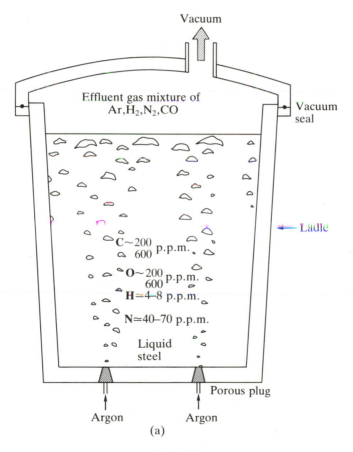

FIG. 5.12(a)

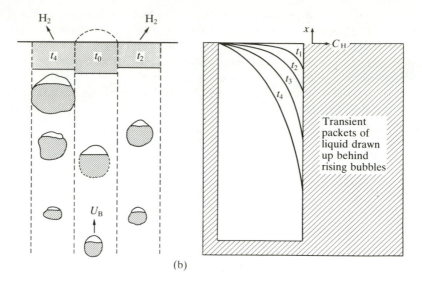

(b)

FIG. 5.12.   (a) Illustration of ladle vacuum degassing operation. (b) Fresh elements of liquid from the bulk of the steel are drawn up behind surfacing bubbles to promote surface evaporation of dissolved metalloids.

while in the bulk, $C_H = C_H^B$. The relevant differential equation describing unsteady-state mass transfer from a stagnant volume element of steel exposed to vacuum, is Fick's second law:

$$\frac{\partial C_H}{\partial t} = D \frac{\partial^2 C_H}{\partial x^2},$$

(5.76)

subject to the boundary conditions:

B.C.1 $x = 0,\ 0 \leqslant t \leqslant t_e,\ C_H = C_H^* = 0;$

B.C.2 $x \to -\infty,\ 0 \leqslant t \leqslant t_e, C_H = C_H^B;$

B.C.3 $x = 0,\ 0 \leqslant t \leqslant t_e,\ \dot{N}_H'' = 2\dot{N}_{H_2}''.$

Once more, the general solution is of the form

$$C_{H,x,t} = A + B\ \text{erf}\left(\frac{x}{\sqrt{4D_H t}}\right),$$

(5.77)

so that

$$C_{H,x,t} = C_H^B\ \text{erf}\left(\frac{x}{\sqrt{4D_H t}}\right),$$

(5.78)

and

$$\dot{N}''_{\mathrm{H}} = \sqrt{\frac{D}{\pi t}}\, C^{\mathrm{B}}_{\mathrm{H}}, \tag{5.79}$$

and

$$\bar{N}''_{\mathrm{H}} = 2\sqrt{\frac{D}{\pi t_{\mathrm{e}}}}\, C^{\mathrm{B}}_{\mathrm{H}}. \tag{5.80}$$

At the end of t_{e} seconds, the exposure time, the concentration profiles that have been set up in the metal adjacent to the interface, are destroyed, as fresh material drawn up in the wake of the bubbles rising from the bulk of the metal, replaces this 'old surface material' and the whole process is repeated.

Inserting values of $D_{\mathrm{H}} = 90 \times 10^{-9}$ m^2 s^{-1} and a surface renewal rate s of 0.5 s^{-1}, the initial rate of surface desorption of hydrogen from a steel containing 8 p.p.m. H would be

$$\bar{N}''_{\mathrm{H}} = 2\left(\frac{90 \times 10^{-9}}{\pi \times \frac{1}{0.5}}\right)^{\frac{1}{2}} \left(\frac{8 \times 10^{-6} \times M_{\mathrm{H}}}{1/7200}\right)$$

$$= 1.37 \times 10^{-5} \text{ kg-atom H} \quad \text{m}^{-2}\text{s}^{-1}$$

or

$$\bar{N}''_{\mathrm{H}_2} = 2.5 \times 10^{-2} \text{ kg-mole m}^{-2}\text{h}^{-1}.$$

Note that if half the surface is renewed every second, then bubbles rising directly behind each other are spaced at 2-second intervals, i.e. $s = 1/t_{\mathrm{e}}$.

(b)   Danckwerts' surface renewal theory

In the case of transfer of solute across a liquid–liquid interface, Danckwerts (1962) proposed a model for predicting mass transfer coefficients similar to that envisaged for bubble-stirred interfaces. He assumed that convection currents within the main bulk of the two liquids were sufficiently strong to ensure uniform bulk solute concentrations, $C^{\mathrm{B}}_{\mathrm{A}}$, and that at the interface, unsteady-state transfer of solute from packets of fluid (eddies) brought randomly to the interface, occurred. This hypothesis lead to the expression

$$\bar{N}''_{\mathrm{A}} = \sqrt{D_{\mathrm{AB}}s}\,(C^{\mathrm{B}}_{\mathrm{A}} - C^{*}_{\mathrm{A}}), \tag{5.81}$$

where $*$ represents interface and s represents the fraction of surface renewed per unit time ($= 1/t_{\mathrm{e}}$) on a random basis. Consequently, predictions from eqn (5.81) and eqn (5.80) differ by a factor of $2/\pi^{1/2}$, or 1.13.

Over the intervening years since this proposition, the difficulty has been to obtain experimental data to confirm his theory. Various other theories for fluid–fluid interactions have been developed along similar lines to Danckwerts' surface renewal theory. In the film-penetration model, a distribution of eddies are visualized to approach within random distances of the interface. Such refinements again generally lack experimental support to the author's knowledge.

(c)   Exercise: estimation of surface renewal times in open hearth operations

During the course of an open hearth operation, (e.g. Szekely 1964), the 100 tonnes of molten metal charge, at 1600°C, contained within a bath of 16 m² cross-sectional area, registered a drop in carbon content of 0.15 wt.% h⁻¹. On the basis of photographic and experimental work, it is reasonable to assume that the carbon is removed as large spherical cap bubbles of CO, evenly distributed over the bath, each having an average volume, at 1600°C, of 15×10^{-6} m³ and rear cross-sectional area of 20×10^{-4} m². Calculate an approximate rate of surface renewal at the slag–metal interface, and the corresponding exposure time, t_e, before 'old surface' is replaced by fresh material from the bulk metal phase.

Calculate appropriate metal phase mass transfer coefficients for Mn and S transport into the slag ($D_S = 3 \times 10^{-9}$, $D_{Mn} = 2 \times 10^{-9}$ m² s⁻¹). What "stagnant boundary layer thickness" must be postulated for such cases?

5.9.   REACTION PRODUCTS IN SINGLE-, TWO-, AND MULTIPHASE SYSTEMS

(a)   Gas–gas interactions

When two gases intermingle, they may sometimes react to form a gaseous, liquid or solid product. For example, the combustion of an oil droplet atomized within an oil-fired furnace involves volatilization of hydrocarbon from the droplet surface, together with simultaneous diffusion and convection of oxygen (or oxidant) towards it. The interpenetration of hydrocarbon vapour and gas leads to combustion, the formation of a flame front, and the effusion of hot, incandescent products of combustion such as CO, CO_2 and H_2O.

A similar phenomenon takes place when metallic vapours evaporate to react with oxygen. An example of fuming is shown schematically in Fig. 5.13. A cloud of fine oxide particles condense around an evaporating droplet, and are convected away into the bulk gas flow. Their homogeneous nucleation from a gas phase inevitably leads to dust particles predominantly in the submicrometre range. While potentially injurious to the body, they are very difficult to filter from exhaust furnace gases.

On a historical note, the demise of open hearth steel-making operations in North America during the last three decades came about through government regulations requiring mandatory upgrading of pollution control equipment:

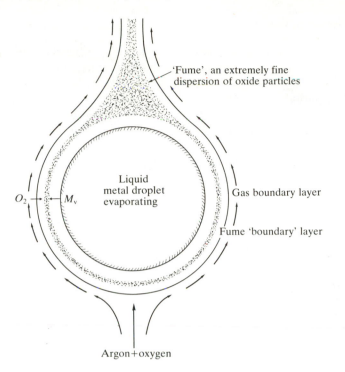

'Fume', an extremely fine
dispersion of oxide particles

Liquid
metal droplet
evaporating

O_2 ←→ M_v

Gas boundary layer

Fume 'boundary' layer

Argon+oxygen

FIG. 5.13.   Illustration of fuming phenomena in metallurgical processes.

top-blown lancing with oxygen was being used to shorten refining times in an attempt to match productivities being achieved by BOF operations. Such practices lead to unacceptable pollution of the environment, and equally unacceptable capital spending programmes to meet the new regulations. Currently, for every 100 tonnes of steel produced in top blown converter operations, approximately 1.5 tonnes are exhausted as iron oxide dusts. About 40 per cent of this derives from fuming. The remainder is the result of splintering and explosion of small steel droplets, generated under the impact zones of the oxygen jets, and carried away in gases exhausting from the converter mouth.

Other fuming phenomena to be noted are related to the extraction of zinc, from a slag bearing zinc oxide, as zinc fume. Similarly, in vacuum refining operations, in-leaking oxygen reacting with impurity metallic vapours (e.g. Bi_v and Pb_v from copper melts), or impurity sulphide vapours from mattes, can form fine condensates on chamber walls, etc. Finally, the decomposition of nickel carbonyl $Ni(CO)_4$ to nickel and carbon monoxide forms the basis of the important Mond process for the refining of nickel, and is again representative of this class of reaction system.

(i)   Kinetics of fume formation

Figure 5.13 illustrates the phenomenon of fuming around a droplet of liquid metal surrounded by oxygen containing gas mixtures. In the absence of any oxygen, (see Fig. 5.14) and depending on the precise geometry and flow conditions around the liquid surface, a boundary layer of metallic vapour, of thickness Δ, will be present, together with a concentration gradient for diffusion of metallic vapour into the bulk gas flow (curve A). If the partial pressure of oxygen within the bulk gas phase is raised, oxygen can combine to form an oxide fume, subject to normal thermodynamic and nucleation constraints, according to the general equation

$$pM + \frac{q}{2}O_2 \rightarrow M_pO_q. \tag{5.82}$$

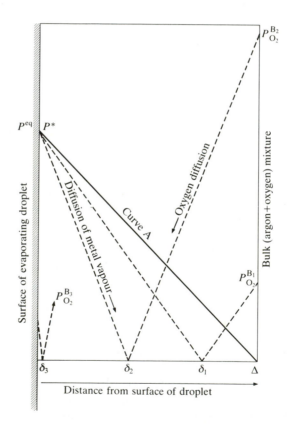

FIG. 5.14.   Plot of vapour pressure distributions arising during fuming of a metallic vapour.

Here M_pO_q may represent FeO, ZnO, As$_2$O$_3$, Li$_2$O, AlO$_2$, etc. Continuity requires that

$$\frac{1}{p}\dot{N}''_M = -\frac{2}{q}\dot{N}''_{O_2}, \tag{5.83}$$

while free energy considerations will normally demand that the activities of the reacting species in the gaseous state are close to zero at the reaction zone, δ, shown in Fig. 5.14.

The condition that $C_M \cong C_{O_2} \sim 0$ (or $P_M \sim P_{O_2} \sim 0$) at δ is met by outward diffusion of metal vapour and inward diffusion of oxygen down concentration gradients of the type shown in the broken curves of Fig. 5.14. For convenience, we can treat Δ as an effective stagnant boundary layer thickness, and write

$$\dot{N}''_M = -\left(\frac{2p}{q}\right)\frac{D_{O_2/A}}{(\Delta-\delta)}\left[\frac{P^B_{O_2}-0.0}{RT}\right]. \tag{5.84}$$

It is evident that a rise in oxygen partial pressure will lead to increases in the inward flux of oxygen. This must be met by an increase in $\dot{N}''_M$ and can occur by the reaction plane moving in towards the droplet surface, so as to steepen the gradient of metal vapour concentration. Clearly, a point may be reached at which the evaporation of metallic vapour can hardly keep pace with the penetrating flux of oxygen, even though eqn (5.84) implies that $\dot{N}''_M$ can become infinite as the inner boundary layer thickness δ approaches zero.

(ii) Limiting rates of evaporation

An equation developed by Langmuir and based on the kinetic theory of gases introduced in Chapter 1, shows that the *nett* rate of evaporation of a gas from a liquid surface per unit time and unit area is equal to

$$\dot{N}''_{vap} = \frac{\alpha(P^{eq} - P^*)}{(2\pi RTM_i)^{1/2}} \tag{5.85}$$

where $\dot{N}''_{vap}$ = flux of evaporating species (kg-mole m^{-2} s^{-1});
   P^{eq} = equilibrium vapour pressure of evaporating species (Pa);
   P^* = interfacial partial pressure above melt (Pa);
   R = gas constant (8314.34 J kg-mole^{-1} K^{-1});
   α = coefficient indicating escape efficiency of molecules (~ 1)
   M_i = molecular weight of evaporating species (kg kg-mole^{-1})

Generally, interfacial kinetics are much faster than transport of the evaporated molecules, and a dynamic equilibrium between evaporating and re-condensing molecules is struck, in which $P^* \to P^{eq}$. However, for the case of a hard vacuum, where $P^* = 0$, or for the situation involving fume formation, where an external demand by oxygen may require a greater efflux of molecules

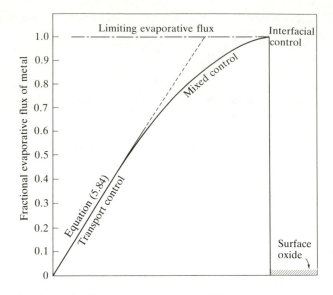

FIG. 5.15.   A schematic diagram showing the effect of oxygen partial pressure on metal vaporization and fuming kinetics.

than can be supplied (i.e. at δ_3 in Fig. 5.14), interfacial kinetics can become rate-controlling. Figure 5.15 summarizes the behaviour of a gas fuming reaction on oxygen pressure, showing how the flux of evaporating species will first rise linearly with oxygen pressure, until coming slowly to a halt once the maximum (or intrinsic) rate of evaporation is exceeded. Beyond this, an oxide of the metal will be formed, and fuming will cease. Alternatively, hydrodynamic instability of the fume layer close to the liquid surface may result in an earlier end to fuming.

Table 5.1 provides data on typical maximum rates of evaporation for a variety of common metals.

TABLE 5.1. *Data on typical maximum rates of evaporation for some common metals*

Metal vapour evaporation, $\dot{N}_{MAX}$	Temperature (K)	Limiting rate of evaporation (kg-mole m^{-2} s^{-1})
Nickel	1873	10
Iron	1873	20
Chromium	1873	70
Copper	1473	1

(b)   Gas–liquid interactions

So far in our treatment of gas–liquid systems, examples have been restricted to situations in which gases have been soluble in the liquid. Equally, however, many metallurgical examples can be cited in which the gas reacts with the liquid, or with solutes within that liquid, to form new phases. We have already noted in Chapter 1 that silicon, manganese, phosphorus, and iron are oxidized to join the slag phase in steel-making operations.

An equivalent example from the aluminium industry would be the purification of molten aluminium from contaminating lithium, sodium, magnesium or iron solutes by fluxing with argon–chlorine gas mixtures, i.e.

$$3\{Cl_2\} \quad + \quad 2(Al) \quad \rightarrow \quad 2\{AlCl_3\}, \tag{5.86}$$

$$\text{gas bubble} \qquad \text{melt} \qquad\qquad \text{gas bubble}$$

followed by

$$2\{AlCl_3\} \quad + \quad 3(Mg)_{Al} \quad \rightarrow \quad 3(MgCl_2) \quad + \quad 2Al. \tag{5.87}$$

$$\text{gas bubble} \qquad\qquad\qquad\qquad\quad \text{droplets} \qquad \text{melt}$$

Such a reaction product, if swept away from the interfaces of rising bubbles, can be expected to generate a fine dispersion of salt particles, which must be removed, or allowed to float out of the melt, before continuous casting operations. Figure 5.16 illustrates the process equipment (a), and likely mechanisms (b), involved in such operations.

Other important reactions of this type include the desulphurization of carbon-saturated iron melts by magnesium vapour according to:

$$\{Mg\}_v \quad + \quad (S)_{Fe} \quad \rightarrow \quad \langle MgS \rangle, \tag{5.88}$$

$$\text{bubble} \qquad\qquad\qquad\qquad \text{film on}$$
$$\text{bubble}$$

$$\{Mg\}_v \quad \rightarrow \quad (Mg)_{Fe} \tag{5.89}$$

$$\text{bubble}$$

and

$$(Mg)_{Fe} \quad + \quad (S)_{Fe} \quad \rightarrow \quad \langle MgS \rangle. \tag{5.90}$$

$$\text{particle,}$$
$$\text{inclusion}$$

Research (Irons and Guthrie 1982), has shown that about 90% of the sulphur removed in magnesium injection operations into a 4.3%C, 0.05–0.07%S iron, takes place according to the second mechanism (eqns (5.89) and (5.90)). In this, dissolved magnesium and dissolved sulphur co-diffuse and precipitate on particles of magnesium sulphide already present and rising out of the melt.

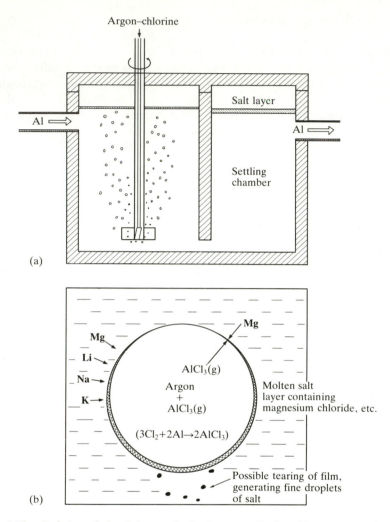

FIG. 5.16.   Refining of aluminium melts by inert gas fluxing: (a) process equipment; (b) proposed mechanisms.

The second phase particles dispersed in the melt come about as the result of initial reaction between dissolving magnesium vapour bubbles and any solute sulphur which has diffused towards the bubble interfaces as illustrated in Fig. 5.17(b). Figure 5.17(a) shows possible process equipment for carrying out this desulphurization.

Evidently, the mathematical treatment of such systems is relatively difficult, since one must know the number of MgS particles being generated at the

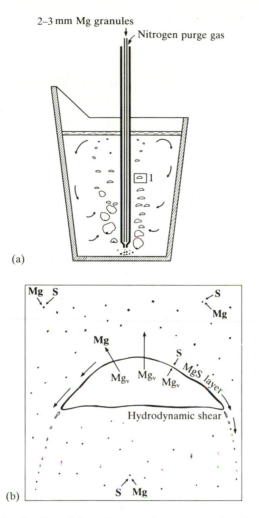

FIG. 5.17. Desulphurization of iron from carbon saturated melts by magnesium granules: (a) process equipment; (b) proposed mechanisms.

bubble interfaces, their size and disposition within the melt, and so on. Coefficients for rate constants are therefore empirical or, at present, simply not available.

Finally, to underscore the fact that similar phenomena are to be observed in hydrometallurgical systems, one can cite the precipitation of nickel powder from ammoniacal aqueous solutions in the Sherritt–Gordon operation, in which the reduction of dissolved nickel by hydrogen at 2 atm and 90 °C takes place according to

$$Ni^{2+} + 2e \rightarrow \langle Ni \rangle, \tag{5.91}$$
$$\text{powder}$$

coupled with the complementary oxidation of hydrogen

$$\{H_2\} \rightarrow 2H^+ + 2e. \tag{5.92}$$

(c)   Gas–solid interactions

There are many metallurgical examples of gas–solid reactions; those involving the iron oxides have already been mentioned in Chapter 1. Apart from these, many other important extraction routes involve similar reduction of metal oxides to metal using carbon and/or carbon monoxide. Those oxides include oxides of zinc, nickel, manganese, chromium, lead, molybdenum, and niobium. The first steps in the extraction of metals from sulphide concentrates of copper, nickel, lead and zinc provide yet other examples of gas–solid interactions. In those cases, the sulphides are first 'roasted' and partially or fully oxidized, prior to reduction with carbon.

The sulphide and oxide ores are typically liberated from siliceous rocks by grinding, crushing and flotation. All these solid particles (which are usually agglomerated, or pelletized) contain impurities. These impurities can greatly influence reduction or oxidation kinetics of the concentrates. For instance, 'green' iron oxide pellet concentrates, containing 2–5 per cent silica, are intentionally fluxed with lime additions during firing at 1200 °C, so as to form a network of silicates. These strengthen a pellet for its passage down the blast furnace. However, an excess can lower a pellet's reducibility to carbon monoxide, if the silicate network blocks passageways between the iron oxide particles. Other impurities such as Na_2O or K_2O, may 'poison' the surfaces of these particles and by preventing general reduction of the oxide surface lead to localized reduction and iron whisker formation, followed by catastrophic swelling and pellet breakdown. Needless to add, this has a disastrous effect on blast furnace operation.

In some instances, a reaction front can be observed when a pellet is being reduced by a gas. When this takes place, the reaction front is clearly delineated and the reaction is said to be topochemical in nature. Figure 5.18 illustrates the principal characteristics of a pellet undergoing a topochemical and generalized reduction scheme. Thus, the iron oxide pellet shown in (b) is reduced by pure hydrogen, of much higher diffusivity and driving potential versus reduction by a CO/CO_2 mixture shown in Figure 5.18(a), (corresponding to a proper blast furnace environment simulation). The difference in behaviour is striking, pellet (a) undergoing general reduction and (b) topochemical reduction (Guthrie, 1979).

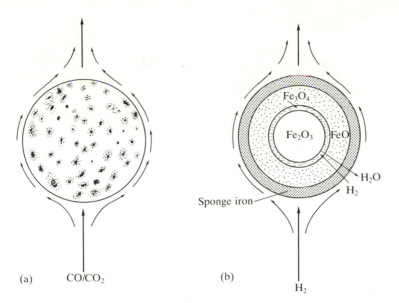

(a)   CO/CO₂

(b)

H₂

Fɪɢ. 5.18.   Illustration of (a) general reduction of an iron oxide pellet by a CO/CO_2 gas mixture and (b) topochemical reduction by an H_2/H_2O gas mixture.

In attempting to analyse the kinetics of gas–solid reactions, adsorption, desorption, and nucleation, as well as chemical and transport kinetics, can all conspire to render *a priori* predictions on industrial systems impractical.

Physical tests have therefore been developed to simulate real environments, to establish rate-controlling features, and to provide specific rate constants for use in global mathematical models of packed-bed reduction processes.

(i)   Kinetics of limestone decomposition

The metallurgical practice of burning limestone to lime, for fluxing silica and other impurities during the pyrometallurgical treatment of ores and concentrates, is shown in Fig. 5.19(a). The example provides a simple illustration of the way gas–solid reactions can be treated. Consider the topochemical decomposition of the pellet of calcium carbonate shown in Fig. 5.19(b) according to:

$$\overset{\text{heat}}{CaCO_3 \rightleftarrows CaO + CO_2}. \tag{5.93}$$

Evidently, individual particles of lime must thermally decompose by nucleation of CaO crystallites, and the carbon dioxide produced must then diffuse away into the bulk gas phase which, for simplicity, is taken to be argon.

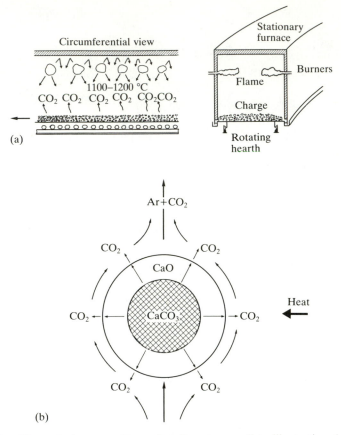

FIG. 5.19. Thermal decomposition of a limestone pellet, illustrating (a) typical process equipment for lump decomposition, and (b) topochemical reaction during pellet decomposition.

Chemical control. Supposing that the nucleation of lime is rate-controlling, and that it can be characterized by first-order kinetics of the form

$$\dot{N}'' = -k_f C, \tag{5.94}$$

where $\dot{N}''$ is the rate of formation of calcium carbonate/unit area/unit time, and C represents the concentration of calcium carbonate per unit volume of pellet. Continuity then requires that

$$\frac{d}{dt}(CV) = -k_f AC. \tag{5.95}$$

Taking $C = $ constant and inserting geometrical factors,

$$C\frac{d}{dt}(\tfrac{4}{3}\pi r^3) = -k_f 4\pi r^2 C,$$

we obtain

$$\frac{dr}{dt} = -k_f. \tag{5.96}$$

This shows that the reaction front should move back at a constant velocity k_f during thermal decomposition, under conditions of chemical control. Taking $r = R_0$ at $t = 0$, and $r = 0$ at $t = T$, we obtain an estimate for the time needed for complete decomposition:

$$T = R_0/k_f. \tag{5.97}$$

Transport control. It is equally feasible that the rate of carbonate decomposition is controlled by the ease with which carbon dioxide can diffuse through the product layer into the bulk gas stream. Under conditions of transport control, we can write:

(1) For CO_2 transport through the gas boundary layer:

$$\dot{N} = 4\pi R^2 k_g (C_g^* - C_g^B), \tag{5.98}$$

where C_g^* and C_g^B represent the concentration of CO_2 at the surface of the pellet and in the bulk gas phase.

(2) For quasi-steady state transport through the CaO product layer, we may similarly write,

$$\dot{N} = 4\pi r^2 \left(-D\frac{dC_g}{dr} \right), \tag{5.99}$$

which may be integrated according to

$$\frac{N}{4\pi} \int_R^r \frac{dr}{r^2} = -D \int_{C_g^*}^{C_g^{eq}} dC_g, \tag{5.100}$$

to give

$$\frac{N}{4\pi}\left(\frac{1}{r} - \frac{1}{R} \right) = D(C_g^{eq} - C_g^*). \tag{5.101}$$

Substituting for C_g^* from eqn (5.98), and re-arranging:

$$\dot{N}_g = (C_g^{eq} - C_g^B) \bigg/ \left\{ \frac{1}{4\pi D}\left(\frac{1}{r} - \frac{1}{R} \right) + \frac{1}{4\pi R^2 k_g} \right\}. \tag{5.102}$$

From stoichiometry, we know that the net efflux of carbon dioxide, $+\dot{N}_g$, must equal the rate of molar decomposition of lime, $-\dot{N}$, which, as before, can be replaced with $d(VC)/dt$ or $C4\pi r^2\, dr/dt$. Consequently,

$$\frac{dr}{dt} = \frac{\dot{N}}{C4\pi r^2} = -\frac{\dot{N}_g}{C4\pi r^2},$$ (5.103)

or

$$\frac{dr}{dt} = \frac{-1}{r^2 C}\left(\frac{P^*}{RT} - \frac{P^B}{RT}\right)\Big/\left\{\frac{1}{k_g R^2} + \frac{1}{D}\left(\frac{1}{r} - \frac{1}{R}\right)\right\}.$$ (5.104)

(ii) Discussion

This expression shows that an increase in k_g, the mass transport coefficient in the bulk gas phase, or D, the effective diffusivity of gas through the pores of the lime, lead to faster rates of topochemical reaction. Figure 5.20 (Hills 1966) shows a dimensionless plot of the fractional increase in the product layer thickness $(R-r)/R$ versus dimensionless time. The important feature to note is the near linearity for wide ranges of $D/k_g R$. Certainly for $D/k_g R \cong 0.5$, deviations from linearity are only observed beyond about 0.75. As this corresponds to 98.5 per cent decomposition by weight, and is beyond the

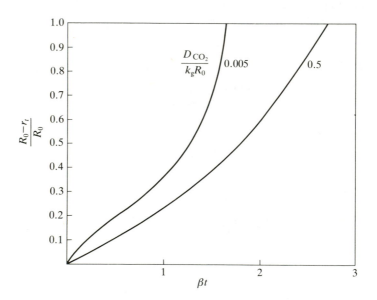

Fig. 5.20.   Plot of fractional increase in product layer thickness versus time for the thermal decomposition of a limestone pellet under conditions of CO_2 mass transport control through product layer and gas phase boundary layer surrounding pellet.

$$\beta = \frac{6 D_{CO_2/CaCO_3}(P^* - P^B)}{RT R_0^2 \rho_{CaO}}.$$

(After Hills 1966.)

normal limits of experimental detection, it is practically impossible to use $[dr/dt]$ versus time curves as a means of deciding whether a reaction is chemically controlled or transport-controlled.

The most useful guideline that can be stated for gas–solid reactions is first to estimate reaction rates on the basis of transport control. If experimental rates are similar, then transport control can be assumed to dominate, whereas if rates are much lower, chemical, or some other controlling feature must be the dominating kinetics. To illustrate the difficulties in analysing an apparently simple reaction such as this, one should note that the structure of lime fired at two different temperatures can be completely different with regards to porosity, specific surface area, etc. Similarly, thermal kinetics plays an important role in this particular reaction, the reaction being highly endothermic, but was not considered, as such.

(d)   Liquid–liquid interactions

The formation of reaction products when two liquids come into contact provides yet other examples of complex mass transport phenomena. As with the other systems, only a few examples can be cited to illustrate the wide range of possibilities.

(i)   Foaming slags in steel-making

When a liquid of high oxygen potential comes into contact with a liquid of low oxygen potential (e.g., containing carbon), explosive formation of gas is possible. For example, droplets of iron containing dissolved carbon can react with the oxidizing slag in a BOF to nucleate carbon monoxide and sustain a foaming slag, according to

$$(C)_{Fe} \quad + \quad (FeO) \quad \rightarrow \quad Fe \quad + \quad \{CO\}. \quad (5.105)$$

$$\text{droplet} \qquad \text{bulk slag} \qquad \text{bulk iron} \qquad \text{bubble}$$

Alternatively, when the concentration of nascent oxygen in the slag rises, intermittent explosive release of gases leads to slopping, and yield loss.

(ii) Aluminium deoxidation of steel

When solid aluminium is added to molten steel, the aluminium melts and pure liquid disperses into the steel bath. The reaction of this aluminium, or of aluminium-rich iron, with solute oxygen dissolved in the steel ladle leads to the nucleation and growth of solid Al_2O_3 particles, according to

$$(Al) \quad + \quad (O)_{Fe} \quad \rightarrow \quad \langle Al_2O_3 \rangle$$

$$\text{liquid} \qquad\qquad\qquad \text{particles} \qquad\qquad (5.106)$$

(e) Liquid–solid interactions

Many hydrometallurgical leaching operations associated with the extraction
of metals from calcined ores result in the formation of a product phase. For
example, acid leaching of zinc from calcined particles of $ZnO.Fe_2O_3$, leaves a
spent product layer. This layer provides hindrance to the transport of chloride
and zinc ions to and from the reacting interface. Similar examples apply in the
extraction of nickel and copper calcines.

Pyrometallurgical examples include ladle metallurgy operations such as the
desulphurization of iron by lime particles, or the modification of inclusions of
Al_2O_3 in steel melts by calcium or rare earth additions. As with other complex
systems, an understanding of the reaction mechanisms, together with empirical rate constants, provides the best approach to their characterization.

5.10. GENERAL EXERCISES INVOLVING INTERPHASE HEAT AND MASS TRANSFER

(1) The lead–zinc splash condenser for the removal of zinc vapour in the ISF process

A spherical drop of initially pure lead at 800 K is formed in a lead splash
condenser, and travels a distance of 0.45 m at a velocity of 0.3 m s^{-1} through a
gas mixture (CO, N_2, Zn vapour) containing 10 per cent by volume zinc
vapour. The rate of zinc pick-up by the drop is controlled by the transport of
zinc vapour from the bulk of the gas phase (at 1200 K) to the gas–liquid
interface at 800 K. Stirring within the drop, and its small size, is sufficient to
eliminate any significant concentration gradients of zinc within the liquid.
Calculate the final concentration of zinc in the drop (kg-atom Zn/m^3 Pb), and
express this as a percentage of the saturation concentration of zinc in lead at
800 K.

Temperature of drop $= 800$ K; temperature of gases $= 1200$ K; Drop radius
$= 0.3$ mm; Total gas pressure $= 1$ atm; mass transfer coefficient of zinc vapour,
$k_G = 1.09$ m s^{-1}.

$$C_{Zn,gases} = \frac{n_{Zn}}{V_{gas}} = \frac{P_{Zn}}{RT},$$

where $C_{Zn,gases}$ is the number of kg-atoms Zn vapour/m^3 gas; P_{Zn} is the partial
pressure Zn vapour, (atm). The concentration of zinc dissolved in lead, in
equilibrium with its vapour, is related by

$$C_{Zn/Pb} = 2.48 \times 10^{-3} P_{Zn} \text{ at (800 K)},$$

where $C_{Zn/Pb}$ is the number of kg-atoms Zn/m^3 liquid metal; P_{Zn} is the partial
pressure of zinc vapour (atm). Maximum solubility of zinc in lead $= 4.7$
$\times 10^{-6}$ kg-atom Zn/m^3 liquid.

(2) The absorption of gases into teeming streams issuing from teeming ladles

The presence of hydrogen in steel, even in very small amounts, can have extremely deleterious effects on its mechanical properties. In the casting of an hydrogen-free steel, with an oxygen content of 0.1 wt.%, it was suggested that no precautions were necessary to avoid hydrogen pick-up from the atmosphere during teeming into the casting. Examine this statement by estimating the hydrogen pick-up (in wt.% or p.p.m.) over the distance H that the stream of molten metal is in contact with the atmosphere. Use the thermodynamic data provided in Appendix III, to show that one can assume

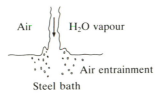

Air   H₂O vapour

Air entrainment

Steel bath

that the kinetics of the reaction are completely controlled by diffusion of water vapour through the atmosphere to the jet surface. In addition, since the reaction is far from equilibrium, it is permissible to assume that the interfacial water vapour concentration is negligible in comparison with the bulk gas water vapour concentration when calculating *the flux* of water vapour to the molten metal surface. The jet may be considered to be in plug flow in the vertical axis, completely mixed in the horizontal plane (i.e., linear reactor), and to have a constant cross-sectional area.

Data.

Mass transfer coefficient of water vapour through gas phase, k_g	$0.1 \ \mathrm{m\,s^{-1}}$
Diameter of jet, d	$0.013 \ \mathrm{m}$
Bulk vapour pressure of water 100% humidity at 90°F	$12.13 \times 10^2 \ \mathrm{N\,m^{-2}}$
Length of jet, H	$0.61 \ \mathrm{m}$
Velocity of jet, U	$6.0 \ \mathrm{m\,s^{-1}}$
Density of iron, ρ_{Fe}	$7.2 \times 10^3 \ \mathrm{kg\,m^{-3}}$
Atomic weight of iron	56
Initial concentration of O in steel, C_O	$0.1 \ \mathrm{wt.\%}$
Initial concentration of H in steel, C_H	$0.001 \ \mathrm{wt.\%}$
Gas constant, R	$83.7 \ \mathrm{J\,kg\text{-}mole^{-1}\,K^{-1}}$
Bulk temperature of gas, θ	$305 \ \mathrm{K}$

Comment

Far more hydrogen than the amount calculated here is absorbed at the junction of the free jet and free metal surface, as a result of massive volumes of air entrainment. More and more companies now shroud all teeming streams and provide argon purging, so as to exclude hydrogen, nitrogen, and oxygen from entering steel delivered to the caster.

(3)   Deoxidation of copper in anode furnace

A copper refinery decided to change its operating practice for the deoxidation of copper in an anode furnace, from 'poling' to natural gas lancing. In this method, the natural gas, essentially comprising methane, is introduced through a pipe (lance) into the bath of molten copper, so as to generate a submerged 'jet' of dispersed bubbles of mean initial volume of 1 cm^3, (i.e., prior to reaction with dissolved oxygen in the copper). The interfacial reaction which occurs can be regarded as an *instantaneous*, irreversible reaction between methane and oxygen to yield CO_2 and H_2O (steam) by the reaction

$$CH_4 + 4(O)_{Cu} = CO_2 + 2H_2O.$$

Using the data given below, calculate the *maximum possible* fluxes of oxygen and methane to the bubble's interface and show that the process is mass-transport-controlled in the liquid phase (for 0.2 wt.% O in Cu) provided the mole fraction of unreacted methane in the bubble is always in excess of 0.15 (i.e., 15% CH_4, 28.3% CO_2 and 56.6% H_2O). Using this fact, calculate the minimum contact time that is required between bubble and liquid to ensure that 85 per cent of the methane is used up before the bubble exits from the liquid copper.

Data:

Temperature of liquid copper	1500 K
Density liquid copper	8000 kg m^{-3}
Bulk oxygen content in copper	0.2 wt.%O
Total gas pressure in bubble	1 atm
Overall liquid phase mass transfer coefficient of O to bubble surface	4×10^{-4} m s^{-1}
Overall gas phase mass transfer coefficient	8.24×10^{-2} m s^{-1}
Initial bubble volume	100 mm^3
Gas constant, R	83.7 J kg-mole^{-1} K^{-1}
Surface area A of bubble is related to bubble volume V by	$A = SV^{2/3}$
Shape factor, S	4.84

(4) Dezincification of lead

In the separation of molten zinc from lead, a mixture containing 10 atom% of zinc in molten lead is pumped from the lead splash condenser into a cooling launder (temperature 700 K). Since the solubility of zinc in lead decreases with temperature, the zincy lead becomes supersaturated with zinc, resulting in the transfer of zinc to form a virtually pure upper phase of zinc ($+1\%$ Pb). The launders used are rectangular in cross-section.

It is desired to calculate (a) the bulk flow velocity of zincy lead, (b) the length of the launder necessary to lower the zinc concentration in the lead by 1 atom%, under the operating conditions below. Consider two cases: the stirred reactor and the linear reactor.

Data:

Throughput of zincy lead	400 tonnes h^{-1}
Height of zincy lead in launder	0.15 m
Width of launder	0.6 m
Mass transfer coefficient of zinc in lead across interface	4×10^{-4} m s^{-1}
Atomic weight of lead	207
Atomic weight of zinc	65.38
Density of lead, ρ_{Pb}	11000 kg m^{-3}
Density of zinc, ρ_{Zn}	6900 kg m^{-3}
Maximum solubility of zinc in lead in equilibrium with zinc phase at 700 K	8 atom%

You may neglect changes in volume of the zincy lead due to zinc removed. In addition, the bulk of the zincy lead phase may be assumed to be sufficiently well-mixed that the vertical bulk concentration distribution of zinc in lead at any given position along the launder is uniform.

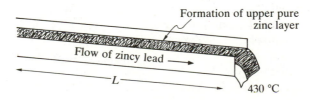

Formation of upper pure zinc layer

Flow of zincy lead →

L

430 °C

(5)   Liquid metal casting and gating practices

A copper casting of an intricate shape is to be manufactured using a sand mould. On the basis of one-dimensional heat flow (i.e. x direction in figure), calculate:

(i) The thickness of the skin of metal formed in the connecting slit (downgate) between the pouring basin and the casting after one minute of pouring;

(ii) the minimum initial thickness, b, of the downgate necessary to ensure that the casting, of volume V, is completely filled with molten metal within three minutes from start of pouring.

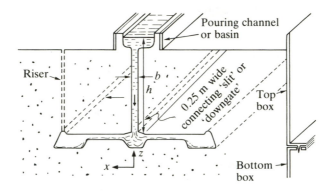

You may assume that the pure molten metal is poured with zero superheat (i.e., at melting point), that the volume of the casting and also the pouring channel is much greater than the volume of the downgate, and that Bernoulli's equation may be applied to calculate the volumetric flow rate of metal through the downgate.

Data:

Width of slit, b	0.25 m
Height of slit, h	0.25 m
Volume of casting	1 m^3
Allowed filling time	3 min
Density of copper	8400 kg m^{-3}
Latent heat of fusion	205 kJ kg^{-1}
Thermal conductivity of sand mould, k	0.33 W m^{-1} s^{-1}
Thermal diffusivity of sand mould, α	12.1 m^2 s^{-1}
Melting point temperature of copper	1353 K
Ambient temperature	293 K
Coefficient of discharge	0.90

(6) Rate of fume formation from metal droplets ejected from baths of molten metal

Calculate the rate of metal vaporization ($\mu g\,s^{-1}$) from a 1 mm diameter droplet of (i) pure copper at 1500 K, and (ii) pure iron at 1900 K travelling at a relative velocity of $3.0\,m\,s^{-1}$ through the following gas mixtures at 1 atm.

(a) pure argon;
(b) argon $+0.1$ vol.% O_2;
(c) argon $+1$ vol.% O_2.

Calculate the equivalent stagnant boundary layer thickness, Δ, and the distances, δ_r, of the fume layer from the droplet surfaces.

Data:
Physical properties for copper at 1500 K and for iron at 1900 K

Saturation vapour pressure of copper at 1500 K	7.7×10^{-6} atm		
Saturation vapour pressure of iron at 1900 K	1.1×10^{-4} atm		
$D_{Cu/A}$	$2.2 \times 10^{-4}\,m^2\,s^{-1}$	$D_{Fe/A}$	$3.1 \times 10^{-4}\,m^2\,s^{-1}$
$D_{O_2/A}$	$3.1 \times 10^{-4}\,m^2\,s^{-1}$	$D_{O_2/A}$	$4.4 \times 10^{-4}\,m^2\,s^{-1}$
v_A	$2.2 \times 10^{-4}\,m^2\,s^{-1}$	v_A	$2.92 \times 10^{-4}\,m^2\,s^{-1}$
v_{O_2}	$3.0 \times 10^{-4}\,m^2\,s^{-1}$	v_{O_2}	$3.36 \times 10^{-4}\,m^2\,s^{-1}$
N_{Sh}	$2 + (0.4\,Re_D^{1/2} + 0.06\,Re_D^{2/3})Sc^{0.4}$,		
or, alternatively,			
N_{Sh}	$2 + 0.6\,N_{Re}^{1/2}N_{Sc}^{1/2}$		

where N_{Sh}, Sherwood number $= kD/D_{AB}$; k is the mass transfer coefficient; v is the kinematic viscosity; N_{Re}, Reynolds number $= Ud/v$; D_{AB} is the diffusion coefficient; N_{Sc}, Schmidt number $= v/D_{AB}$; U is the relative velocity.

You may assume instantaneous 100% conversion of oxygen and metal vapours to the oxide forms (Cu_2O and FeO) at δ_r.

(7) Thermal limits of hot rolling mill capacity

A hot rolling mill superintendent considered the possibility of increasing his mill's tonnage capacity by increasing the heights of the ingots charged to the soaking pits. Following the soaking operation, these ingots are rolled down to 32-mm-thick slabs in the rougher stand, exit at 1590 K, and then pass along an entry (or holding) table into the finishing stands. Slab temperature rundown during this period is an important constraint on the finishing operation; in order to avoid overloading the electric motors running the finishing stands,

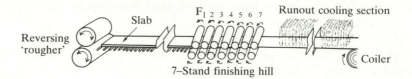

the slab temperature must never drop below 1422 K on entry into stand Fl (see figure).

Assuming the critical conditions of a minimum lag of 5 s between each slab and a coiling speed of $20 \, \text{m s}^{-1}$ on 2.3 mm gauge material, calculate the maximum height of ingot that could be handled by the operation, and the new theoretical annual mill capacity (i.e. tonnes/annum on a 24 hour basis of operation with no shut-downs producing 2.3 mm gauge material).

Data:

Density of slab	$7450 \, \text{kg m}^{-3}$
Thickness of slab	32 mm
Temperature of slab at rougher	1590 K
Minimum temperature of slab entering finishing stands	1422 K
Ingot thickness	0.61 m
Width of ingot	1.21 m
Thickness of slab	32 mm
Thickness of strip	2.3 mm
Coiling speed (speed of strip)	$20 \, \text{m s}^{-1}$
View factor of slab, $F_{\text{slab}/\infty}$	1
Emissivity of slab	0.8
Stefan–Boltzmann constant	$5.67 \times 10^{-8} \, \text{W m}^{-2} \text{K}^{-4}$

For the purposes of this calculation, ignore temperature gradients across the slab, natural convection from the slab surfaces, conduction into rollers and, finally, back-radiation from the plant.

(8) Surface evaporation of magnesium

Magnesium vapour is bubbled into an inductively stirred bath of molten pig iron at 1473 K, at a constant flow rate. Although the bath is protected with a layer of graphite chips and the bath itself is carbon-saturated, the possibility of oxygen dissolution and diffusion into the bath cannot be ignored. Using the penetration theory of mass transfer, estimate the net molar inflow of gaseous oxygen into the system and the percentage of magnesium vapour entering the

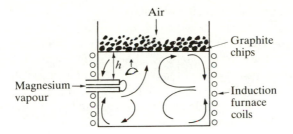

bath that will be used up in reacting with dissolved oxygen to form MgO. You should assume good mixing in the bulk of the steel (and air).

Data:

C (wt.%) + O (wt.%) = CO (atm) $K^{eq} = 500 = \dfrac{P_{CO}}{(\text{wt.\% C})(\text{wt.\% O})}$

2C(s) + O$_2$ = 2CO (atm) $\Delta G_T^0 = 231 \text{ kcal mole}^{-1} \text{ O}_2$

2Mg + O$_2$ = 2MgO $\Delta G_T^0 = -204 \text{ kcal mole}^{-1} \text{ O}_2$

Solubility of Mg in iron (4.3% C, 0.1% Si) = 0.7 wt.% for iron in contact with a partial pressure of 1 atm of magnesium vapour at 1523 K.

Gaseous oxidizing atmosphere above melt	air (1 atm)
Diameter of crucible	0.2 m
Melt weight, W	62.2 kg
Bubble depth, h	0.134 m
Density of iron melt	7000 kg m^{-3}
Magnesium vapour flowrate from nozzle	0.38 litre s^{-1}
Diffusion coefficient of oxygen in carbon-saturated iron, $D_{O/Fe}$	12 × 10^{-9} m^2 s^{-1}
Carbon content of melt	4.3 wt.%
Activity of dissolved magnesium in melt	0.1
Mass transfer coefficient of gaseous oxygen to melt surface, k_g	0.05 m s^{-1}
Mass transfer coefficient of dissolved oxygen, k_O	$2\left(\dfrac{D_{O/Fe}\,\bar{U}_s}{\pi R}\right)^{1/2}$
Mean surface velocity of inductively stirred melt, $\bar{U}_s$	0.13 m s^{-1}

---

(9)   A problem in Imperial (British) units

The economic and technical feasibility of a rival company producing a specialty steel using a low-capital-cost annealing furnace is to be estimated. It is known that the heat treating line is composed of a 40-foot heating section, followed immediately by a 60-foot cooling section. The strip to be processed is 0.160 inch thick by 4 feet wide and the processing cycle requires heating the material to 1450°F followed by a minimum 30-second soak at 1450°F and subsequent cooling to a line exit temperature of 500°F. The cooling rate of the strip must not exceed 30°F per second, or buckling will occur.

Determine the maximum speed at which such material could be run (in ft/min) knowing that radiant energy losses from the strip to the surrounding water-cooled inside walls of the furnace is the rate-limiting step in the heating/cooling program. Make reasonable assumptions as you see fit and justify or explain.

Data:

Strip width	4 feet
Strip gauge	0.16 inches
Strip density	465 lb ft.$^{-3}$
Strip heat capacity	0.15 BTU lb.$^{-1}$°F^{-1}
Strip emmissivity	0.4
Strip thermal conductivity	20 BTU h^{-1} ft^{-1}°F^{-1}
Temperature of strip entering cooling section	1450°F
Temperature of furnace walls	60°F
Stefan–Boltzmann constant	0.174 × 10^{-8} BTU h^{-1} ft^{-2} R^{-4}

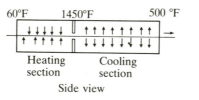

Heating section   Cooling section

Side view   End view

(10)   Fume in BOF dusts

The primary refining action in BOF steel-making arises as a result of reaction between entrained metal droplets and a foaming slag, the foam being generated by reaction of dissolved carbon in the metal droplets and oxygen absorbed at their surfaces. However, industrial research experiments reveal that smaller droplet sizes than those compatible with the hydrodynamic action of sonic oxygen jets impacting on liquid steel form a major fraction of the entrained droplets in these slags.

If we postulate that some of the larger steel drops formed by the oxygen jets become so supersaturated with oxygen that they explode, and go on to estimate that this fragmentation will occur when the internal build-up of carbon monoxide pressure exceeds outside ambient and surface tension pressures by a factor of 50, derive an expression for the residence time in the slag before such droplets spontaneously shatter.

You should show and discuss how droplet diameter, slag oxygen activity, droplet carbon content and other relevant factors affect explosion times T_E. Do you suppose, based on your results, that an optimum size range exists which will yield a maximum in drop disintegration rates?

Calculate, in particular, the time for a steel droplet of 1 mm diameter, containing 1 wt.% carbon, to explode given the following data:

Surface tension of steel (oxygen saturated), σ	800 dynes cm^{-1}
Density of steel, ρ	7000 kg m^{-3}
Droplet diameter, d	1 mm
Diffusion coefficient of oxygen in iron, $D_{O/Fe}$	10×10^{-9} m^2 s^{-1}
Equilibrium constant for	
$\dfrac{(\text{wt.\% C})(\text{wt.\% O})}{P_{CO}\,(\text{atm})} = K$	0.002 (wt.%)^2 atm^{-1}
Ambient static pressure of CO in furnace	1 atm
Activity of oxygen in steelmaking slags	0.7
Solubility limit of oxygen in steel	0.23 wt.%
Sherwood number, $Sh_l = k_l d / D_{O/Fe}$	2.0

For the purposes of this simplified calculation, you can make a good approximation by assuming transport control through a stagnant boundary layer inside the metal droplet such that the internal mass transfer coefficient $k_L = \dfrac{D_{O/Fe}}{r}$ where r = radius of droplet. Similarly, ignore CO formation, but explain how you think this would alter your results.

(11)   Removal of tramp elements from liquid steel

In a feasibility study of copper removal from molten steel scrap, it was decided to estimate the vacuum level required so as to reduce the interfacial vapour pressure of dissolved copper, P^*, to half its equilibrium value of P^{eq}. Given the information below, and first ignoring the possibility of liquid phase control, calculate:

(1)   the necessary gas phase mass transfer coefficient;
(2)   the necessary vacuum level (in $\mu m = 10^{-4} \, cm \, Hg$);
(3)   the molar efflux of copper (g-atom $m^{-2} s^{-1}$).

If the liquid phase mass transfer coefficient for transport of dissolved copper is $3 \times 10^{-5} \, m \, s^{-1}$, roughly estimate how P^*, the interfacial vapour pressure, might be affected and draw your conclusion as to whether the interfacial step can be rate-controlling.

Data:

Temperature of steel	1600 C
Density of steel	$7000 \, kg \, m^{-3}$
Density of copper	$8200 \, kg \, m^{-3}$
Molecular weight of steel	56
Molecular weight of copper	63.5
Concentration of dissolved copper in contaminated liquid steel	1 wt.%
Raoultian activity of copper in steel at 1600°C	8.6
Gas phase mass transfer coefficient (at 1 atm press), k_G^0	$0.04 \, m \, s^{-1}$
Gas phase mass transfer coefficient (at reduced pressure), k	k_G^0/P (atm)

Maximum rate of evaporation of copper is given by the Langmuir equation,

$$\dot{N}'' = \frac{P^{eq}}{\sqrt{2\pi RTM}},$$

where R is the gas constant $= 8.31432 \times 10^7 \, g \, cm^2 \, s^{-2} \, °K^{-1} \, g\text{-mole}^{-1}$, M is the molecular weight of copper impurity (g/g mole), $P^{eq} =$ vapour pressure of impurity in $g \, cm^{-1} s^{-2}$.

Vapour pressure of pure liquid copper is given by

$$\log (P) = -A/T + B + C \log (T), \text{ (mm Hg)},$$

where $A = 17,650.0$, $B = 13.39$, $C = -1.273$.

(12)   Heat extraction by laminar cooling header boxes for steel strip

A new cooling bank system has been designed to cool steel strip produced in a hot strip rolling mill, prior to coiling (see figure). The novel stands incorporate

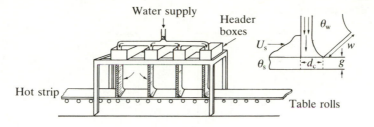

four header boxes filled with water and designed so as to produce a maintenance-free, non-clogging system, in which four curtains of water successively quench the top surface of the strip. Between quenches, the strip, being thin, rapidly re-establishes thermal equilibrium across its thickness.

Develop an expression for the total heat that will be extracted by such a system, and go on to derive an analytical expression for the overall bank heat transfer coefficient. You should assume that the effective impact width d is equal to *twice* the impact thickness of the water blades, and that the *surfaces* of strip under the jets are instantaneously quenched to water temperature. You can ignore the strip's cooling by radiant heat losses and by conduction to the table rolls, as these are of secondary importance and effect.

Given the data below, calculate:

(i) the exit temperature of strip from the stand;
(ii) the overall bank heat transfer coefficient h_B, where

$$h_B = \frac{\Sigma Q_{\text{extracted}}}{L W}(h \text{ in cal cm}^{-2}\text{s}^{-1}\text{°C}^{-1}).$$

Data:

Strip gauge, g	2.5 mm
Strip width, w	1 m
Strip temperature at inlet, θ_i	860 °C
Strip velocity, U_s	10 m s^{-1}
Strip density, ρ_s	7.4 g cm^{-3}
Strip heat capacity, C_s	0.15 cal g^{-1}°C^{-1}
Strip thermal conductivity, k_s	0.11 cal cm^{-1}°C^{-1}s^{-1}
Water temperature, θ_w	30 °C
Bank length, L	18 m
No. of headers, n	4
Thickness of water blade just prior to impact, d_i	1 cm

(13)   Zinc fuming

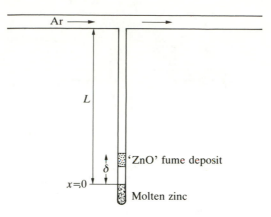

In a study of the mechanism of ZnO fume formation, molten zinc was contained in an alumina tube as shown in diagram, and air was passed over the top of the column. Since the diameter of the alumina tube was only 1 mm, any longitudinal mixing resulting from possible turbulence was eliminated as a complicating factor. Experiments showed that a very fine condensate of white ZnO precipitated out on the sidewalls. Using the data below, calculate how far up the tube you would expect the zinc vapour to oxidize.

Data:

Gas diffusion length of alumina tube, L	10 cm
Diffusion coefficient of oxygen in nitrogen, D_{O_2/N_2}	$20 \times 10^{-4}\,\mathrm{m^2\,s^{-1}}$
Diffusion coefficient of zinc in nitrogen, D_{Zn/N_2}	$20 \times 10^{-4}\,\mathrm{m^2\,s^{-1}}$
Operating temperature, T (K) of apparatus	1115K (842 C)
$Zn(g) + \frac{1}{2}O_2\,(g) = ZnO(s);\ \Delta G_T^0 = 115,420 + 10.35T\log(T) - 82.38T$	

Vapour pressure of zinc is given by:

$$\log(P_{Zn})\,(\mathrm{mm\,Hg}) = \frac{-6\,620}{T} - 1.255\log(T) + 12.34.$$

(14)   Thin-strip casting of metals

There are various methods by which molten metals can be cast directly into the form of thin strip from a melt. These techniques include vertical twin roll

'Bessemer' casting, in which liquid metal is poured into the space between two water-cooled rolls, contra-rotating, so as to expel solidified strip from the 'nip' or separating space, between the two rolls. An alternative is to feed liquid metal between two contra-rotating water cooled belts (i.e. Hazelett twin belt casters), whose separation distance, and belt velocities, are adjusted so as to ensure complete solidification of the strip prior to exit.

The figure illustrates a possible metal delivery system to a water-cooled belt, block, or wheel caster arrangement.

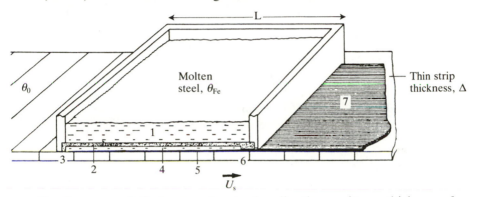

Develop an analytical expression to describe the maximum thickness of solidified shells, that can be formed as a function of t, the time of contact of solidifying metal with a substrate of infinite thermal capacity and conductivity. It should take the form

$$\Delta = k \, t^{1/2}$$

where k is a constant depending on the physical and thermal properties of the metal solidifying to this strip. A typical value for k would be 20 mm/min$^{1/2}$ for steel on a water cooled substrate. (Hint: assume linear temperature gradients within the freezing strip).

Given a contact length L of 1 m, and a substrate velocity, U_s, of $1 \, \text{m s}^{-1}$, deduce the final thickness, Δ, of solid strip exiting from the delivery system.

(15) Enhanced corrosion of galvanized steel by acid rain

When atmospheric oxygen dissolves in acidified water, zinc is subject to corrosive attack according to the cathodic reactions:

$$O_2 + 2H_2O + 4e \rightarrow 4(OH)^-,$$

$$2H^+ + 2e \rightarrow H_2$$

which causes a corresponding anodic dissolution of zinc

$$Zn \rightarrow Zn^{2+} + 2e.$$

Consider the case of a galvanized steel tank for collecting rainwater, whose rate of corrosion is controlled by *cathodic concentration polarization* (i.e. the transport of dissolved O_2 and H^+ ions through a mass transfer boundary layer adjacent to the tank's inside surface governs corrosion kinetics). Under normal conditions, the water is neutral at pH 7, but acid rain (H_2SO_4) then reduces this to 4 (North Eastern U.S. 1988). Estimate the respective corrosion rate of zinc ($mg/Dm^2/day$ and ipy (inches penetration per year)), given the following information:

Data

Density of zinc, Atomic weights (see Appendix)
Mass transfer coefficient for O_2 transport to tank walls $= 3.3 \times 10^{-6}\,m\,s^{-1}$
Mass transfer coefficient for H^+ transfer to tank walls $= 5.0 \times 10^{-5}\,m\,s^{-1}$
$pH = -\log(C_{H^+})$, C_{H^+} (g-ions/litre); pH $= 4$ (acidic).
Oxygen content of rainwater $= 2.68 \times 10^{-4}$ molar

5.11. CONCLUSIONS

This chapter has concerned itself with interphase heat and mass transfer between solids and fluids, liquids and gases, liquids and liquids, gases and gases, when the fluids are in motion. Obviously, the few examples and exercises presented can only illustrate a tiny fraction of situations in which interphase transport processes occur in metallurgical operations. Nevertheless, it is hoped that a sufficient component of the basic techniques and approaches for analysing such phenomena has been provided and will be helpful. Much of the research literature of recent years has been concerned with such phenomena, and an enormous number of publications have been presented, in periodicals including:

Archiv. für das Eisenhuttenwesen of V.D.eh
Canadian Metallurgical Quarterly, Transactions of C.I.M.
Iron and Steel Institute of Japan, Transactions of I.S.I.J.
Ironmaking and Steelmaking, Transactions of The Metals Society
Metallurgical Transactions, Sections A & B of A.I.M.E.
Scandinavian Journal of Metallurgy, Jernkonteret
Stahl and Eisen of V.D.eh

Subjects include gas flows through: pellets, packed beds, sinter, and sinter strands, blast furnaces, kilns, and fluidized beds; gas flows through liquids in reactor vessels; pneumatic conveying of solid powder into liquid metals, flows in metal casting operations, gating practices; electromagnetic stirring, metal/

alloy solidification processes in static and continuous casting operations, and a host of others concerned with metal finishing operations.

FURTHER READING

1. Holman, J. P. (1981). *Heat transfer* (5th edn). McGraw-Hill, New York.
2. Kreith, F. and Black, W. (1980). *Basic heat transfer*. Harper and Row, New York.
3. Bird, R., Stewart, W., and Lightfoot, E. (1966). *Transport phenomena*. Wiley, New York.
4. Darken, L. S. and Gurry R. W. (1953). *Physical chemistry of metals*. McGraw-Hill, New York.
5. Welty, J. R., Wicks, C. E., and Wilson, R. E. (1976). *Fundamentals of momentum, heat and mass transfer* (2nd edn). John Wiley, New York.

6

NUMERICAL TECHNIQUES AND COMPUTER APPLICATIONS

6.0. INTRODUCTION

Numerical methods for solving differential equations have become a very powerful weapon in the armoury of the practising engineer, following the widespread availability of high-speed digital computers since the early 1960s. Prior to that time, realistic solutions to commonplace engineering problems involving heat, mass, or momentum transfer were not easily obtained, except in those particular cases in which the boundary conditions were 'well set' and analytical solutions were available. Some of these, such as the error function solutions and series solutions have already been presented in preceding chapters.

The purpose of the present chapter is to illustrate the basic methods involved in the numerical solution of complex fluid/heat/mass transfer problems and to demonstrate the relative ease with which solutions may often be obtained with a minimum of programming difficulty.

Following a brief guide to the subject, therefore, some typical metallurgical case histories drawn from the author's areas of research are included to demonstrate the value these techniques have for the present and future design of metallurgical processing vessels and processing operations.

6.1. BASIC CONCEPTS

(a) Taylor's series approach

If a differential equation describing a physical phenomenon is of the form (say)

$$\frac{dy}{dx} = f(x), \tag{6.1}$$

then one can approximate this numerically by the equation

$$\frac{\Delta y}{\Delta x} \simeq f(x), \tag{6.2}$$

in which the continuous function dy/dx has been *discretized* between the interval Δx according to

$$y^{+} - y = f(x)\Delta x. \tag{6.3}$$

Here y^+ represents the new value of y being sought at $x + \Delta x$, while y represents the 'old' value of y at x.

If we know our starting point values (x_0, y_0) for x and y, eqn (6.3) may be used on an iterative basis to calculate values for y at each successive increment in Δx. This procedure is termed numerical integration.

To illustrate the point, we will take a specific example. Consider the first-order ordinary differential equation

$$\frac{dy}{dx} = A + Bx, \tag{6.4}$$

for which we wish to obtain a numerical integration over the range $x = 0$ to $x = 5$.

Its approximate form for numerical integrations can be written as

$$\Delta y = (A + Bx)\Delta x, \tag{6.5}$$

or, as in eqn (6.3) as

$$y^+ - y = (A + Bx)\Delta x. \tag{6.6}$$

Taking the simplest case of $A = B = 1$, for ease of computation and numerical integration,

$$y^+ = y + (1 + x)\Delta x. \tag{6.7}$$

Using eqn (6.6) on an iterative basis, we can use our simple discretization formula to write

$$
\begin{array}{ll}
y_1 = y_0 + (A + Bx_0)\Delta x, & x_1 = x_0 + \Delta x, \\
y_2 = y_1 + (A + Bx_1)\Delta x, & x_2 = x_1 + \Delta x, \\
y_3 = y_2 + (A + Bx_2)\Delta x, & x_3 = x_2 + \Delta x, \\
y_4 = y_3 + (A + Bx_3)\Delta x, & x_4 = x_3 + \Delta x, \\
y_5 = y_4 + (A + Bx_4)\Delta x, & x_5 = x_4 + \Delta x, \\
\quad\vdots & \quad\vdots \\
y_{i+1} = y_i + (A + Bx_i)\Delta x, & x_{i+1} = x_i + \Delta x.
\end{array}
$$

We may now proceed to a numerical solution of eqn (6.6), choosing first an incremental step of $\Delta x = 1$. Thus, rewriting our general equation with $A = B = 1$, we have for the particular case in hand,

$$y_{i+1} = y_i + (1 + x_i)\Delta x \qquad x_{i+1} = x_i + \Delta x, \qquad \Delta x = 1.0.$$

Adopting our starting conditions and coefficients: $y_0 = 0$ $x_0 = 0$, $A = 1, B = 1$, we can construct Table 6.1 proceeding in a stepwise manner from $x = 0$ to $x = 5$. Taking a smaller increment in Δx, we can prepare another table showing the iterative calculations performed. These are shown in Table 6.2. These 'hand' solutions are shown in Table 6.3 for comparison with the

TABLE 6.1. *Numerical integrations of eqn (6.6), taking discretization interval $\Delta x = 1$*

Iteration number	Value y	Value x
1	$y^+ = \ 0 + (1+0)1 \ \to \ 1$	$x^+ = 0.0 + 1.0 \to 1.0$
2	$y^+ = \ 1 + (1+1)1 \ \to \ 3$	$x^+ = 1.0 + 1.0 \to 2.0$
3	$y^+ = \ 3 + (1+2)1 \ \to \ 6$	$x^+ = 2.0 + 1.0 \to 3.0$
4	$y^+ = \ 6 + (1+3)1 \ \to 10$	$x^+ = 3.0 + 1.0 \to 4.0$
5	$y^+ = 10 + (1+4)1 \ \to 15$	$x^+ = 4.0 + 1.0 \to 5.0$

TABLE 6.2. *Numerical integrations of eqn (6.6), taking discretization interval $\Delta x = 0.5$*

Iteration number	Value y	Value x
1	$y^+ = \ 0.00 + (1+0)0.5 \ \ \to \ \ 0.50$	$x^+ = 0.0 + 0.5 \to \ 0.5$
2	$y^+ = \ 0.5 \ +(1+0.5)0.5 \ \to \ \ 1.25$	$x^+ = 0.5 + 0.5 \to \ 1.0$
3	$y^+ = \ 1.25 + (1+1)0.5 \ \ \to \ \ 2.25$	$x^+ = 1.0 + 0.5 \to \ 1.5$
4	$y^+ = \ 2.25 + (1+1.5)0.5 \ \to \ \ 3.50$	$x^+ = 1.5 + 0.5 \to \ 2.0$
5	$y^+ = \ 3.50 + (1+2)0.5 \ \ \to \ \ 5.00$	$x^+ = 2.0 + 0.5 \to \ 2.5$
6	$y^+ = \ 5.0 \ +(1+2.5)0.5 \ \to \ \ 6.75$	$x^+ = 2.5 + 0.5 \to \ 3.0$
7	$y^+ = \ 6.75 + (1+3.0)0.5 \ \to \ \ 8.75$	$x^+ = 3.0 + 0.5 \to \ 3.5$
8	$y^+ = \ 8.75 + (1+3.5)0.5 \ \to 11.00$	$x^+ = 3.5 + 0.5 \to \ 4.0$
9	$y^+ = 11.00 + (1+4)0.5 \ \ \to 13.50$	$x^+ = 4.0 + 0.5 \to \ 4.5$
10	$y^+ = 13.50 + (1+4.5)0.5 \ \to 16.25$	$x^+ = 4.5 + 0.5 \to \ 5.0$

TABLE 6.3. *Solution of eqn (6.4) via numerical and analytical routes—summary of results*

x	y, ($\Delta x = 1$)	y, ($\Delta x = 0.5$)	y, Analytical
0	0	0	0
1	1	1.25	1.5
2	3	3.5	4.0
3	6	6.75	7.5
4	10	11.0	12.0
5	15	16.25	17.5

N.B. $y_i - y(x_i) = \varepsilon_i = $ local truncation error.

analytical solution of

$$\frac{dy}{dx} = 1 + x,$$

this being, of course,

$$y = x + x^2/2, \tag{6.8}$$

with $C = 0$ ($y = 0$ at $x = 0$).

We see immediately from Table 6.3 that the finer the increment in Δx, the more closely our numerically estimated y values approach their analytical counterparts. Since hand calculations using smaller Δx's than those chosen become too time-consuming, it would be appropriate at this stage to write a small computer program to calculate y values for Δx increments of 0.001.

The reason for the improved accuracy as integration step length Δx is decreased is evident from Fig. 6.1(a). There, the functional relationship between y and x has been graphed, and the curve's representation by a series of straight line extrapolations between each discretization interval Δx is reproduced according to the values appearing in Tables 6.1 and 6.2:

Ideally, we want our numerical predictions to fall exactly on the curve representing the analytical solution as shown in Fig. 6.1(a). The fact that the discrete y values do not generally do so is the result of approximating the slope dy/dx by $\Delta y/\Delta x$ the first term of a Taylor's expansion.

The error between y_i and $y(x_i)$ is known as the local truncation error ε_i, which must be minimized as far as possible.

The method just outlined represents a primitive *explicit* scheme, known as Euler's method, in which the forward value of y is explicitly defined in relation to known values of x and $\hat{y}$ at the previous integration step. Its accuracy can be improved upon if second and higher order derivatives are incorporated into the solution algorithm. For example, if we expand the Taylor's series approximating $f(x)$, we can write, in general, where h is the incremental step change

$$y(x_0 + h) = y(x_0) + hf(x_0, y(x_0)) + \frac{h^2}{2!}f'(x_0, y(x_0)) + \frac{h^3}{3!}f''(x_0, y(x_0)) + O(h^4)$$

$$\tag{6.9}$$

or in the notation used for eqn (6.3),

$$y^+ - y = f(x)\Delta x + f'(x)\frac{\Delta x^2}{2} + f''(x)\frac{\Delta x^3}{6} + O(\Delta x^4) \tag{6.10}$$

where

$$f'(x) = \frac{d^2y}{dx^2}, \quad f''(x) = \frac{d^3y}{dx^3}, \quad \text{etc.}$$

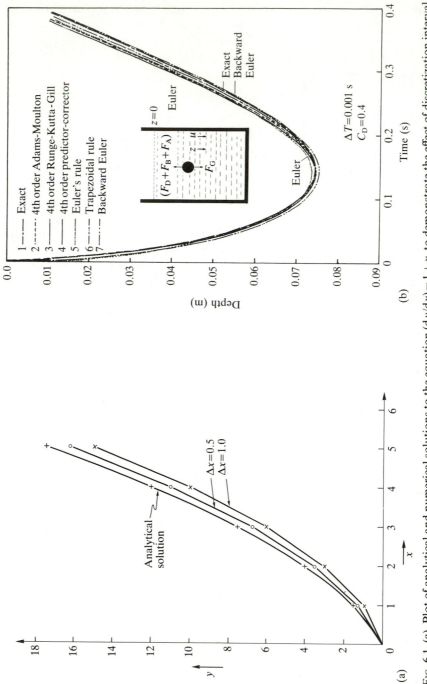

FIG. 6.1 (a) Plot of analytical and numerical solutions to the equation $(dy/dx) = 1 + x$, to demonstrate the effect of discretization interval Δx. (b) Performance of various algorithms (explicit and implicit) in the numerical solution to the particle trajectory equations.

Provided Δx is very small, these second, third, and fourth order terms become negligible as we have just demonstrated. Furthermore, the algorithms needed to estimate second and higher order derivatives lead to increasingly cumbersome solution procedures (algorithms). For instance, in evaluating d^2y/dx^2, the slopes at the start and end of the integration interval are needed; i.e.

$$\frac{d^2 y}{dx^2} \simeq \frac{\Delta(\Delta y/\Delta x)}{\Delta x} = \frac{\left(\dfrac{\Delta y}{\Delta x}\right)^+ - \left(\dfrac{\Delta y}{\Delta x}\right)}{\Delta x} \tag{6.11}$$

Similarly, for a third-order Euler technique, one would need to choose three points over the integration step, in order to estimate d^3y/dx^3. Thus the solution of a differential equation by direct Taylor's expansion of the object function becomes impractical for all but the simplest of equations (i.e. eqn (6.4)), if higher order derivatives are called for.

(b) Runge–Kutta techniques

Fortunately, it is possible to develop one-step procedures which involve only first order slope evaluations, but which are also as accurate as higher order Taylor formulae. These algorithms are known as the Runge–Kutta methods and have the form

$$y^+ = y + \Delta x \phi(x, y, \Delta x) \tag{6.12}$$

where ϕ is known as the increment function. The most widely used is the fourth order Runge–Kutta algorithm available in most computer program libraries. For completeness, the algorithm and its constants are given below

$$y^+ = y + \frac{\Delta x}{6} [k_1 + 2k_2 + 2k_3 + k_4] \tag{6.13a}$$

where

$$k_1 = f[x, y], \quad k_2 = f\left[x + \frac{\Delta x}{2}, \quad y + \frac{\Delta x}{2} k_1\right] \tag{6.13b}$$

$$k_3 = f\left[x + \frac{\Delta x}{2}, \quad y + \frac{1}{2} \Delta x k_2\right] \tag{6.13c}$$

and

$$k_4 = f[x + \Delta x, \quad y + \Delta x k_3] \tag{6.13d}$$

As Runge–Kutta formulae can be used to solve simultaneous differential equations, they can be applied to higher order differential equations, since one can make 'n' first-order differential equations from a single nth-order differential equation. For example, in the second order differential equation

$$\frac{d^2 y}{dx^2} = f\left(x, y, \frac{dy}{dx}\right) \tag{6.14}$$

one can let

$$z = \frac{dy}{dx} \qquad (6.15)$$

so that

$$\frac{dz}{dx} = \frac{d^2y}{dx^2} \qquad (6.16)$$

and the equivalent two first-order differential equations become

$$\frac{dz}{dx} = f(x, y, z) \quad \text{and} \quad \frac{dy}{dx} = z = g(x, y, z) \qquad (6.17a, b)$$

A typical metallurgical example illustrating the solution of a first-order ODE (ordinary differential equation), and for which a variety of algorithms has been tested, is shown in Fig. 6.1(b). There, the depth of immersion of a sphere (see worked example in Section 3.6) projected vertically into a steel bath is plotted as a function of time. Provided a sufficiently small time-step is taken, we see that the simple Euler method can essentially match higher-order techniques (Mazumdar and Guthrie 1988).

(c) Example: program for laminar boundary layer flows

Develop a program to solve the differential equations describing the boundary layer over a flat plate using the Runge–Kutta numerical integration procedure.

Solution

For thin boundary layers (see Figure 6.2) at high Reynolds values, the shear stresses within the boundary layer are closely approximated by $\mu(\partial U/\partial y)$, while the normal shear stress terms σ_{xx} and σ_{yy} are equal and opposite to the local static pressure, P. Further, pressure gradients normal to the plate are much less than those along the length of the plate and can be ignored. When

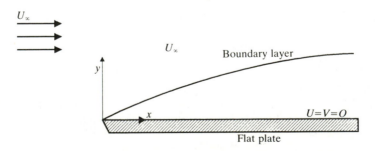

FIG. 6.2. Developing boundary layer for laminar flow of liquid over a flat plate.

these simplifications are made to the general Navier–Stokes equations, the resultant differential equation for steady, laminar incompressible boundary-layer flow parallel to a flat plate, can be written as

$$U\frac{\partial U}{\partial x} + V\frac{\partial U}{\partial y} = v\frac{\partial^2 U}{\partial y^2} \tag{1}$$

which must be solved in conjunction with the appropriate continuity equation,

$$\frac{\partial U}{\partial x} + \frac{\partial V}{\partial y} = 0 \tag{2}$$

subject to the boundary conditions that $U = V = 0$ at the plate surface, and that: $U = U_\infty$ at $y \to \infty$. Since the flow is two dimensional, it is convenient to introduce the stream function Ψ, thereby satisfying continuity automatically, as discussed in Chapter 2. Further, Blasius (1908) showed that by transforming the independent x, y variables to the dimensionless distance $\eta = (y/2)(U_\infty/vx)^{1/2}$ and the dependent variables U, V contained within $\Psi(x, y)$ to the dimensionless equivalent $f(\eta)$, where $f(\eta)$, for $f = \Psi/(vxU_\infty)^{1/2}$, then:

$$U = \frac{\partial \Psi}{\partial y} = \frac{U_\infty}{2}f' \tag{3}$$

$$V = -\frac{\partial \Psi}{\partial x} = \frac{1}{2}\left(\frac{vU_\infty}{x}\right)^{1/2}(\eta f' - f) \tag{4}$$

$$\frac{\partial U}{\partial x} = -\frac{U_\infty \eta}{4x}f'', \tag{5}$$

$$\frac{\partial U}{\partial y} = \frac{U_\infty}{4}\left(\frac{U_\infty}{vx}\right)^{1/2}f'' \tag{6}$$

and

$$\frac{\partial^2 U}{\partial y^2} = \frac{U_\infty}{8}\frac{U_\infty}{vx}f''' \tag{7}$$

Substitution for U, $\partial U/\partial x$, V, $\partial U/\partial y$, and $\partial^2 U/\partial y^2$ in eqn (1) yields a third-order, non-linear, ordinary differential equation (ODE),

$$f''' - ff'' = 0, \tag{8}$$

subject to the boundary conditions that $f = f' = 0$ at $\eta = 0$ (the plate's surface) and $f' = 2$ at $\eta \to \infty$. (This condition derives from eqn (3) where we see $f' = 2$ when $U = U_\infty$). Equation (8) and its boundary conditions form a two point boundary value problem which can be solved via a standard finite difference scheme, or can be treated as an initial value problem. Owing to the difficulty of setting up a suitable grid of appropriate spacing and extent for a finite difference scheme, the latter approach is favoured and is treated here.

Addressing the problem set then, eqn (8) must be solved via the Runge–Kutta method. Converting eqn (8) into three first-order ODEs, in which we define $y_1 = f$, $y_2 = f'$, and $y_3 = f''$ where f is a function of η, we have:

$$\frac{dy_1}{d\eta} = f' = y_2 \tag{9}$$

$$\frac{dy_2}{d\eta} = f'' = y_3 \tag{10}$$

and

$$\frac{dy_3}{d\eta} = f''' = -ff'' = -y_1 y_3 \tag{11}$$

These three coupled ODEs must be solved simultaneously, subject to $y_1(0) = 0$, $y_2(0) = 0$, and $y_3(0) = f''(0) =$ unknown and $y_2(\infty) = 2$.

The flow chart of the program 'Laminar boundary layer', written to solve this set of equations as a function of η (or X in program), is provided in Table 6.4 and listed in Table 6.5.

To start the integration, we make an initial guess, g, at the value of $f''(0)$. Thus we know that f' varies between 0 and 2 between the plate surface and edge of the boundary layer, and by experience we learn that the boundary layer would probably be contained within the range of $\eta = 0$ to $\eta = 3$. Assuming a linear relationship between f' and η we see the slope, $f'' \sim (2.0)/3 \simeq 0.66$. In the program, g was set equal to 0.8 as a first guess.

We then solve the set of equations to see if the condition at $\eta \to \infty$, i.e. $f' = 2$, is satisfied with the desired degree of accuracy (ERROR). Assuming not, we must re-guess $f''(0)$. Assuming the error (r) is a monotonous function of the guessed value:

$$r = r_1 + \left(\frac{\partial r}{\partial g}\right) \Delta g, \tag{12}$$

then we can drive the error towards zero ($r \to 0$) by incrementing g, according to

$$g_2 = g_1 - r_1/(\partial r/\partial g). \tag{13}$$

In the program Y3IS and Y3I represent successive values of our guess for $f''(0)$ while Y2ERR1, and Y2ERR represent corresponding values of the error r in the η_∞ boundary condition where $f'(\infty) = 2$. Further details of the program can be inspected in Table 6.5. A typical output is given in Table 6.6 for the input data noted. There it is seen that a converged solution satisfying the boundary conditions $f = f' = 0$ at $\eta = 0$ and $f' = 2$ at $\eta \to \infty$ was reached after six iterations around $\eta = 0$, following which values for X were incremented from zero to 8 over intervals of 0.5 and corresponding values for Y1, Y2, and Y3 provided.

Recalling that a momentum boundary layer thickness, δ, is defined such that $U(x, y) = 0.99 U_\infty$, eqn (3) shows us that $f' = 2 \times 0.99 U_\infty / U_\infty = 1.98$. Referring to Table 6.5's output data (Table 6.6), we see that f' (or y_2) = 1.98 at $X = 2.5$.

Remembering $X = (y/2)(U_\infty / vx)^{1/2}$, we can then solve for δ according to

$$2.5 = \frac{\delta}{2} \sqrt{(U_\infty / vx)} \tag{14}$$

or

$$\delta = 5 \sqrt{(vx/U_\infty)} \quad \text{or} \quad \delta / x = 5 \sqrt{(v / U_\infty x)} = 5(\text{Re}_x)^{-1/2} \tag{15}$$

This result has already been quoted in Chapter 5 (eqn (5.5)). The solution also allows us to deduce wall shear stresses and velocity profiles.

(d) Exercise: thermal boundary layer analysis

As an exercise, extend the similarity solutions for the laminar momentum boundary layer developed above, to the treatment of thermal boundary layers, in which the plate is at a uniform temperature, θ_0, and the bulk liquid, flowing at a velocity of U_∞, is at a uniform temperature, θ_∞. The relevant form of the energy equation (where viscous energy dissipation is ignored) is:

$$U \frac{\partial \theta}{\partial x} + V \frac{\partial \theta}{\partial y} = \alpha \frac{\partial^2 \theta}{\partial y^2}, \tag{16}$$

subject to the boundary conditions:

$$\theta(x, 0) = \theta_0, \quad \theta(0, y) = \theta_\infty, \quad \theta(x, \infty) = \theta_\infty.$$

Normalizing temperature according to $\Theta = (\theta - \theta_0)/(\theta_\infty - \theta_0)$, and adopting the variables for dimensionless distance $\eta = (y/2)(U_\infty / vx)^{1/2}$, and $f = \Psi / (vx U_\infty)^{1/2}$, show that eqn (16) can be reduced to:

$$\Theta'' + \text{Pr} \cdot f \cdot \Theta' = 0, \tag{17}$$

with the corresponding boundary conditions:

$$\Theta(0) = 0 \text{ at } \eta = 0, \text{ and } \Theta(\infty) = 1 \text{ at } \eta = \infty,$$

where $\text{Pr}(= v/\alpha)$ is the Prandtl number.

Hints. The major problem in solving eqn (2) is that

$$\left. \frac{d\theta}{dy} \right|_{y=0} \quad \{\equiv \theta'(0)\}$$

is unknown and must be solved for during the numerical integration procedures. Recognize that the influence of the momentum boundary layer appears through the presence of f and that $f(\eta)$ and its derivatives have already been solved.

Therefore extended the 'boundary layer program' to include the solution of eqn (17), using Prandtl values of 0.001, 0.01, 0.1, and 1.0. One can also show

TABLE 6.4. *Flow chart of 'Laminar boundary layer' program.*

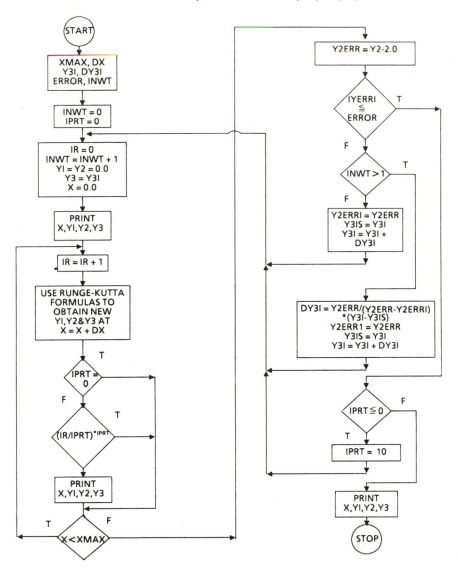

TABLE 6.5. *Listing of a program named 'Laminar boundary layer' for solving the boundary layer flow equations via the Runge–Kutta numerical integration technique*

```
C BLASIUS SOLUTION FOR LAMINAR FORCED CONVECTIVE FLOW OVER A FLAT
C PLATE. THE RESULTANT THIRD ORDER NONLINEAR ORDINARY
C DIFFERENTIAL EQUATION IS SOLVED AS AN INITIAL VALUE PROBLEM USING A
C FOURTH-ORDER RUNGE-KUTTA SCHEME. A SHOOTING SCHEME, WHICH IN TURN
C USES NEWTON'S SCHEME, IS USED FOR UPDATING INITIAL GUESS VALUES
C DURING ITERATIONS.
C
C K1, K2, K3, K4 ; L1, L2, L3, L4 ; M1, M2, M3, M4 ; REPRESENT
C      RUNGE-KUTTA VARIABLES.
C INWT - NUMBER OF ITERATIONS REQUIRED FOR CONVERGENCE
C IPRT - ITERATION COUNTER FOR PRINT
C IR   - NUMBER OF INTEGRATION STEPS IN RUNGE-KUTTA SCHEME
C X    - INDEPENDENT VARIABLE
C XMAX - MAXIMUM VALUE OF THE INDEPENDENT VARIABLE
C DX   - INCREMENT OF X FOR RUNGE-KUTTA SCHEME
C Y1   - DEPENDENT VARIABLE
C Y2   - FIRST DERIVATIVE OF Y1
C Y3   - FIRST DERIVATIVE OF Y2 OR SECOND DERIVATIVE OF Y1
C Y3I  - UPDATED VALUES OF Y3 AT X = 0 DURING ITERATIONS TOWARDS
C               A CONVERGED SOLUTION
C DY3I - INCREMENT FOR Y3I NEEDED FOR NEWTON'S METHOD
C ERROR - TOLERABLE ERROR FOR Y2ERR
C Y2ERR - DIFFERENCE BETWEEN THE GIVEN BOUNDARY VALUE AT
C        XMAX FOR Y2 AND THE CALCULATED VALUE
C Y2ERR1 - STORED Y2ERR REQUIRED FOR NEWTON'S METHOD
C
        REAL K1,K2,K3,K4,L1,L2,L3,L4,M1,M2,M3,M4
        READ(5,*) DX, XMAX, Y3I, DY3I, ERROR, INWT
        INWT = 0
        IPRT = 0
        WRITE(6,1)
1    FORMAT('1',6X,'INITIAL VALUES DURING ITERATION UNTIL CONVERGED',/
     1/9X,'X',11X,'Y1',11X,'Y2',11X,'Y3',//)
2    IR = 0
        INWT = INWT + 1
        Y1  = 0.0
        Y2  = 0.0
        Y3  = Y3I
        X   = 0.0
C
C PRINT THE RESULTS WHEN CONVERGED ( IPRT = 1 )
        WRITE(6,4) X, Y1, Y2, Y3
4    FORMAT(5X, F8.3, 3(2X,F10.5),//)
C START INTEGRATION OF THE EQUATIONS
C
8   IR = IR + 1
C START OF FOURTH-ORDER RUNGE-KUTTA SCHEME
C
        K1 = Y2
        L1 = Y3
        M1 = -Y1*Y3
C
```

(Continued)

TABLE 6.5. (*Continued*)

```
        K2 = Y2 + L1*DX/2.0
        L2 = Y3 + M1*DX/2.0
        M2 = -(Y1 + K1*DX/2.0)*(Y3 + M1*DX/2.0)
C
        K3 = Y2 + L2*DX/2.0
        L3 = Y3 + M2*DX/2.0
        M3 = -(Y1 + K2*DX/2.0)*(Y3 + M2*DX/2.0)
C
        K4 = Y2 + L3*DX
        L4 = Y3 + M3*DX
        M4 = -(Y1 + K3*DX)*(Y3 + M3*DX)
C
        Y1 = Y1 + (K1 +2.0*K2 + 2.0*K3 + K4)/6.0*DX
        Y2 = Y2 + (L1 + 2.0*L2 + 2.0*L3 + L4)/6.0*DX
        Y3 = Y3 + (M1 + 2.0*M2 + 2.0*M3 + M4)/6.0*DX
        X = X + DX
C
        IF(IPRT .EQ. 0) GO TO 10
        IF((IR/IPRT)*IPRT .NE. IR) GO TO 10
        WRITE(6,4) X, Y1, Y2, Y3
10    IF( X .LT. XMAX) GO TO 8
C
C CHECK BOUNDARY CONDITION AT XMAX FOR ERROR=Y2ERR
        Y2ERR = Y2 - 2.0
        IF(ABS(Y2ERR) .LE. ERROR) GO TO 50
C
C STORE VALUES FOR NEWTON ITERATION
C
        IF(INWT .GT. 1) GO TO 30
        Y2ERR1 = Y2ERR
        Y3IS = Y3I
        Y3I = Y3I + DY3I
        GO TO 2
C
C NEWTON STEP FOR CALCULATING IMPROVED MISSING VALUE OF Y3I
C
 30   DY3I = -Y2ERR/(Y2ERR - Y2ERR1)*(Y3I - Y3IS)
        Y2ERR1 = Y2ERR
        Y3IS = Y3I
        Y3I = Y3I + DY3I
        GO TO 2
C
C ITERATION LOOP ENDS. FOR PRINT OUT INTEGRATION IS AGAIN CARRIED OUT.
C
 50   IF(IPRT) 55, 55, 60
 55   IPRT = 10
        WRITE(6,58)
 58   FORMAT(/5X,'AFTER CONVERGENCE RESULTS FROM X = 0 TO X = XMAX',//)
        GO TO 2
 60   WRITE(6,70) INWT
 70   FORMAT(//,6X,'NUMBER OF ITERATIONS = ',I3)
        STOP
        END
```

TABLE 6.6. *Output data generated by 'Laminar boundary layer' program.*

$DATA
0.050 5.0 0.8 0.20 1.E-06 10

INITIAL VALUES DURING ITERATIONS UNTIL CONVERGED

X	Y1	Y2	Y3
0.000	0.00000	0.00000	0.80000
0.000	0.00000	0.00000	1.00000
0.000	0.00000	0.00000	1.30145
0.000	0.00000	0.00000	1.32707
0.000	0.00000	0.00000	1.32826
0.000	0.00000	0.00000	1.32827

CONVERGED RESULTS FROM X = 0 TO X = XMAX

0.000	0.00000	0.00000	1.32827
0.500	0.16558	0.65958	1.29206
1.000	0.65004	1.25956	1.00702
1.500	1.39683	1.69212	0.64544
2.000	2.30578	1.91106	0.25693
2.500	3.28331	1.98310	0.06363
3.000	4.27966	1.99796	0.00961
3.500	5.27928	1.99985	0.00088
4.000	6.27925	2.00000	0.00005
4.500	7.27925	2.00000	0.00000
5.000	8.27924	2.00000	0.00000

analytically, through integration of eqn (17), that

$$\Theta'(0) = 1 \bigg/ \int_0^\infty \left(\frac{f''}{1.328}\right)^{\mathrm{Pr}} \mathrm{d}\eta$$

and this should be compared with your numerical solutions for various Pr.

6.2. INTEGRAL CONTROL VOLUME APPROACH AND CONSERVATION

An alternative route to the discretization of differential equations and their numerical integration via a Taylor's series expansion, is the control, or integral, volume approach. This discretization technique is more favoured by engineers, in that it is based on physical reasoning and is relatively easy to apply to the complex partial differential equations typically needed to describe systems of engineering interest.

Consider therefore Fig. 6.3(a), where a series of control volumes have been drawn across the central region of a cooling ingot. Nodal points have been drawn at the centroids of these elements. The faces of the elements lie midway between the nodal points at $i \pm \frac{1}{2}$. Recalling the governing partial differential equation (PDE) for transient heat conduction in the presence of a heat source, or sink, (for example):

$$\rho C_P \frac{\partial \theta}{\partial t} = \nabla(k\nabla\theta) + \dot{q}''', \tag{6.18}$$

then the integral form of this equation can be written for the control volume element, of volume δV, as:

$$\int_{\delta V} \rho C \frac{\partial \theta}{\partial t} = \int_{\delta V} \nabla \cdot (k\nabla\theta)\,\mathrm{d}V + \int_{\delta V} \dot{q}'''\,\mathrm{d}V. \tag{6.19}$$

Reversing the order of differentiation and integration in the first term, assuming ρC to be constant, and using Gauss's theorem, leads to

$$\rho C \frac{\partial}{\partial t} \int_{\delta V} \theta\,\mathrm{d}V = \int_{\delta S} (k\nabla\theta) \cdot \mathrm{d}S + \int_{\delta V} \dot{q}'''\,\mathrm{d}V, \tag{6.20}$$

where δS denotes the bounding surface of the volume element. Provided the volume element is sufficiently small one can ignore variations in temperature within its volume and suppose that θ represents the average or mean temperature, and that $\dot{q}'''$ represents the mean rate of heat generation, within the node. Similarly, for the case illustrated in Fig. 6.3(a), we have one-dimensional heat flow in the x direction, so that the surface area normal to heat flow is, (for unit width) $\Delta y(1)$ for conduction into, and out of, the volume

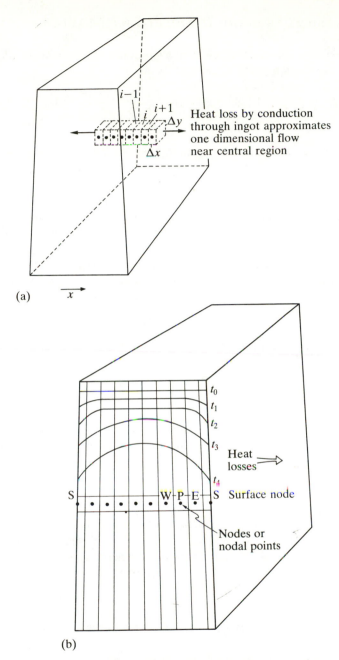

(a)

(b)

FIG. 6.3. (a) Nodal elements of interest in ingot cooling: approximate one-dimensional conduction of heat: owing to the much larger width and height of the ingot, the dominant heat flow is through the thickness of the ingot. (b) Temperature–time distributions within ingot, showing choice of nodal points and their associated differential volume elements.

element; no other parts of the element's surfaces need be considered. Under these conditions, eqn (6.20) would reduce to:

$$\rho C \Delta x \Delta y(1) \frac{\partial \theta}{\partial t} = -\int_{i-\frac{1}{2}, \Delta y(1)} k \frac{\partial \theta}{\partial x} \cdot dy + \int_{i+\frac{1}{2}, \Delta y(1)} k \frac{\partial \theta}{\partial x} dy + \dot{q}''' \Delta x \, \Delta y(1).$$

$$(6.21)$$

The resultant two surface integrals are formed in such a way that the heat flux crossing each face is conserved in transport between volumes. Evidently therefore, the heat flux entering node i through the $\Delta y(1)$ surface at $i-\frac{1}{2}$ must be evaluated, numerically, in exactly the same way as the heat flux leaving the i volume element through the $i+\frac{1}{2}$ surface. This dual aspect of conservation in transport and of conservation within *a control volume element* is a key feature of the control volume approach. In the example taken above, the surface integrals would be estimated numerically, by estimating temperature gradients on the basis of adjacent nodal point temperatures, according to

$$\frac{\partial \theta}{\partial t}\bigg|_{i-\frac{1}{2}} \simeq \frac{\theta_i - \theta_{i-1}}{\Delta x} \quad \text{and} \quad \frac{\partial \theta}{\partial x}\bigg|_{i+\frac{1}{2}} \simeq \frac{\theta_{i+1} - \theta_i}{\Delta x}.$$

The final discretization formula would therefore read

$$\rho C \Delta x \Delta y(1) \frac{\Delta \theta}{\Delta t} = -k \Delta y(1) \frac{(\theta_{i-1} - \theta_i)}{\Delta x} + k \Delta y(1) \frac{(\theta_{i+1} - \theta_i)}{\Delta x} + \dot{q}''' \Delta x \Delta y(1).$$

$$(6.22)$$

As mentioned, the control-volume approach is based on physical reasoning and is generally easy to apply. It is the recommended approach when finite difference approximations (FDA) of partial differential equations met in transient heat and mass or fluid flow become difficult to construct. It is also useful for non-uniform mesh spacings and the staggered grid arrangement for fluid flow analyses discussed later in this chapter. Similarly, the nodal points can be placed on the surfaces of the control volume, rather than at their centroids, and this again will be shown to be helpful in treating heat and mass flow through interfaces. The remaining examples in this chapter are devoted to an illustration of the control volume approach.

6.3. TRANSIENT HEAT AND MASS TRANSFER

(a) The explicit-integral control volume approach

In order to illustrate the set-up of an integral control volume method for numerical computations, we can continue with the example of the cooling ingot. Thus, referring to Fig. 6.3(a), in which we wish to predict the rate of temperature decline within a cooling ingot as a result of radiative and conductive heat losses from the ingots' surfaces, we saw that the ingot had been cut into a number of slices, or elements, and a series of prominent *nodal points* placed at their centres for elements inside the ingot, and at the surfaces for half-volume elements adjacent to the ingot's surfaces. The section of interest to us is at the ingot's centre, so that only heat conduction across the thickness of the ingot, i.e., in the x direction, is of any interest or significance (vertical and lateral temperature gradients will be effectively zero). Similarly, heat generation terms caused by phase transformations will be ignored.

Carrying out a local heat balance on an internal volume element at P (see Fig. 6.3(b)), as influenced by its cardinal point neighbours W and E, we can then write,

$$\begin{pmatrix} \text{Heat conducted into P} \\ \text{from the east} \end{pmatrix} - \begin{pmatrix} \text{heat conducted out of} \\ \text{P to the west} \end{pmatrix} = \begin{pmatrix} \text{rate of heat gain} \\ \text{in central volume} \\ \text{element} \end{pmatrix}$$

i.e.

$$k_{EP}\frac{(\theta_E - \theta_P)}{\Delta x} - k_{PW}\frac{(\theta_P - \theta_W)}{\Delta x} = \rho C \Delta x (1 \times 1)\frac{\theta_P^+ - \theta_P}{\Delta t} \qquad (6.23)$$

Here we have assumed unit cross-section area $(\Delta y, \Delta z)$ for the control volume faces through which heat is conducted. Similarly, assuming $k \neq f(\theta)$ for simplicity of representation, we obtain

$$\frac{\alpha \Delta t}{\Delta x^2}(\theta_E - \theta_P) - \frac{\alpha \Delta t}{\Delta x^2}(\theta_P - \theta_W) + \theta_P = \theta_P^+,$$

or

$$\theta_P^+ = \theta_P(1 - 2M) + M\theta_E + M\theta_W. \qquad (6.24)$$

Equation (6.24) represents a numerical approximation to our partial differential equation for transient heat conduction. 'θ_P^+' represents the new temperature of the centre node P at time t_2 following heat conduction into and out of the element over a time period Δt (i.e. $t_2 = t_1 + \Delta t$). We see that θ_P^+ is stated in *explicit form*, since all the quantities appearing on the right of eqn (6.24) are known.

This numerical algorithm can therefore be used in an iterative mode corresponding to that demonstrated for eqn (6.7) to seek a solution to the problem $\theta = f(x,t)$ at $t = t_f$. Provided therefore that the initial conditions $\theta_0 = f(x,0)$ are specified, and the boundary conditions are well set, numerical integration of eqn (6.24) for each internal nodal point can proceed, taking suitable integration time intervals Δt.

In this respect, the time interval which can be taken is not entirely arbitrary. We see from eqn (6.24) that a term $(1 - 2M)$ appears. Numerical stability of the equation, as well as basic laws of thermodynamics, require that the term always remain positive or zero. If $(1-2M)$ is allowed to become negative, one finds, for instance, that a nodal element can be losing heat to its westerly and easterly neighbours while simultaneously heating up in the absence of any heat source! Consequently, one requires that

$$M \leqslant \tfrac{1}{2}, \text{ or } \frac{\alpha \Delta t}{\Delta x^2} \leqslant \tfrac{1}{2}, \text{ or } \Delta t \leqslant \frac{1}{2}\frac{\Delta x^2}{\alpha}. \tag{6.25}$$

Note that the modulus M is equivalent to a local Fourier modulus, $\alpha \Delta t/\Delta x^2$ or Fo, its macroscopic equivalent, $\alpha t/L^2$, being presented in Chapter 4, in connection with Hessler charts, as well as in Table 3.2.

This represents a limitation of the explicit technique, since excessive computer time may be necessary in some cases in which extended heating or cooling times are of interest. For example, if we briefly consider the example presented in Section 4.11, in which we were interested in the transient temperature profiles developed within an ingot of stainless steel cooling under ambient conditions; the ingot was numerically sectioned, or discretized, into slabs of 12 mm width. Given $\alpha = 5 \times 10^{-6} \text{ m}^2 \text{ s}^{-1}$, eqn (6.25) requires that

$$\Delta t \leqslant \frac{1}{2}\frac{(12 \times 10^{-3})^2}{5 \times 10^{-6}},$$

i.e., $\Delta t \leqslant 14.4$ s.

Should one be interested in an operation lasting 10 hours for example, in which transient temperature profiles were needed simultaneously in 10 ingots within a soaking pit, one can see how the number of interations can rapidly escalate ($\sim 10 \times 3600/15 \sim 2400$ time iterations per ingot). For such problems, an alternative scheme, known as the implicit technique, is available.

Returning, for the moment, to the explicit scheme, one needs to develop the numerical equivalents of the boundary condition equations or statements. Consider a surface node, S, a heat balance requires that

Heat loss from surface volume element by convection and radiation	+	Heat input to surface element by conduction	=	rate of heat accumulation

or

$$-h_{\mathrm{t}}(1 \times 1)(\theta_{\mathrm{S}}-\theta_{\infty}) \quad +k(1 \times 1)\frac{(\theta_{\mathrm{S}-1}-\theta_{\mathrm{S}})}{\Delta x} \;=\; \rho C(1 \times 1)\frac{\Delta x}{2}\frac{(\theta_{\mathrm{S}}^{+}-\theta_{\mathrm{S}})}{t_2-t_1},$$

$$\tag{6.26}$$

where we note that the surface node's volume is half that of an internal node. Dividing by $\rho C\Delta x/2$, and rearranging terms, we can then provide an explicit expression for the new surface temperature θ_S^+, following a time interval Δt. Consequently, θ_S^+ can be computed in terms of the known temperatures in the ingot at the previous time t_1. Thus,

$$\theta_S^+ = \theta_S(1 - 2M - 2H) + 2M\theta_{S-1} + 2H\theta_\infty,$$

where θ_∞ is the ambient temperature, $M = (k/\rho C)(\Delta t/\Delta x^2)$, the Fourier modulus Fo, and $H = h_t\Delta t/\rho C\Delta x$, the product of the Fourier and Biot moduli, (Fo)(Bi). Under these circumstances, we see that stability requires

$$(M + H) \leqslant \frac{1}{2}.$$

It should now be evident how this scheme can be built up to consider quite complex heat (or mass) transfer problems of a one-, two- or three-dimensional character.

For a 2D problem:

$$M \leqslant \frac{1}{4} \text{ and } \theta_P^+ = \theta_P(1 - 4M) + M\theta_W + M\theta_E + M\theta_N + M\theta_S. \quad (6.27)$$

For a 3D problem:

$$M \leqslant \frac{1}{6} \text{ and } \theta_P^+ = \theta_P(1 - 6M) + M\theta_W + M\theta_E + M\theta_N + M\theta_S + M\theta_{high} + M\theta_{low}.$$
$$(6.28)$$

Phase transformations, heat sources, thermal contact between different phases, variable interfacial resistances, variable thermal properties and heat flux conditions, cylindrical or spherical geometries, etc., are all relatively easily incorporated into explicit numerical schemes.

As a final note, nodes adjacent to interfaces are best placed at, or on, the interface itself, since better numerical approximations of the governing boundary conditions or equations are thereby ensured.

(b) Example: temperature–time profiles in rectangular slabs

Figure 6.4 provides an example of a typical solution that was obtained with a computer program written to simulate an alternative firing practice under consideration by a North American steel company faced with a downturn in steel demand. Since British Thermal Units are currently used in industrial practice there, Figs 6.4 and 6.5 have been plotted with this in mind. Figure 6.4 shows the temperature at the surface and centreplane for a 27-inch-thick ingot of steel being heated in a soaking pit at a firing rate of 26×10^6 British Thermal Units per hour until the surface of the ingot reaches 2420°F. At that point, a control thermocouple decreases the fuel rate of flow, so as to maintain constant ingot surface temperature.

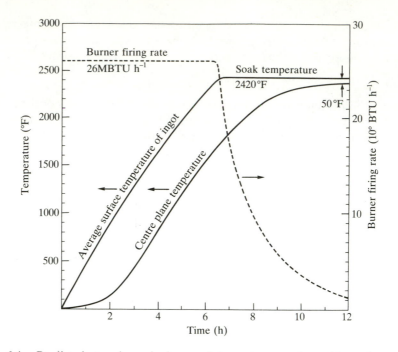

FIG. 6.4. Predicted step-down in burner firing rates, together with surface and internal ingot temperatures during soaking in a pit, prior to hot rolling operations.

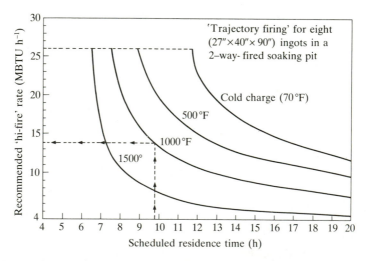

FIG. 6.5. 'Trajectory firing': recommended burner settings as a function of ingot preheat temperatures and time available in soaking pits.

Sometime later, heat conducted to the centre of the ingot raises its internal temperature to within 50°F of 2450°F, the soak temperature. At this point, the ingot is judged ready to roll. These computations were made using an iterative procedure based on equations of the type illustrated in eqns (6.24) to (6.27). In the solutions shown, a modulus value of 0.1 was taken. The inherent simplicity of such algorithms (see Table 6.8) and their ready physical interpretation lead to minimal programming difficulties, and make them popular candidates for engineering simulations, as already noted. Figure 6.5 summarizes the set of firing practices to be recommended for the particular steelworks soaking pit that was operating at reduced capacity. It shows minimum firing rates needed for ingots to be ready, given specified pit residence times and initial superheat temperatures.

(c) The implicit-integral control volume approach

Returning to eqn (6.23), which is the numerical equivalent of our transient heat conduction problem (Fourier's second law) we may write it in an apparently identical, but in fact, very different form of the integral control volume algorithm, i.e.,

$$M(\theta_E - \theta_P) - M(\theta_P - \theta_W) = \theta_P - \theta_P^-, \tag{6.29}$$

Here we say that the heat being conducted through the solid depends upon temperature differences between the nodal points at the new or forward time step, i.e., at t_2 rather than t_1. Rather than write superscripts θ^+ on all the temperatures appearing on the left-hand side of eqn (6.29), we can write instead θ^- on the temperature at node P corresponding to the old, or known, temperature prior to the integration step. Expanding eqn (6.29), we have

$$(1 + 2M)\,\theta_P - M\theta_W - M\theta_E = \theta_P^- \tag{6.30}$$

Equation (6.30) is known as the *implicit form* of the discretization formula, since the desired value of θ_P at the forward time requires a knowledge of the forward-time temperatures of its nearest neighbours. While this leads to some extra difficulty in seeking a solution, the great advantage of eqn (6.30) is the form of the $(1 + 2M)$ bracket, which places no upper limit on the time-step interval, remaining as it does, positive at all times.

Returning to our cooling ingot example, therefore, we can write equivalent expressions for the two surface nodes located at nodal points 1 and N respectively.

$$(1 + 2M + 2H_1)\theta_1 - 2M\theta_2 = \theta_1^- + 2H_1\theta_\infty, \tag{6.31}$$

and

$$(1 + 2M + 2H_N)\theta_N - 2M\theta_{N-1} = \theta_N^- + 2H_N\theta_\infty. \tag{6.32}$$

Referring to eqns (6.30) through to (6.32), one should note that they take the form of a tridiagonal matrix:

$$
\begin{array}{llll}
b_1\theta_1 + c_1\theta_2 & & & = d_1 \qquad (6.33) \\
a_2\theta_1 + b_2\theta_2 + c_2\theta_3 & & & = d_2 \\
\quad a_3\theta_2 + b_3\theta_3 + c_3\theta_4 & & & = d_3 \\
\qquad a_4\theta_3 + b_4\theta_4 + c_4\theta_5 & & & = d_4 \\
\qquad\quad a_i\theta_{i-1} + b_i\theta_i + c_i\theta_{i+1} = d_i & & & \qquad (6.34) \\
\qquad\qquad a_N\theta_{N-1} + b_N\theta_N = d_N & & & \qquad (6.35)
\end{array}
$$

The equations imply that θ_1 is known in terms of θ_2. The equation for $i = 2$ is a relation between θ_1, θ_2, and θ_3. However, since θ_1, can be expressed in terms of θ_2, using eqn (6.33), this second relation reduces to a relation between θ_2 and θ_3. In other words, θ_2 can be expressed in terms of θ_3. This process of substitution can be continued until θ_N is expressed in terms of θ_{N+1}. Since θ_{N+1} has no real existence, we can obtain the numerical value for θ_N at this stage, enabling us to begin a back-substitution process in which θ_{N-1} is obtained from θ_N, θ_{N-2} from θ_{N-1}, etc., down to θ_1 from θ_2. This, then, is the essence of the tridiagonal matrix algorithm, which becomes a most efficient technique when programmed on a high-speed digital computer. The system of equations can be summarized according to

$$
\theta_i = \gamma_i - \frac{c_i}{\beta_i}\theta_{i+1}, \qquad (6.36)
$$

where

$$
\beta_i = b_i - \frac{a_i c_{i-1}}{\beta_{i-1}} \text{ for } i = 2, 3, 4, \ldots, N, \qquad (6.37)
$$

and

$$
\beta_1 = b_1, \qquad (6.38)
$$

$$
\gamma_i = d_i - \frac{a_i \gamma_{i-1}}{\beta_i}. \qquad (6.39)
$$

(d) Example: a two-dimensional heat transfer program for predicting temperature–time profiles in batch-annealing operations

Table 6.7 provides a flow chart and Table 6.8 provides a listing of the main components of a computer program TWIN, written in FORTRAN IV code, that was developed for predicting the temperature–time distributions within a multicoil stack placed in a batch annealing furnace (Figure. 6.6(a)) during heating and cooling over a 5-day period. A typical coil, together with the notation used in constructing the finite difference scheme, is given in Fig. 6.6(b). Note that nodal points were placed at the surfaces of the coil, so that the

TABLE 6.7. *Flow chart of 'Twin' program.*

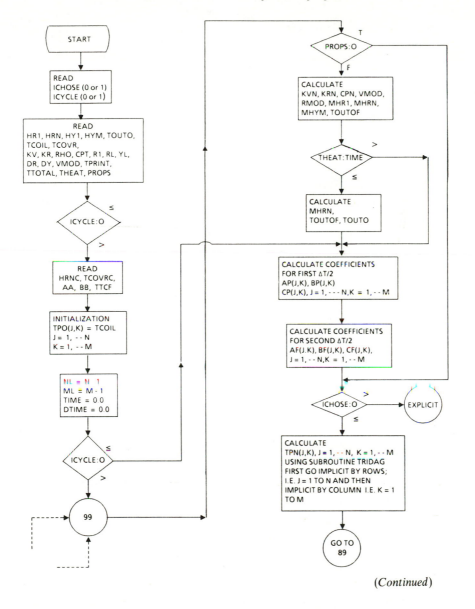

(*Continued*)

TABLE 6.7. (*Continued*)

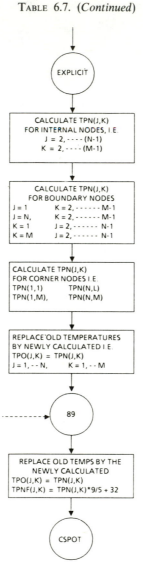

(*Continued*)

Table 6.7. (*Continued*)

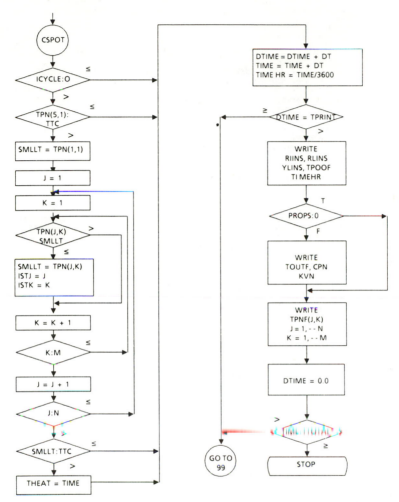

surface control volume element 1 in Figure 6.6(b) is half the volume of an internal element at that radius, while a corner element is a little less than a quarter as large.

Since the program required two-dimensional treatment of the heat flow over extended time periods, use was made of the implicit alternating direction approach (see Carnahan *et al.* 1969) and provided as an option in Table 6.8(b). This was first introduced by Peaceman and Rachford (1955), and involves writing eqns (6.30)–(6.32) in terms of a half-time-step interval $\Delta t/2$ and

TABLE 6.8. *Listing of a program named 'TWIN' for simulating the heating or cooling of steel coils in a batch anneal furnace. Implicit and explicit solving routines are included for comparison*

```
C     PROGRAM "TWIN" IS A HEAT TRANSFER SIMULATION DEVELOPED FOR
C     STUDYING BATCH ANNEALING.   IN ITS PRESENT FORM, IT RELATES
C     SPECIFICALLY TO PREDICTING THE TEMPERATURE TIME DISTRIBUTION
C     INSIDE A COIL HEATED ON ALL FOUR SIDES BY A COMBINATION OF
C     RADIATIVE HEAT TRANSFER FROM THE INSIDE WALLS OF THE BATCH
C     ANNEAL FURNACE AND THE CONVECTIVE HEAT INPUT FROM THE
C     RECIRCULATING GASES.
C
C     THE PROGRAM PROVIDES TWO ALTERNATIVE METHODS OF COMPUTATION.
C     THE IMPLICIT METHOD IN WHICH NO RESTRICTION IS PLACED ON THE
C     TIME STEP ,DT, (OR VMOD+RMOD)AND THE EXPLICIT METHOD WHERE
C     0.5 REPRESENTS THE UPPER LIMIT  ON (VMOD+RMOD)
C
C     THE FOLLOWING PARAMETERS ARE CONSIDERED
C     HR1   =  HEAT TRANSFER COEFFICIENT FOR INSIDE VERTICAL SURFACE
C              OF COIL
C     HRN   =  HEAT TRANSFER COEFFICIENT FOR OUTSIDE FLAT SURFACE OF COIL
C     HY1   =  HEAT TRANSFER COEFFICIENT FOR BOTTOM FLAT SURFACE OF COIL
C     HYM   =  HEAT TRANSFER COEFFICIENT FOR TOP FLAT SURFACE OF COIL
C     R1    =  INSIDE RADIUS OF COIL
C     RL    =  RADIAL WIDTH OF COIL (OUTER-INNER RADII)
C     YL    =  HEIGHT OF COIL
C     DR    =  INTER-NODAL DISTANCE IN RADIAL DIRECTION
C     DY    =  INTER-NODAL DISTANCE IN VERTICAL DIRECTION
C     KV    =  VERTICAL THERMAL CONDUCTIVITY OF COIL
C     KR    =  RADIAL THERMAL CONDUCTIVITY OF COIL
C     VMOD   =  VERTICAL MODULUS
C     TPRINT  =  PRINTOUT TIME
C     RHO   =  DENSITY OF STEEL
C     CPT   =  SPECIFIC HEAT CAPACITY OF STEEL
C     TOUT  =  AMBIENT GAS TEMPERATURE
C     TPN(J,K)  =  NEW TEMPERATURE OF NODAL POINT (J,K) AT END
C                  OF TIME INTERVAL, DT
C     TPO(J,K)  =  TEMPERATURE OF NODAL POINT (J,K) AT START
C                  OF TIME INTERVAL, DT
      REAL HR1,HRN,HY1,HYM,MHR1,MHYM,MHY1,MHYM,KV,KR
      REAL KVN,KRN,KVO,MHR1O,MHRNO,MHY1O,MHYMO
      DIMENSION TPN(25,25),TPO(25,25),TPNF(25,25),TSTAR(25,25)
      DIMENSION AP(25,25),AF(25,25),BP(25,25),BF(25,25),CP(25,25),
     $CF(25,25),A(25),B(25),C(25),D(25),TDAG(25)
      WRITE(6,88)
88    FORMAT(1H ,'ENTER CHOICE OF METHOD; IMPLICIT=0, EXPLICIT = 1',//)
      READ(5,*) ICHOSE
      WRITE(6,87)
87    FORMAT(1H ,'ENTER CHOICE OF CONDITIONS+; HEATING OR COOLING=0',
     $',COUPLED =1'///)
      READ(5,*) ICYCLE
      WRITE(6,77)
77    FORMAT (1H, 'ENTER INSIDE,OUTSIDE,BOTTOM AND TOP HT. TRANSFER',
     $'COEFFICIENTS,GAS,COIL,COVER TEMPS,VERTICAL AND RADIAL CONDS.',/
     $'DENSITY,HT. CAPACITY,INSIDE RADIUS,RADIAL WIDTH,COIL HT.',
     $ ',RADIAL AND VERTICAL NODAL SPACING,VERT. MODULUS,TPRINT.',//,
     $'TOTAL TIME,HEATING TIME   ,VARIABLE THERMAL PROPERTIES=1 ',//)
      READ(5,*)HR1,HRN,HY1,HYM,TOUTO,TCOIL,TCOVR,KV,KR,RHO,CPT,R1,RL,YL,
     $DR,DY,VMOD,TPRINT,TTOTAL,THEAT,PROPS
      IF(ICYCLE) 66,66,65
65    WRITE(6,67)
67    FORMAT(1H ,'ENTER SURFCE HEAT TRANSFER COEFF FOR COOLING(F),THOOD'
     $'FIRST AND SECOND ORDER COEFFICIENTS,COLD SPOT TEMPERATURE',//)
      READ(5,*) HRNC,TCOVRC,AA,BB,TTCF
66    CONTINUE
      TPRINT = TPRINT*3600.
```

(Continued)

TABLE 6.8 (*Continued*)

```
        TTOTAL = TTOTAL*3600.
        DT=VMOD*RHO*CPT*DY**2.0/KV
        THEAT = THEAT*3600.
        RMOD=KR*DT/(RHO*CPT*DR**2.0)
        MHR1=HR1*DT/(RHO*CPT*DR)
        MHRN=HRN*DT/(RHO*CPT*DR)
        MHY1=HY1*DT/(RHO*CPT*DY)
        MHYM=HYM*DT/(RHO*CPT*DY)
        KVO = KV
        VMODO=VMOD
        RMODO=RMOD
        MHR1O=MHR1
        MHRNO=MHRN
        MHY1O=MHY1
        MHYMO=MHYM
        TOUTI = TOUTO
        TOUTAV = (TOUTI+TOUTO)/2.0
        TOUT = TOUTAV
        N=RL/DR+1.1
        M=YL/DY+1.1
        DO 210 J=1,N
        DO 220 K=1,M
        TPO(J,K)=TCOIL
        TPOO=TPO(1,1)
220     CONTINUE
210     CONTINUE
        TPOOF = 9.0*TPOO/5.0 + 32.
        TTC = (TTCF - 32.0)*5.0/9.0
        RIINS = R1/2.54
        RLINS = RL/2.54
        YLINS = YL/2.54
        NL=N-1
        ML=M-1
        TIME=0.0
        DTIME = 0.0
        IF (ICHOSE) 151,151,99
99      CONTINUE
        IF(PROPS.EQ.0.0) GO TO 98
C   CALCN OF NEW THERMAL PROPS FOR K AND C AND AMBIENT GAS TEMPS
        KVN=KVO - 0.0000548*(TPO(5,3) - TPOO)
        KRN=(KR/KV)*KVN
        CPN = 3.04+ 0.00758*(TPO(5,3)+273.)+60000./(TPO(5,3)+273.)
        $**2.
        CPN=CPN/53.85
        VMOD=VMODO*(KVN/KV)*(CPT/CPN)
        RMOD=VMOD*(RMODO/VMODO)
        MHR1=MHR1O*(CPT/CPN)
        MHRN=MHRNO*(CPT/CPN)
        MHY1=MHY1O*(CPT/CPN)
        MHYMO=MHYMO*(CPT/CPN)
        TOUTOF=(TCOVR+0.43*(TPO(1,5)*(9.0/5.0)+32.00))/1.43
        IF(THEAT - TIME ) 68,68,69
68      MHRN = MHRN*(HRNC/HRN)
        TOUTOF = (TCOVRC + 0.43*(TPO(1,5)*(9.0/5.0) + 32.))/1.43
        CONTINUE
69      TOUTO=5./9.*(TOUTOF-32.0)
        TOUTI=TOUTO
        TOUTAV=(TOUTI+TOUTO)/2.0
        TOUT = TOUTAV
C
C
C                   IMPLICIT   METHOD
C
C
C   SET TIME INDEPENDENT COEFFICIENTS FOR FIRST HALF OF TIME INTERVAL.
C   FOR CORNERS
151     AP(N,1)= -RMOD*((R1 + (N-1.5)*DR)/(R1 + (N -1.0)*DR))
        AP(N,M) = -RMOD*((R1 + (N-1.5)*DR)/(R1 + (N-1.0)*DR ))
```
(*Continued*)

TABLE 6.8 (*Continued*)

```
      AP(1,1) = 0.0
      AP(1,M) = 0.0
C
      BP(1,1) = ( 1 + MHR1 + RMOD*(1 + DR/(2.0*R1)))
      BP(1,M) = BP(1,1)
      BP(N,1) = (1 + MHRN + RMOD*((R1 + (N-1.5)*DR)/(R1 + (N-1.0)*DR)))
      BP(N,M) = BP(N,1)
C
      CP(1,1) = -RMOD*(1 + DR/(2.0*R1))
      CP(1,M) = CP(1,1)
      CP(N,1) = 0.0
      CP(N,M) = 0.0
C
C
C
C SET COEFFICIENTS FOR TOP AND BOTTOM SURFACES FOR FIRST HALF
C  OF TIME INTERVAL J =2,NL
C
      DO 1 J=2,NL
      AP(J,1) = -0.5*RMOD*((R1+(J-1.5)*DR)/(R1 + (J-1.0)*DR))
      AP(J,M) = AP(J,1)
      BP(J,1) = (1.0 + .5*RMOD*(R1 + (J-1.5)*DR)/(R1 + (J-1.0)*DR)
     $+0.5*RMOD*(R1 + (J-0.5)*DR)/(R1 + (J-1.0)*DR))
      BP(J,M) = BP(J,1)
      CP(J,1) = -0.5*RMOD*((R1+(J-0.5)*DR)/(R1 + (J-1.0)*DR))
1     CP(J,M) = CP(J,1)
      CONTINUE
C
C
C SET COEFFICIENTS FOR INNER AND OUTER VERTICAL SURFACES;
C     K=2,ML AND J =1 OR N
      DO 2 K = 2,ML
      AP(1,K) = 0.0
      AP(N,K) = -RMOD*((R1 + (N-1.5)*DR)/(R1 + (N-1.0)*DR))
      BP(1,K) = 1+MHR1 + RMOD*(1 + DR/(2.0*R1))
      BP(N,K) = 1 + MHRN + RMOD*((R1 + (N-1.5)*DR)/(R1 + (N-1.0)*DR))
      CP(1,K) = -RMOD*(1 + DR/(2.0*R1))
2     CP(N,K) = 0.0
       CONTINUE
C
C
C  SET COEFFICIENTS FOR INSIDE NODAL POINTS; J=2,NL; K=2,ML
      DO 3 J = 2,NL
      DO 4 K = 2,ML
      AP(J,K) = -0.5*RMOD*(R1 + (J-1.5)*DR)/(R1 + (J-1.0)*DR)
      BP(J,K) = 1 + RMOD
      CP(J,K) = -0.5*RMOD*(R1 + (J-0.5)*DR)/(R1 + (J-1.0)*DR)
4     CONTINUE
3     CONTINUE
C
C
C
C  SET TIME INDEPENDENT COEFFICIENTS FOR SECOND HALF OF TIME INCREMENT
C
C     FOR CORNERS
C
      AF(1,1) = 0.0
      AF(1,M) = -VMOD
      AF(N,1) = 0.0
      AF(N,M) = -VMOD
      BF(1,1) = 1 + MHY1 + VMOD
      BF(1,M) = 1+MHYM + VMOD
      BF(N,1) = 1 + MHY1 + VMOD
      BF(N,M) = 1 + MHYM + VMOD
      CF(1,1) = -VMOD
      CF(1,M) = 0.0
      CF(N,M) = 0.0
      CF(N,1) =-VMOD
```

(*Continued*)

TABLE 6.8 (*Continued*)

```
C
C
C
C   SET COEFFICIENTS FOR TOP AND BOTTOM SURFACES FOR SECOND HALF
C   OF TIME INTERVAL; J=2,NL   K=M AND 1
C
C
        DO 5   J = 2,NL
        AF(J,1) = 0.0
        AF(J,M) = -VMOD
        BF(J,1) = 1 + MHY1 +VMOD
        BF(J,M) = 1 + MHYM + VMOD
        CF(J,1) = -VMOD
        CF(J,M) = 0.0
5       CONTINUE
C
C
C   SET COEFFICIENTS FOR INNER AND OUTER VERTICAL SURFACES
C      K=2,ML AND J = 1 OR N
C
        DO 6 K = 2,ML
        AF(1,K) = -0.5*VMOD
        AF(N,K) = -0.5*VMOD
        BF(1,K) = 1 + VMOD
        BF(N,K) = 1 + VMOD
        CF(1,K) = -0.5*VMOD
        CF(N,K) = -0.5*VMOD
6       CONTINUE
C
C
C   SET COEFFICIENTS FOR SECOND HALF OF TIME INCREMENT FOR INSIDE NODAL
C   POINTS ; J = 2,NL  ; K = 2,ML
        DO 7 J = 2,NL
        DO 8 K = 2,ML
        AF(J,K) = -0.5*VMOD
        BF(J,K) = 1 + VMOD
        CF(J,K) = -0.5*VMOD
8       CONTINUE
7       CONTINUE
C
98      IF (ICHOSE) 150,150,160
150     CONTINUE
C
C   CALCULATION OF NEW TEMPERATURES INSIDE COIL ,TPN(J,K)
C    THE COMPUTATION OF TEMPERATURE NOW PROCEEDS FOR THE FIRST HALF TIME
C   INTERVAL,GOING IMPLICIT BY ROWS;  I.E.  J=1 TO J= N , AND RISING
C   VERTICALLY FROM K=1 TO K= M FOLLOWING EACH SCAN
        DO 9 K = 2,ML
        A(1)= AP(1,K)
        B(1) = BP(1,K)
        C(1) = CP(1,K)
        D(1) = MHR1*TOUT + 0.5*VMOD*(TPO(1,K+1) + TPO(1,K-1)) +TPO(1,K)
     $*(1-VMOD)
C
        DO 10 J=2,NL
        A(J) = AP(J,K)
        B(J) = BP(J,K)
        C(J) = CP(J,K)
10      D(J) = 0.5*VMOD*(TPO(J,K+1) + TPO(J,K-1)) + (1-VMOD)*TPO(J,K)
C
        A(N) = AP(N,K)
        B(N) = BP(N,K)
        C(N) = CP(N,K)
        D(N) = MHRN*TOUT + 0.5*VMOD*(TPO(N,K+1) + TPO(N,K-1))
     $+TPO(N,K)*(1-VMOD)
        CALL TRIDAG(1,N,A,B,C,D,TDAG)
        DO 11 J = 1,N
```

(*Continued*)

Table 6.8 (*Continued*)

```
11      TSTAR(J,K) = TDAG(J)
9       CONTINUE
C
C
C   SOLVE INTERMEDIATE TEMPERATURES FOR BOTTOM AND TOP SURFACES OF
C   COIL ; (IMPLICIT BY ROW)
C   K=1; BOTTOM SURFACE
        A(1) = AP(1,1)
        B(1) = BP(1,1)
        C(1) = CP(1,1)
        D(1) = MHR1*TOUT + MHY1*TOUT + VMOD*TPO(1,2) + TPO(1,1)*
     $(1-MHY1-VMOD)
        A(N) = AP(N,1)
        B(N)= BP(N,1)
        C(N) = CP(N,1)
        D(N) = MHRN*TOUT + MHY1*TOUT + VMOD*TPO(N,2) + TPO(N,1)*(1-MHY1
     $-VMOD)
C
C
        DO 13 J = 2,NL
        A(J) = AP(J,1)
        B(J) = BP(J,1)
        C(J) = CP(J,1)
13      D(J) = MHY1*TOUT + VMOD*TPO(J,2) + (1-MHY1 -VMOD)*
     $TPO(J,1)
        CALL TRIDAG(1,N,A,B,C,D,TDAG)
        DO 14 J= 1,N
14      TSTAR(J,1) = TDAG(J)
C
C
C       K=M; TOP SURFACE
C
        A(1) = AP(1,M)
        B(1) = BP(1,M)
        C(1) = CP(1,M)
        D(1)= MHR1*TOUT + MHYM*TOUT + VMOD*TPO(1,M-1) +(1-MHYM-VMOD)
     $*TPO(1,M)
        A(N) = AP(N,M)
        B(N) = BP(N,M)
        C(N) = CP(N,M)
        D(N) = MHRN*TOUT + MHYM*TOUT + VMOD*TPO(N,M-1)
     $+(1-MHYM-VMOD)*TPO(N,M)
        DO 16 J=2,NL
        A(J) = AP(J,M)
        B(J) = BP(J,M)
        C(J) = CP(J,M)
16      D(J) = MHYM*TOUT + VMOD*TPO(J,M-1)+(1-MHYM-VMOD)
     $*TPO(J,M)
        CALL TRIDAG(1,N,A,B,C,D,TDAG)
        DO 17 J=1,N
17      TSTAR(J,M) = TDAG(J)
C
C
C
C   COMPUTATION OF TEMPERATURE FOR SECOND HALF OF TIME INTERVAL,GOING
C   IMPLICIT BY COLUMN; K=1 TO K=M AND MOVING HORIZONTALLY FROMJ=1 TO N
C   AFTER EACH SUCCESSIVE SCAN
C
C
        DO 18 J = 2,NL
        A(1) = AF(J,1)
        B(1) = BF(J,1)
        C(1) = CF(J,1)
        D(1) = (1-0.5*RMOD*(R1+(J-1.5)*DR)/(R1+(J-1.0)*DR)
     $-0.5*RMOD*(R1+(J-0.5)*DR)/(R1+(J-1.0)*DR))*TSTAR(J,1)
     $+MHY1*TOUT + 0.5*RMOD*((R1+(J-1.5)*DR)/(R1 +(J-1.0)*DR))*TSTAR(J-1
     $,1)     +0.5*RMOD*((R1+(J-0.5)*DR)/(R1+(J-1.0)*DR))*TSTAR(J+1,1)
```

(*Continued*)

TABLE 6.8 (*Continued*)

```
        DO 19 K=2,ML
        A(K) = AF(J,K)
        B(K) = BF(J,K)
        C(K) = CF(J,K)
19      D(K) = (1-0.5*RMOD*(R1+(J-1.5)*DR)/(R1+(J-1.0)*DR)
      $-0.5*RMOD*(R1+(J-0.5)*DR)/(R1+(J-1.0)*DR))*TSTAR(J,K)
      $+0.5*RMOD*((R1+(J-1.5)*DR)/(R1+(J-1.0)*DR))*TSTAR(J-1,K)
      $+0.5*RMOD*((R1+(J-0.5)*DR)/(R1+(J-1.0)*DR))*TSTAR(J+1,K)
C
C
        A(M) = AF(J,M)
        B(M) = BF(J,M)
        C(M) = CF(J,M)
        D(M) = (1-0.5*RMOD*(R1+(J-1.5)*DR)/(R1+(J-1.0)*DR)
      $-0.5*RMOD*(R1+(J-0.5)*DR)/(R1+(J-1.0)*DR))*TSTAR(J,M)
      $+MHYM*TOUT + 0.5*RMOD*((R1+(J-1.5)*DR)/(R1+(J-1.0)*DR))*TSTAR
      $(J-1,M)    +0.5*RMOD*((R1+(J-0.5)*DR)/(R1+(J-1.0)*DR))*
      $TSTAR(J+1,M)
C
C
        CALL TRIDAG(1,M,A,B,C,D,TDAG)
C
        DO 20 K = 1,M
20      TPN(J,K) = TDAG(K)
18      CONTINUE
C
C   SOLVE SECOND HALF TEMPERATURES FOR THE INNER AND OUTER
C VERTICAL SURFACES OF THE COIL (IMPLICIT BY COLUMN)
C     J= 1; INNER SURFACE OF COIL
C
C
        A(1) = AF(1,1)
        B(1) = BF(1,1)
        C(1) = CF(1,1)
        D(1) = MHY1*TOUT + MHR1*TOUT + RMOD*(1+DR/(2.0*R1))*TSTAR(2,1)
      $+(1-MHR1 -RMOD*(1+DR/(2.0*R1)))*TSTAR(1,1)
C
        A(M) = AF(1,M)
        B(M) = BF(1,M)
        C(M) = CF(1,M)
        D(M) = MHYM*TOUT + MHR1*TOUT + RMOD*(1+DR/(2.0*R1))*TSTAR(2,M)
      $+(1-MHR1 - RMOD*(1 + DR/(2.0*R1)))*TSTAR(1,M)
        DO 21 K = 3,ML
        A(K) = AF(1,K)
        B(K) = BF(1,K)
        C(K) = CF(1,K)
21      D(K) = MHR1*TOUT + RMOD*(1 + DR/(2.0*R1))*TSTAR(2,K)
      $+(1-MHR1 - RMOD*(1 + DR/(2.0*R1)))*TSTAR(1,K)
        CALL TRIDAG (1,M,A,B,C,D,TDAG)
C
        DO 22 K = 1,M
22      TPN(1,K) = TDAG(K)
        CONTINUE
C
C
C     J = N ; OUTER SURFACE OF COIL
C
        A(1) = AF(N,1)
        B(1) = BF(N,1)
        C(1) = CF(N,1)
        D(1) = MHY1*TOUT + MHRN*TOUT + RMOD*((R1 +(N-1.5)*DR)/(
      $R1 + (N-1)*DR))*TSTAR(N-1,1) + (1-MHRN - RMOD*(R1 + (N-1.5)*DR)/
      $(R1 + (N-1.0)*DR))*TSTAR(N,1)
C
C
        A(M) = AF(N,M)
        B(M) = BF(N,M)
```

(*Continued*)

TABLE 6.8 (*Continued*)

```
      C(M) = CF(N,M)
      D(M) = MHYM*TOUT + MHRN*TOUT + RMOD*((R1+(N-1.5)*DR)/
     $(R1+(N-1)*DR))*TSTAR(N-1,M) +(1-MHRN -RMOD*(R1 + (N-1.5)*DR)/
     $(R1 + (N-1.0)*DR))*TSTAR(N,M)
      DO 23 K = 2,ML
      A(K) = AF(N,K)
      B(K) = BF(N,K)
      C(K) = CF(N,K)
23    D(K) =MHRN*TOUT + RMOD*((R1 +(N-1.5)*DR)/(R1 +(N-1.)*DR))
     $*TSTAR(N-1,K)+(1-MHRN-RMOD*(R1+(N-1.5)*DR)/(R1+(N-1)*DR))*TSTAR
     $(N,K)
C
      CALL TRIDAG(1,M,A,B,C,D,TDAG)
      DO 24 K = 1,M
24    TPN(N,K) = TDAG(K)
      GO TO 89
C
C
C                      EXPLICIT   METHOD
C
C
160   CONTINUE
      DO 30 J=2,NL
      DO 40 K=2,ML
      TPN(J,K)=TPO(J,K)*(1.0-2.0*(VMOD+RMOD))+VMOD*(TPO(J,K+1)+
     @TPO(J,K-1))+TPO(J-1,K)*RMOD*(R1+(J-1.5)*DR)/(R1+(J-1)*DR)+
     @TPO(J+1,K)*RMOD*(R1+(J-0.5)*DR)/(R1+(J-1)*DR)
40    CONTINUE
30    CONTINUE
C     CALCULATION OF NEW TEMPERATURES ON INSIDE VERTICAL SURFACE OF COIL
      J=1
      DO 51 K=2,ML
      TPN(J,K)=TPO(J,K)*(1.0-2.0*MHR1-RMOD*(2.0+(DR/R1))-2.0*VMOD)+
     @VMOD*TPO(J,K+1)+VMOD*TPO(J,K-1)+RMOD*(2.0+(DR/R1))*TPO(J+1,K)+
     @TOUTI*2.0*MHR1
51    CONTINUE
C     CALCULATION OF NEW TEMPERATURES ON OUTER VERTICAL SURFACE OF COIL
      J=N
      DO 52 K=2,ML
      TPN(J,K)=TPO(J,K)*(1.0-2.0*MHRN-2.0*VMOD-2.0*RMOD*(R1+(J-1.5)*DR
     @)/(R1+(J-1.0)*DR))+VMOD*(TPO(J,K+1)+TPO(J,K-1))+TPO(J-1,K)*2.0*
     @RMOD*(R1+(J-1.5)*DR)/(R1+(J-1.0)*DR)+TOUTO*2.0*MHRN
52    CONTINUE
C     CALCULATION OF NEW TEMPERATURES ON BOTTOM SURFACE OF COIL
      K=1
      DO 53 J=2,NL
      TPN(J,K)=TPO(J,K)*(1.0-2.0*(MHY1+VMOD)-2.0*RMOD*(R1+(J-1)*DR)/
     @(R1+(J-1.5)*DR))+2.0*MHY1*TOUTAV+2.0*VMOD*TPO(J,K+1)+
     @RMOD*TPO(J-1,K)+RMOD*((R1+(J-0.5)*DR)/(R1+(J-1.5)*DR))*TPO(J+1,K)
53    CONTINUE
C     CALCULATION OF NEW TEMPERATURES ON TOP SURFACE OF COIL
      K=M
      DO 54 J=2,NL
      TPN(J,K)=TPO(J,K)*(1.0-2.0*MHYM-2.0*VMOD-2.0*RMOD*(R1+(J-1)*DR)/
     @(R1+(J-1.5)*DR))+2.0*MHYM*TOUTAV+2.0*VMOD*TPO(J,K-1)+RMOD*
     @TPO(J-1,K)+RMOD*((R1+(J-0.5)*DR)/(R1+(J-1.5)*DR))*TPO(J+1,K)
54    CONTINUE
C     BOTTOM INSIDE EDGE OF COIL
      TPN(1,1)=TPO(1,1)*(1.0-MHY1*(2.0+DR/(2.0*R1))-MHR1*2.0-VMOD*
     @(2.0+DR/(2.0*R1))-RMOD*(2.0+DR/(2.0*R1)))+TOUTAV*MHY1*(2.0+
     @DR/(2.0*R1))+MHR1*2.0*TOUTI+VMOD*(2.0+DR/(2.0*R1))*TPO(1,2)+
     @RMOD*(2.0+DR/(2.0*R1))*TPO(2,1)
C     TOP INSIDE EDGE OF COIL
      TPN(1,M)=TPO(1,M)*(1.0-MHYM*(2.0+DR/(2.0*R1))-MHR1*2.0-VMOD*(2.0+
     @DR/(2.0*R1))-RMOD*(2.0+DR/(2.0*R1)))+TOUTAV*MHYM*(2.0+DR/(2.0*R1))
     @+TOUTI*MHR1*2.0+TPO(1,M-1)*VMOD*(2.0+DR/(2.0*R1))+
     @TPO(2,M)*RMOD*(2.0+DR/(2.0*R1))
```

(*Continued*)

TABLE 6.8 (*Continued*)

```
C     TOP OUTSIDE EDGE OF COIL
      TPN(N,M)=TPO(N,M)*(1.0-MHYM*((2.0*R1+2.0*DR*N-2.5*DR)/(R1+(N-1)*
     @DR))-MHRN*((2.0*R1+2.0*DR*N-DR)/(R1+(N-1)*DR))-VMOD*
     @(2.0*R1+2.0*DR*N-2.5*DR)/(R1+(N-1)*DR)-RMOD*(2.0*R1+2.0*N*DR-
     @3.0*DR)/(R1+(N-1)*DR))+TOUTAV*MHYM*(2.0*R1+2.0*N*DR-2.5*DR)/(R1+
     @(N-1)*DR)+MHRN*TOUTO*((2.0*R1+2.0*N*DR-DR)/(R1+(N-1)*DR))+
     @TPO(N,M-1)*VMOD*(2.0*R1+2.0*N*DR-2.5*DR)/(R1+(N-1.0)*DR)+
     @TPO(N-1,M)*RMOD*(2.0*R1+2.0*N*DR-3.0*DR)/(R1+(N-1.0)*DR)
C     BOTTOM OUTSIDE EDGE OF COIL
      TPN(N,1)=TPO(N,1)*(1.0-MHY1*(2.0*R1+2.0*N*DR-2.5*DR)/(R1+(N-1.0)*
     @DR)-MHRN*(2.0*R1+2.0*N*DR-DR)/(R1+(N-1.0)*DR)-VMOD*(2.0*R1+
     @2.0*N*DR-2.5*DR)/(R1+(N-1.0)*DR)-RMOD*(2.0*R1+2.0*N*DR-3.0*DR)/
     @(R1+(N-1)*DR))+TOUTAV*MHY1*(2.0*R1+2.0*N*DR-2.5*DR)/(R1+(N-1)*
     @DR)+MHRN*TOUTO*(2.0*R1+2.0*N*DR-DR)/(R1+(N-1)*DR)+TPO(N,2)*VMOD*
     @((2.0*R1+2.0*N*DR-2.5*DR)/(R1+(N-1.0)*DR))+TPO(N-1,1)*RMOD*((
     @2.0*R1+2.0*N*DR-3.0*DR)/(R1+(N-1)*DR))
89    DO 100 J=1,N
      DO 200 K=1,M
      TPO(J,K)=TPN(J,K)
      TPNF(J,K) = TPN(J,K)*9.0/5.0   + 32.
200   CONTINUE
100   CONTINUE
C
C
C
C     SEQUENCE FOR SEEKING COLD SPOT TEMPERATURE
C
      IF(ICYCLE) 70,70,71
71    IF(TPN(5,1)-TTC) 72,72,73
73    CONTINUE
      SMLLT = TPN(1,1)
      DO 101 J= 1,N
      DO 201 K = 1,M
      IF(TPN(J,K)  .GT.  SMLLT)  GO  TO 201
      SMLLT  =   TPN(J,K)
      ISTJ   =  J
      ISTK   =  K
201   CONTINUE
101   CONTINUE
      IF (SMLLT - TTC) 76,76,78
78    THEAT = TIME
76    CONTINUE
72    CONTINUE
70    CONTINUE
      DTIME = DTIME + DT
      TIME=TIME+DT
      TIMEHR = TIME/3600.
      IF(DTIME.GT.TPRINT)GO TO 170
      GO TO 99
170   WRITE(6,300) RIINS,RLINS,YLINS,TPOOF,TIMEHR
300   FORMAT(1H,'TEMPERATURE ARRAY IN COIL',F8.1,'INS.INNER RADIUS',F8.1
     $,'INS.RADIAL WIDTH AND',F8.1,'INS. HIGH',//
     $,1H ,'HAVING AN INITIAL TEMPERATURE OF',F8.1,'DEGREES FAHRENHEIT',
     $///, 1H ,'TIME(HOURS)',F8.3,/)
      IF(PROPS.EQ.0.0) GO TO 55
      WRITE(6,154) TOUTOF,CPN,KVN
154   FORMAT(1H,'GAS TEMP F',F10.1,'HT CAPACITY',F6.2,'THERMAL CONDUCTIV
     $ITY',F9.3,//)
55    CONTINUE
      DO 310 I=1,M
      K = M+1-I
      WRITE(6,155)(TPNF(J,K),J=1,N)
155   FORMAT(2X,13F7.1,/2X,13F7.1,/2X,13F7.1,/2X,13F7.1,/////////)
310   CONTINUE
      DTIME = 0.0
      IF(TIME.LT.TTOTAL) GO TO 99
      CONTINUE
      STOP
      END
```

(*Continued*)

TABLE 6.8 (*Continued*)

```
C  TRIDAG IS A SUBROUTINE FOR SOLVING A SET OF LINEAR SIMULTANEOUS
C  EQUATIONS HAVING A TRIDIAGONAL COEFFICIENT MATRIX USING A RECUR-
C SIVE METHOD OFSOLUTION.
C        THE EQUATIONS ARE NUMBERED FROM IF THROUGH L ,AND THEIR
C    SUB-DIAGONAL ,DIAGONAL ,AND SUPRA-DIAGONAL COEFFICIENTS
C    ARE STORED IN THE ARRAYS A,B,C.   THE COMPUTED TEMPERATURE SET
C    Y(IF) ............Y(L) IS STORED IN THE ARRAY Y
C
         SUBROUTINE TRIDAG(IF,L,A,B,C,D,Y)
C  ....COMPUTE INTERMEDIATE ARRAYS BETA AND GAMMA
         DIMENSION A(25),B(25),C(25),D(25),Y(25),BETA(25),GAMMA(25)
C
         BETA(IF) = B(IF)
         GAMMA(IF) = D(IF)/BETA(IF)
         IFP1 = IF + 1
         DO 1 I = IFP1,L
         BETA(I) = B(I) - A(I)*C(I-1)/BETA(I-1)
1        GAMMA(I) = (D(I) - A(I)*GAMMA(I-1))/BETA(I)
C  COMPUTE THE SOLUTION FOR TEMPERATURE AT Y(L)
         Y(L) = GAMMA(L)
         DO 31 I = IF,L
C        WRITE(6,32) GAMMA(I),A(I),B(I),C(I),D(I)
32       FORMAT(3X,5F10.5,//)
31       CONTINUE
C  EMPLOY THE RECURSIVE FORMULA
C
         LAST = L - IF
         DO 2 K = 1,LAST
         I = L-K
         Y(I) = GAMMA(I) - C(I)*Y(I+1)/BETA(I)
C        WRITE(6,*) Y(I)
2        CONTINUE
         RETURN
C
         END
```

updating temperatures by scanning across the vertical column of nodes from 1 to N. This is then followed by an equivalent horizontal scan from bottom to top, i.e., from nodes 1 to M over the next half-time-step. The summation of the two sets of temperature increments at each nodal point then provides the new temperature profile being sought.

Thus, referring to an interior element whose nodal point is located at (J, K) within the coil, we can discretize the partial differential equation, also shown in Fig. 6.6(b), as follows;

$$+\left\{\begin{matrix}\text{Radial heat into}\\\text{element } (J, K) \text{ from}\\\text{element } (J-1, K)\end{matrix}\right.\quad\begin{matrix}\text{Radial heat out of}\\-\text{ element } (J, K) \text{ into}\\\text{element } (J+1, K)\end{matrix}\qquad\begin{matrix}\text{Rate of heat accumu-}\\\text{lation in volume}\\=\text{element } (J, K) \text{ over}\end{matrix}$$

$$+\quad\begin{matrix}\text{Vertical heat into}\\\text{element } (J, K) \text{ from}\\\text{element } (J, K-1)\end{matrix}\quad\left.\begin{matrix}\text{Vertical heat out}\\-(J, K) \text{ into element}\\(J, K+1)\end{matrix}\right\}\quad\begin{matrix}\text{half-time interval,}\\\Delta t/2,\end{matrix}$$

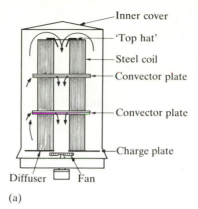

(a)

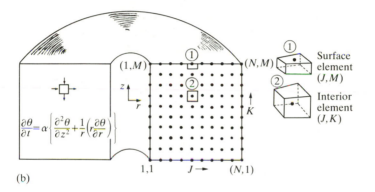

(b)

FIG. 6.6. (a) A schematic of typical batch-anneal process equipment for annealing steel coils for good deep drawing properties. (b) Cross-section of coil in a batch anneal furnace, showing discretization procedure and nomenclature adopted. Also shown are typical surface and internal volume elements.

or

$$
k_r A_J \left(\frac{\theta^*_{J-1,K} - \theta^*_{J,K}}{\Delta R} \right) - k_r A_{J+1} \left(\frac{\theta^*_{J,K} - \theta^*_{J+1,K}}{\Delta R} \right)
$$
$$
+ k_v A_K \left(\frac{\theta^-_{J,K-1} - \theta^-_{J,K}}{\Delta Z} \right) - k_v A_{K+1} \left(\frac{\theta^-_{J,K} - \theta^-_{J,K+1}}{\Delta Z} \right) = \rho C \Delta V_{J,K} \frac{\theta^*_{J,K} - \theta^-_{J,K}}{\Delta t/2}.
$$

$$(6.40)$$

Dividing by $(\rho C \Delta V_{J,K})/(\Delta t/2)$ and denoting $\alpha_r \Delta_t/\Delta R^2$ by RMOD and $\alpha_z \Delta t/\Delta Z^2$ by VMOD, we obtain:

$$
-\frac{\text{RMOD}}{2} \left(\frac{A_J \Delta R}{\Delta V_{J,K}} \right) \theta^*_{J-1,K} + \left\{ 1 + \frac{\text{RMOD}}{2} \left(\frac{A_J \Delta R}{\Delta V_{J,K}} + \frac{A_{J-1} \Delta R}{\Delta V_{J,K}} \right) \right\} h^*_{J,K}
$$

$$-\frac{\text{RMOD}}{2}\left(\frac{A_{J+1}\Delta R}{\Delta V_{J,K}}\right)\theta^*_{J+1,K}=\left\{1+\text{VMOD}\left(\frac{A_K\Delta Z}{\Delta V_{J,K}}\right)\right\}\theta^-_{J,K}$$

$$-\frac{\text{VMOD}}{2}\frac{A_K\Delta Z}{\Delta V_{J,K}}\theta^-_{J,K-1}-\frac{\text{VMOD}}{2}\frac{A_{K+1}\Delta Z}{\Delta V_{J,K}}\theta^-_{J,K+1}. \tag{6.41}$$

We see that this equation takes the tridiagonal form:

$$a_i\theta_{i-1}+b_i\theta_i+c_i\theta_{i+1}=d_i, \tag{6.42}$$

the various coefficients being functions of geometric factors, thermal properties and integration time-step Δt. Note that a total of 72 coefficients (4 per equation $\times$ 2 half time steps $\times$ (4 edges + 4 exterior surfaces + 1 internal) have to be specified, making the programming task rather laborious, in comparison with the explicit technique, which would require only 27 coefficients. This can be seen by referring to the TWIN program listing, where the more compact nature of the explicit method is evident.

Table 6.9 gives a direct comparison of data provided to the implicit and explicit components of the TWIN program, while Fig. 6.7 shows computed temperature profiles across a diagonal of a coil following 4 h of heating. Input boundary conditions were expressed in terms of effective heat transfer coefficient bulk gas temperatures to the four surfaces of the coil. One sees that consistent results can be obtained by taking M up to a value of 8.0 without incurring any significant penalties for expanding the time interval. By contrast, the maximum Fourier modulus for the explicit technique was essentially 0.25, above which solutions became unstable. The nett result was a program capable of running about twenty times faster on a given machine. Figure 6.8 shows some typical heating rate curves for various points within a coil during heating in a batch anneal furnace; Fig. 6.8(d) provides a comparison of predicted and observed temperature–time plots for a three-coil configuration using an extended version of the TWIN program (see Harries and Guthrie 1982; Guthrie and Perrin 1982; and Perrin *et al.* 1987).

6.4. COMPUTATIONAL FLUID FLOW ANALYSES

Similar computational schemes to those just presented can, and have, been developed for fluid flow problems. These can again be divided into two classes, depending on whether an explicit or implicit numerical scheme is adopted. A further subdivision, into parabolic and elliptic flows is common, depending on whether downstream flow affects the character of the upstream flow. Typical elliptic flows involve zones of recirculation, and these are a common feature in many metallurgical systems. Flows which readily come to mind include electromagnetic stirring in continuous casting, fluid flow in filling ladles and gas driven flows in batch processing operations.

TABLE 6.9. *Comparison of implicit and explicit computational efficiency*

	Explicit	Implicit
Maximum Fourier modulus	0.25	8
Maximum time-step ($M = \alpha \Delta t / \Delta R^2$)	375 s	12 000 s
Total iterations required for 30-h	288	9
Estimated ratio of execution times	19	1

Data for Table 6.9 and Figures 6.7 and 6.8

Inside coil heat transfer coefficient	4.18 W m^{-2} K^{-1}
Outside coil heat transfer coefficient	41.8 W m^{-2} K^{-1}
Top surface of coil heat transfer coefficient	41.8 W m^{-2} K^{-1}
Bottom surface of coil heat transfer coefficient	41.8 W m^{-2} K^{-1}
Inside ambient gas temperature (constant)	1273 K
Outside ambient gas temperature (constant)	1273 K
Vertical thermal conductivity	4.18×10^{-3} W m^{-1} K^{-1}
Radial thermal conductivity	4.18×10^{-4} W m^{-1} K^{-1}
Density	7000 kg m^{-3}
Heat capacity	6.27×10^{-4} J kg^{-1} K^{-1}
Inside radius of coil	0.25 m
Radial width of coil	0.50 m
Coil height	1.0 m
Radial nodal spacing	50 mm
Vertical nodal spacing	100 mm
Vertical modulus	variable

Although the numerical methods for solving flow problems are conceptually equivalent to those just presented, it is more difficult to obtain satisfactory (converged) solutions, since computations require the simultaneous solution of a number of highly non-linear equations. For example, in order to solve the turbulent axisymmetric flow problem of a vertical coherent jet entering a ladle, shown in Fig. 6.9(a, b), the following time-averaged differential equations must be discretized and solved (Salcudean and Guthrie 1979), (see Chapter 2 for equivalent laminar flow equations, and note overbars on U and V to denote time-averaged quantities are omitted below for clarity of text),

Continuity equation:

$$\frac{\partial U}{\partial r} + \frac{\partial V}{\partial z} + \frac{U}{r} = 0. \qquad (6.43)$$

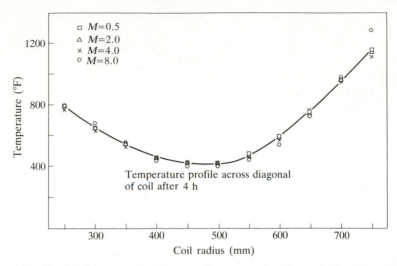

FIG. 6.7.   Predicted temperature–time profiles across the diagonal of a steel coil after four hours of heating in the batch-anneal furnace, showing the effect of local Fourier modulus M on accuracy of numerical solutions.

Transport equation for U

$$\frac{1}{r}\left\{\frac{\partial}{\partial z}(\rho r VU)+\frac{\partial}{\partial r}(\rho r U^2)-\frac{\partial}{\partial z}\left(r\mu_{\text{eff}}\frac{\partial U}{\partial z}\right)-\frac{\partial}{\partial r}\left(r\mu_{\text{eff}}\frac{\partial U}{\partial r}\right)\right\}$$

$$-\frac{\partial P}{\partial r}-\frac{\partial}{\partial z}\left(\mu_{\text{eff}}\frac{\partial V}{\partial r}\right)-\frac{1}{r}\frac{\partial}{\partial r}\left(\mu_{\text{eff}}\,r\,\frac{\partial U}{\partial r}\right)+\mu_{\text{eff}}\frac{U}{r^2}=0. \qquad (6.44)$$

Transport equation for V

$$\frac{1}{r}\left\{\frac{\partial}{\partial z}(\rho r V^2)+\frac{\partial}{\partial r}(\rho r VU)-\frac{\partial}{\partial z}\left(r\mu_{\text{eff}}\frac{\partial V}{\partial z}\right)-\frac{\partial}{\partial r}\left(r\mu_{\text{eff}}\frac{\partial V}{\partial r}\right)\right\}$$

$$-\frac{\partial p}{\partial z}-\frac{\partial}{\partial z}\left(\mu_{\text{eff}}\frac{\partial V}{\partial z}\right)-\frac{1}{r}\frac{\partial}{\partial r}\left(r\mu_{\text{eff}}\frac{\partial U}{\partial z}\right)=0. \qquad (6.45)$$

in conjuction with a turbulence model for μ_{eff}. Adopting the popular two-equation k–ε model of Spalding and Launder (1974), the relevant differential equations are:

Transport equation for k

$$\frac{1}{r}\left\{\frac{\partial}{\partial z}(\rho r Vk)+\frac{\partial}{\partial r}(\rho U\mu k)-\frac{\partial}{\partial z}\left(\frac{r\mu_{\text{eff}}}{\sigma_k}\frac{\partial k}{\partial z}\right)-\frac{\partial}{\partial z}\left(\frac{r\mu_{\text{eff}}}{\sigma_k}\frac{\partial k}{\partial r}\right)\right\}+\mu_t G-C_D\rho\dot\varepsilon=0,$$

$$G=2\left\{\left(\frac{\partial V}{\partial z}\right)^2+\left(\frac{\partial U}{\partial r}\right)^2+\left(\frac{U}{r}\right)^2\right\}+\left(\frac{\partial V}{\partial r}+\frac{\partial U}{\partial z}\right)^2. \qquad (6.46)$$

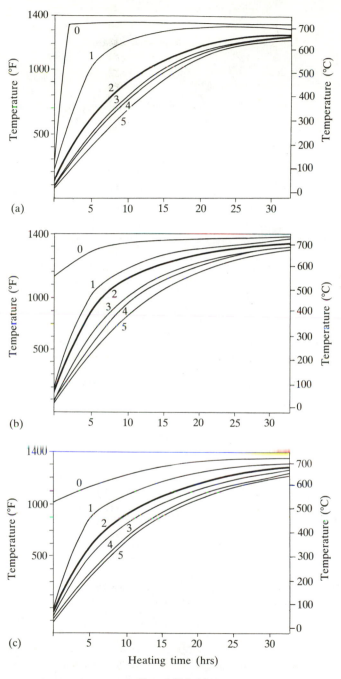

(a)

(b)

(c)

Heating time (hrs)

FIG. 6.8(a)–(c)

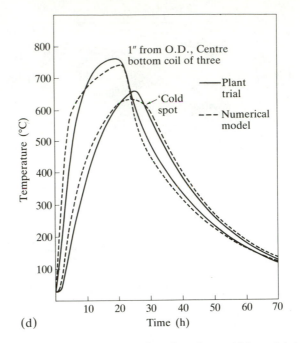

FIG. 6.8. Predicted temperatures at various locations within a tightly wound coil during heating: (a) observed, (b) predicted, (c) tuned, and (d) during total heat–cool cycle. 0, Gas stream; 1, 1 inch from O.D.; 2, 19 inches from O.D.; 3, 10 inches from O.D., 2 inches from top and bottom; 4, 7 inches from O.D.; 5, 14 inches from O.D.

Transport equation for $\dot{\varepsilon}$

$$\frac{1}{r}\left\{\frac{\partial}{\partial z}(\rho r V \dot{\varepsilon}) + \frac{\partial}{\partial r}(\rho r U \dot{\varepsilon}) - \frac{\partial}{\partial z}\left(\frac{r\mu_{\text{eff}}}{\sigma_{\dot{\varepsilon}}}\frac{\partial \dot{\varepsilon}}{\partial z}\right) - \frac{\partial}{\partial r}\left(\frac{r\mu_{\text{eff}}}{\sigma_{\dot{\varepsilon}}}\frac{\partial \dot{\varepsilon}}{\partial r}\right)\right\}$$

$$-C_1 \dot{\varepsilon}\frac{\mu_t G}{k} + \frac{C_2 \rho \dot{\varepsilon}^2}{k} = 0, \qquad (6.47)$$

$$\mu_t = (\mu_{\text{eff}} - \mu) = C_\mu \rho \frac{k^2}{\dot{\varepsilon}}. \qquad (6.48)$$

Efficient computational schemes developed for solving this set of equations rely on the fact that they can be written in a general form

(Convection = diffusion terms + source term).

For instance, replacing our temperature θ with the general variable ϕ, which can now represent temperature, 'concentration', axial and radial velocity components, turbulence energy k, and rate of turbulence energy dissipation $\dot{\varepsilon}$, a general equation for fluid/heat/mass transport processes can be written in the form

$$\frac{\partial}{\partial t}(\rho\phi) + \text{div}(\rho u\phi) = \text{div}(\Gamma_\phi \text{ grad } \phi) + S_\phi, \tag{6.49}$$

where Γ_ϕ grad ϕ represents the diffusive flux of ϕ and S_ϕ is a 'source term' which essentially involves those residual elements which do not fit into the other three groupings. Discretization of this formula, together with a one-dimensional TDMA (of the type just presented), and a Gauss–Siedel routine, linking adjacent lines of nodes, results in an efficient line-by-line computational scheme for numerical integration.

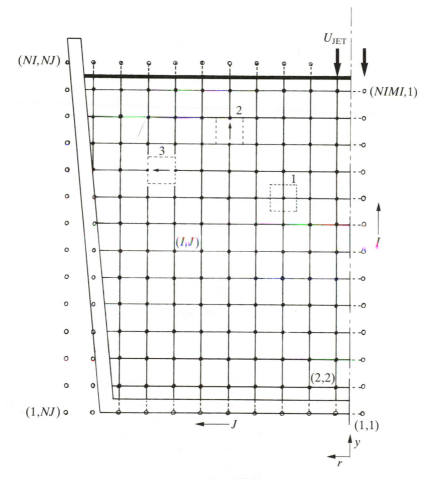

FIG. 6.9(a)

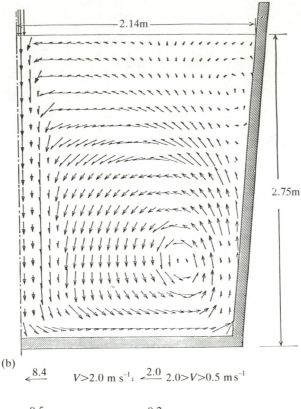

(b)

$\xleftarrow{8.4}$ $V>2.0$ m s^{-1}; $\xleftarrow{2.0}$ $2.0>V>0.5$ m s^{-1}

$\xleftarrow{0.5}$ $0.5>V>0.2$ m s^{-1}, $\xleftarrow{0.2}$ 0.2 m s$^{-1}>V$

FIG. 6.9. (a) Schematic of idealized axisymmetric flow system for simulating furnace tapping operations into teeming ladles, illustrating the set up of the scalar grid arrangement, together with typical volume elements for scalar and vector quantities. 1, scalar control volume; 2, vector U control volume; 3, vector V control volume. (b) Typical predicted flow fields for a jet entering a 250 tonne ladle vertically at a velocity of 10 m s^{-1}, in the absence of air entrainment.

(a) Discretization of fluid flow equations

Before reviewing some of the applications of these, and other, techniques to flows of metallurgical interest, it is worth acquainting the reader with some of the intricacies associated with current computational methods in fluid mechanics. Thus, the discussion has so far only dealt with numerical schemes incorporating diffusive fluxes; nothing has been said regarding the special problems of treating convective flows of liquid passing through volume elements of the type illustrated in the ladle-filling problem in Fig. 6.9a.

There, in similar fashion to the transient heat conduction examples introduced earlier, a grid network is constructed, nodal points being located at points of intersection. At these nodes, scalar properties—pressure P, temperature θ, concentration C_i, turbulence kinetic energy, k, and rate of turbulence energy dissipation $\dot{\varepsilon}$, etc.—acquire values which represent corresponding averages within the fluid bounded by the imagined volume element surrounding each node.

Evidently, liquid will flow from one such volume element into another at a rate dictated by the continuity and momentum equations and imposed boundary conditions. The aim of computational fluid mechanics is then to predict the flows and associated phenomena that would take place: in this case, when the coherent jet of steel enters a ladle in the absence of any air entrainment.

Predictions for a typical industrial case are shown in Fig. 6.9(b), where one sees that a strong recirculating flow deep in the ladle might help to entrain buoyant alloy additions and promote alloy dispersion and mixing. To illustrate the way such convective flows are to be treated, let us now consider a simple example, in which heat is being transported from a westerly volume element into the central element by a combination of steady-state diffusion and convection. This is shown in Fig. 6.10(a). The general ϕ form of the differential energy equation describing this situation would, in Cartesian coordinates, be

$$\frac{d}{dx}(\rho U \phi) - \frac{d}{dx}\left(\Gamma \frac{d\phi}{dx}\right) = 0, \tag{6.50}$$

where $\phi = \theta$ and $\Gamma = \Gamma_t + \Gamma_1 \simeq (k_t + k_1)/C$. Γ, the effective diffusive flux coefficient for conduction of heat, will contain a (normally) dominant turbulent component Γ_t, and the familiar laminar flow component, Γ_1. (For liquid metals, thermal conductivities k are high, and the effect of turbulence in enhancing the total diffusive heat flux is not so marked).

Referring to the nodal points W, P, and E and their associated volume elements in Fig. 10(a), we can apply the familiar statement of heat conservation and say:

Nett rate of heat transport from volume element P across its easterly face by convection and diffusion

= nett rate of heat transport into volume element P across its westerly face by convection + diffusion

or

$$(\rho U \phi)_e - \left(\Gamma \frac{d\phi}{dx}\right)_e = (\rho U \phi)_w - \left(\Gamma \frac{d\phi}{dx}\right)_w. \tag{6.51}$$

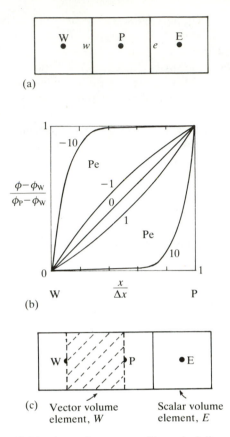

FIG. 6.10. (a) Three fluid volume elements used in typical discretization and numerical solution procedures. Nodes, at their respective centres, are attributed with the mean scalar values (e.g. θ, P, C, ...) of the control volume. (b) Normalized plot of ϕ versus distance between the westerly (W), and central node (P) in (a), to illustrate the role of the convective/diffusive flow ratio, as defined by the Peclet number, $Pe = UL/D_e$. (c) The 'staggered grid' arrangement for vector and tensor quantities (U, τ_{xx}, ...), this being displaced a half-cell distance with respect to the scalar grid in (b).

If we take $\Delta x_w = \Delta x_e$ and assume (as before), a roughly linear change in ϕ between nodal points, then

$$\phi_e = \frac{\phi_E + \phi_P}{2} \quad \text{and} \quad \phi_w = \frac{\phi_P + \phi_w}{2},$$

giving

$$\frac{(\rho U)_e}{2}(\phi_E + \phi_P) - \frac{(\rho U)_w}{2}(\phi_P + \phi_w) = \Gamma_e \frac{(\phi_E - \phi_P)}{\Delta x} - \Gamma_w \frac{(\phi_P - \phi_w)}{\Delta x}. \qquad (6.52)$$

Collecting terms:

$$\left(\frac{(\rho U)_e}{2} + \frac{\Gamma_e}{\Delta x} - \frac{(\rho U)_w}{2} + \frac{\Gamma_w}{\Delta x} \right) \phi_P =$$

$$\left(\frac{\Gamma_e}{\Delta x} - \frac{(\rho U)_e}{2} \right) \phi_E + \left(\frac{\Gamma_w}{\Delta x} + \frac{(\rho U)_w}{2} \right) \phi_W. \tag{6.53}$$

Replacing ρU by F (convective flow flux) and $\Gamma/\Delta x$ by D (diffusive flux)

$$\left(\frac{F_e}{2} + D_e - \frac{F_w}{2} + D_w \right) \phi_P = \left(D_e - \frac{F_e}{2} \right) \phi_E + \left(D_w + \frac{F_w}{2} \right) \phi_W, \tag{6.54}$$

or, more generally,

$$a_P \phi_P = a_E \phi_E + a_W \phi_W. \tag{6.55}$$

It is clear how this sort of discretization can be built up for two-, and three-dimensional steady flows through similar analyses.

For two dimensions:

$$a_P \phi_P = a_E \phi_E + a_W \phi_W + a_N \phi_N + a_S \phi_S. \tag{6.56}$$

For three dimensions:

$$a_P \phi_P = a_E \phi_E + a_W \phi_W + a_N \phi_N + a_S \phi_S + a_T \phi_T + a_B \phi_B. \tag{6.57}$$

More generally, for those differential equations containing a source term (e.g., the U, V, W, momentum equations, and the C, θ, k and $\dot{\varepsilon}$ equations), one can include the coefficient b:

$$a_P \phi_p = a_E \phi_E + a_W \phi_W + a_N \phi_N + a_S \phi_S + a_T \phi_T + a_B \phi_B + b, \tag{6.58a}$$

and condense this into the generalized algebraic form

$$a_P \phi_p = \sum a_{nb} \phi_{nb} + b, \tag{6.58b}$$

where the subscript nb represents nearest neighbour points, and subscript p denotes the node whose ϕ value is currently under determination, i.e.,

$$\phi_p^+ = \frac{\sum a_{nb} \phi_{nb} + b}{a_p}. \tag{6.59}$$

Returning to the one-dimensional convective heat flow example, continuity requires that

$$(\rho U)_e = (\rho U)_w. \tag{6.60}$$

Consequently, denoting $\rho U \Delta x / \Gamma$ by Pe (a Peclet modulus), one can recast eqn (6.54) in the form

$$\phi_P(4) = (2 - \text{Pe}) \phi_E + (2 + \text{Pe}) \phi_W. \tag{6.61}$$

This discretization approach, termed the central difference scheme, suffers from an instability problem associated with the $(2 - \text{Pe})$ bracket. Thus, if the convective flux through the east face is significantly greater than the diffusive flux term, the Peclet modulus (representing the convective-to-diffusive flux ratio) is large, and a_E becomes negative. Take, for instance, $\phi_W = 20°C$, $\phi_E = 10°C$ and Pe $= 10$, then $\phi_P = \{(-8) \times 10 + 12 \times 20\}/4 = 40°C$. Evidently, ϕ_P should lie between 10 and $20°C$ and certainly, from physical arguments, could not exceed 20, which corresponds to a 100 per cent convection-dominated process in which the temperature of the central node, ϕ_P, equals that of the 'upwind' westerly face.

This provides a convenient introduction to the *upwind* scheme, which represents an early, and effective, way to resolve such difficulties. It is also referred to as the upwind-difference scheme, the upstream difference, and donor-cell method. In this procedure, the value of ϕ at a volume element interface is not averaged, but is set equal to (or adjusted towards) the value of ϕ at the grid point on the *upwind* side of the face. Thus

$$\phi_w = \phi_W \text{ if } (\rho U_w) > 0,$$

$$\phi_w = \phi_P \text{ if } (\rho U_w) < 0.$$

This procedure was first proposed by Courant *et al.* (1952). Alternative, and more recent, methods include the exponential, hybrid, and power-law schemes (see Patankar 1980).

The analytical solution to eqn (6.50), taking $\phi = \phi_w$ at $x = 0$ and $\phi = \phi_P$ at $x = \Delta x$, is

$$\frac{\phi - \phi_W}{\phi_P - \phi_W} = \frac{\exp(\text{Pe}\, x/\Delta x) - 1}{\exp(\text{Pe}) - 1}. \tag{6.62}$$

This is graphed in Fig. 6.10(b): the spatial variation of ϕ between the westerly and central nodal points separated by a length equal to Δx is shown for various Peclet moduli, Pe $= \rho U \Delta x/\Gamma$. It confirms that for a straight diffusion problem, in which D_E, or k_E, is constant, a linear profile in ϕ (i.e., C or θ) applies. It shows that for Pe $= 1$, the error is not excessive if we average the values of ϕ_P and ϕ_W, in calculating combined convective and diffusive heat fluxes (say) into the westerly face of the central volume element. However, for Pe $= 10$, one sees that a major fraction of the space between nodes W and P is dominated by the value of ϕ at W for a positive (W to E) flow and by ϕ at P for a negative (E to W) flow, where Pe $= -10$. For Pe $= \pm 100$, a not uncommon situation, these effects are even more pronounced, with ϕ values at $\Delta x/2$ essentially defined by upwind convective flow phenomena, as one might expect.

The analysis demonstrates why the upwind scheme has achieved success in numerical computations at high Reynolds numbers, whereas computational efforts with the central-difference scheme were limited to low Reynolds number flows.

(b) Computational procedures for solving the momentum and continuity equations

The previous discussion tacitly assumes that the velocity components at the interfaces of the scalar volume elements are already known. In reality this requires that the turbulent forms of the Navier–Stokes and continuity equations, in conjunction with equations modelling turbulence, also be discretized and solved.

Again, special problems arise, the major one being the fact that the momentum equations incorporate unspecified pressure gradient terms (e.g. $\partial P/\partial r$ and $\partial P/\partial z$). These can hardly be embedded into the source term of the common ϕ equation for subsequent TDMA solution, since pressure fields are not arbitrary but directly affect flow fields. Indeed, local continuity will only be satisfied for a given flow problem, provided the pressure field has been correctly calculated (i.e., the pressure field is indirectly specified via the continuity equation).

As a consequence, special numerical procedures have been developed in which the continuity and momentum equations are solved by coupling them via a pressure correction equation. Thus, an initial pressure field is first guessed and the implied flow field is then deduced from the discretized momentum equations. The discretized continuity equation is then tested with the proposed velocity field. If the sum of the mass inflows to a given volume flow element exceeds the nett mass outflow, then (for incompressible flows of the type under discussion), the pressure within that elemental volume of fluid is raised (i.e., corrected upwards) using a pressure correction equation to reduce the excess fluid entering. Adjustments are therefore made throughout the flow domain and the momentum equation is re-solved. Again, the resulting flows are checked for local continuity, etc. These procedures are successively repeated. If all goes well during iterations, residual flows into each volume element converge towards zero, and a unique solution, satisfying both the continuity and momentum equations, is obtained.

Two of the 'sub-algorithms' for solving the momentum-continuity combination are SIMPLE (semi-implicit method for pressure linked equations) and SIMPLER (semi-implicit method for pressure linked equations-revised). The interested reader is referred to a text by S. V. Patankar listed at the end of this chapter for further details of these algorithms.

(c) Staggered grids

Referring to Fig. 6.9(a, b), showing the ladle filling problem, it will be remembered that the nodal points locate the scalar quantities P, θ, C, $\dot{\varepsilon}$ and k, whereas the vectors U and V are located on the cell interfaces (as opposed to centres).

Consider once more the case of steady one-dimensional flow, for which the continuity equation is simply

$$\frac{dU}{dx} = 0. \tag{6.63}$$

If we integrate this over the equivalent Cartesian control volume to that shown in Fig. 6.10(a), we obtain

$$U_e - U_w = 0. \tag{6.64}$$

Averaging velocities to correspond to the midway initial faces of the central main control volume, we have

$$\frac{U_P + U_E}{2} - \frac{U_w + U_P}{2} = 0,$$

or

$$U_E - U_w = 0. \tag{6.65}$$

This shows that a discretized continuity equation whose volume elements coincide with the scalar elements requires equal velocities at alternate, rather than adjacent nodes. If so, one can imagine wavy velocity profiles such as

$$U \rightarrow 100 \quad 300 \quad 100 \quad 300 \quad 100 \quad 300$$

$$\text{——o——o——o——o——o——o——},$$

which are clearly unrealistic but could, if present, persist during numerical computations towards a converged solution.

We see a similar problem with the $(-dP/dx)$ term in the momentum equation. Discretization over the same control volume, i.e. integration, gives

$$\frac{P_w - P_e}{\Delta x} = \left(\frac{P_w + P_P}{2} - \frac{P_P + P_E}{2} \right) \frac{1}{\Delta x} = \frac{P_w - P_E}{2\Delta x}.$$

Again, this means that the discretized form of the momentum equation will contain the pressure differences between two *alternate* grid points rather than *adjacent* ones. While this implies a coarser effective grid for the previous field, it would, more seriously, also allow horror pressure fields such as

$$P \rightarrow 100 \quad 1000 \quad 100 \quad 1000 \quad 100 \quad 1000$$

$$\text{——o——o——o——o——o——o——}$$

to persist during interactions towards a 'converged' solution.

Harlow and Welch (1965) in their MAC (marker and cell) method were the first to propose a staggered grid arrangement to resolve these difficulties. This approach is now *derigueur* among practitioners for the type of flow schemes under discussion. Thus, by off-setting the control volumes for V, U, W, to the main volume (scalar) elements, by placing them midway, we see that

$$U_e - U_w \equiv U_P - U_W \tag{6.66}$$

and

$$-\frac{dP}{dx} \simeq \frac{P_w - P_e}{\Delta x} = \frac{P_P - P_E}{\Delta x}. \tag{6.67}$$

This procedure resolves both issues and has the further advantage that the main nodal point pressures P_P and P_E act on the surfaces of the velocity volume element. The velocity volume element is illustrated in Fig. 6.10(c) and, for the ladle filling problem, in Fig. 6.9(a) where U and V control volumes, and a typical scalar cell (I, J), and scalar control volume are illustrated.

(d) The Gauss–Siedel point-by-point method

The reader by now will have appreciated the techniques by which the various differential equations can be discretized. All that remains, once these have been formulated, and written into a computer program, are the solutions! Various approaches are possible, one of the easiest methods being the Gauss–Siedel point-by-point method. Thus, we have a series of discretized equations, of the form

$$\phi_P^+ = \frac{\Sigma a_{nb} \phi_{nb}}{a_p} + \frac{b}{a_p} \tag{6.68}$$

to solve simultaneously, for U, V, C, θ, k, ε, etc., at each of the nodal points on the scalar and vector grid point arrangement chosen.

We will choose a simple example to show how one might tackle such a task. Consider the solution of two equations having the generalized form just noted

$$\theta_1 = 0.5\,\theta_2 + 1 \tag{6.69}$$

and

$$\theta_2 = \theta_1 + 4. \tag{6.70}$$

Analytical solution shows that $\theta_1 = 6$, and $\theta_2 = 10$. Let us start our iterative solution by guessing (arbitrarily) that $\theta_1 = \theta_2 = 0$, and then solving the two equations in sequence in an iterative fashion in which we continually update the values of θ_1 and θ_2 with the most recent deduced:

Iteration No.	0	1	2	3	4	5	6	7	8	...	∞
θ_1	0	1	3.5	4.75	5.38	5.69	5.84	5.92	5.96	...	6
θ_2	0	5	7.5	8.75	9.38	9.69	9.84	9.92	9.96	...	10

We see how this iterative procedure slowly converges towards the correct solution, and how it might be extended to solve hundreds of such algebraic equations. There is however a catch, apart from the slow convergence considerations. Let us re-arrange the two equations, and write them in the form

$$\theta_2 = 2\theta_1 - 2 \tag{6.71}$$

and

$$\theta_1 = \theta_2 - 4. \tag{6.72}$$

Again, guessing $\theta_1 = \theta_2 = 0$ as starting values:

Iteration no.	0	1	2	3	...	...	...	∞
θ_2	0	-2	-14	-38	...	...	...	$-\infty$
θ_1	0	-6	-18	-42	...	...	...	$-\infty$

This time, the iteration procedure is widely divergent! Scarborough (1958) has shown that the Gauss–Siedel method is only convergent provided

$$\frac{\Sigma |a_{nb}|}{a_p} \begin{cases} \leq 1 & \text{for all equations} \\ < 1 & \text{for at least one equation} \end{cases} \tag{6.73a}$$
$$\tag{6.73b}$$

One sees that eqns (6.69), (6.70) satisfied these criteria, whereas eqns (6.71), (6.72), did not. A major disadvantage of the Gauss–Siedel method in solving flow problems is that the method only transmits boundary condition information to the bulk flow field nodes at a rate of one grid interval per iteration. On the other hand, the TDMA method does not suffer from this restriction, but becomes inconvenient to use for three-dimensional problems in view of the number of coefficients needed, and its general level of complexity.

A convenient compromise for solving sets of highly non-linear equations describing flow is a line-by-line technique, in which the direct method (TDMA) for a one-dimensional situation is combined with the Gauss–Siedel method. The procedure can be imagined by referring to Fig. 6.9. Suppose that the chosen line for TDMA is half-way up the ladle; then one can use the Gauss–Siedel routine in iterative sweeps from the top to the bottom of the ladle on a line-by-line basis. Thus, inserting the most recent values for ϕ for corresponding adjacent southerly and northerly nodes, the one-dimensional form of the TDMA is used to solve and update all ϕs along the chosen line. By this means, one can quickly bring information from all boundaries into the interior, speeding up solution procedures significantly.

Other issues

Other issues to which the reader is referred to the specialist literature include the construction and use of variable grids, the phenomenon of 'false diffusion', over-relaxation and under-relaxation techniques for numerical convergence, harmonic mean Γ's and cell blockage procedures for introducing obstacles to the flow or handling irregular boundary zones (such as the tapered ladle in Fig. 6.9(c), the treatment of two-phase flow systems and magnetohydrodynamics. The text by S. V. Patankar provides an excellent introduction to many of these matters.

(f) Alternative numerical approaches and techniques

Many alternate numerical procedures and techniques have been used since the start of computational efforts in fluid mechanics, beginning in the early 1960s with the advent of high-speed digital computers. A popular approach, which was much in vogue into the last decade, was the stream function/vorticity method borrowed from potential flow theory, and well described by Gosman *et al.* (1969). While eliminating the pressure field problem, it was only applicable to two-dimensional flows, hastening a return to the so-called 'primitive variables' of U, V, and W in the late 1970s. The first three-dimensional flow example of metallurgical interest appearing in the literature describes flow generated by a plunging jet of liquid metal into a rectangular cavity (Salcudean and Guthrie 1979). The solving routines were written in accordance with the finite-difference MAC procedures of the Los Alomos Group (Amsden and Harlow, 1970).

The techniques described in this chapter are based on the finite-difference technique. An alternative approach gaining some popularity is the finite-element method, in which the calculation domain is typically subdivided into triangular elements. These are certainly more convenient than the regular grids of the finite-difference approach when fitting complex boundary shapes. The discretization equations are usually derived by the use of a variational principle or by the Galerkin method, which is a special case of the method of weighted residuals. The variational formulation does not have an easy physical interpretation, which has tended to damp its acceptance by fluid-flow, heat transfer, practitioners. Progress has also been blocked until recently (Baliga and Patankar 1983) by the lack of equivalents to either the upwind formulation for treating convection or the staggered grid procedure. Also, the fact that the 'SIMPLE type' procedures have to be replaced with direct simultaneous solution of the continuity and momentum equations calls for much larger computer storage space requirements, making three-dimensional computations restrictive.

Yet another procedure for solving the continuity/Navier–Stokes equations is the spectral element method. Similar to the finite element approach, the method involves the expansion of the solution to a differential equation in a high-order orthogonal expansion, the coefficients of which are determined by a weighted-residual projection technique. The schemes are 'infinite'-order accurate if the expansion functions are chosen properly (Patera 1984).

6.5.   SOME CASE STUDIES

To familiarize the reader with the potential uses of such models in systems of metallurgical interest, a few brief examples are presented before concluding this chapter.

(a) Bubble growth and jet behaviour

Recent work by process metallurgists has shown that significant differences can exist between the injection of gas into liquid metals versus the equivalent chemical engineering counterpart of air into water. As part of those studies, a transient, two-phase solution to predict the growth and size of bubbles forming at nozzles in liquid metals and aqueous counterparts allowed for a quantitative understanding of the phenomenon. By solving the Navier–Stokes equations for the gas and liquid phases simultaneously, it was possible to check the importance of gas jet momentum, surface tension, and liquid density. Figures 6.11(a) and (b) give a comparison of results; argon bubbles growing in molten pig iron at an injection rate of 0.1 litres s^{-1} are shown to grow about three times larger than air bubbles in water ($\sim 12 \text{ cm}^3$ versus 4 cm^3) before releasing (Sahai and Guthrie 1982), in agreement with experimental data (Irons and Guthrie 1978). These results have a significant bearing on process kinetics and melt refining operations.

(b)   Fluid flow patterns generated in ladles by submerged gas injection

Figure 6.12(a) illustrates a commercial CAS ladle station for the controlled addition of ferro-alloys and deoxidizers into a central, slag-free region of steel, the centrally baffled region over a rising plume of steel and gas.

To solve the fluid flow problem, allowance for the lower density region within the plume is included in the computations, together with a model of gas/liquid coupling. Fig. 6.12(b) shows predicted flow, and effective viscosity, fields while 6.12(c) illustrates the trajectories that would be followed by alloy additions that are lighter, heavier or neutrally buoyant when added to the centrally baffled region.

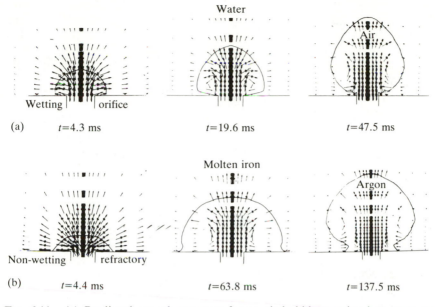

FIG. 6.11.   (a) Predicted growth sequence for an air bubble growing into stagnant water through an orifice of 0.635 cm diameter at a flowrate of $100 \, \text{cm}^3 \, \text{s}^{-1}$. (b) Predicted growth sequence for an argon bubble growing into stagnant liquid iron at 1250°C through an orifice diameter of 0.635 cm at an argon flow rate of $100 \, \text{cm}^3 \, \text{s}^{-1}$.

In the absence of any baffles, radial outflows across the top surface are naturally much higher, as illustrated in Fig. 6.12(d).

The influence of bath hydrodynamics, alloy melting/dissolution rates, and subsequent intermixing, are finally combined to give the 'mixing-in' curves shown in Fig. 6.12(e) (Mazumdar and Guthrie 1987).

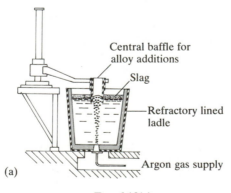

FIG. 6.12(a)

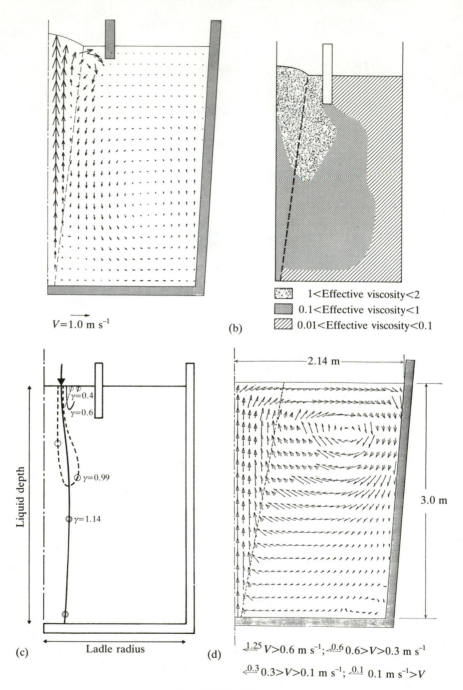

$V=1.0 \text{ m s}^{-1}$

(b)

⬚ 1<Effective viscosity<2
▨ 0.1<Effective viscosity<1
▨ 0.01<Effective viscosity<0.1

$\gamma=0.4$
$\gamma=0.6$
$\gamma=0.99$
$\gamma=1.14$

Liquid depth

(c)   Ladle radius

2.14 m

3.0 m

(d) $^{1.25}$ $V>0.6 \text{ m s}^{-1}$; $^{0.6}$ $0.6>V>0.3 \text{ m s}^{-1}$
$^{0.3}$ $0.3>V>0.1 \text{ m s}^{-1}$; $^{0.1}$ $0.1 \text{ m s}^{-1}>V$

Fig. 6.12(b)–(d)

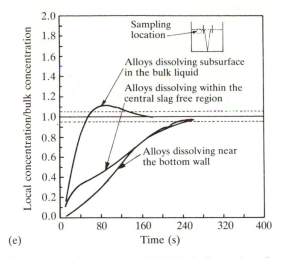

(e)

FIG. 6.12. (a) Illustration of a commercial CAS ladle station, for the controlled injection of ferro-alloys and deoxidizers into a central, slag-free region of the steel, within the central baffle. (b) Predicted flow and effective viscosity fields in a 150-tonne ladle, stirred centrally at a flowrate of $1.13 \, m^3 \, min^{-1}$. The effect of a cylindrical axisymmetrically placed baffle on flow patterns (i.e., CAS method) is illustrated. (Ladle radius 1.82 m, depth of liquid 3.04 m, depth of reference cylinder 0.18 m, radius of refractory cylinder 0.6 m). (c) Predicted trajectories of spherical alloy additions, in a 150 tonne ladle incorporating the C.A.S. technique for alloy additions. (d) Predicted liquid steel flows generated by injection of argon into a slag free 250-tonne cylindrical, tapered ladle at a flow-rate of $0.25 \, Nm^3 \, min^{-1}$ through a centrally placed nozzle, porous plug, or other proprietary element. (e) Predicted changes in steel chemistry with time, following the introduction of buoyant (i.e., 50%, Fe–Si, Al), neutrally buoyant (e.g. Fe–Mn), and more dense (i.e., 60%, Fe–Nb) additions to the centrally baffled region. Note the dimensionless concentration profiles refer to a sampling location outside the baffled region.

(c) Flow and particle behaviour in a blast furnace clarifier system

Figure 6.13(b, c) shows predicted flow patterns generated when water flows out radially from a central feedwell area into a large circular clarifier tank (~ 50 m diameter), illustrated in Fig. 6.13(a). It carries with it iron oxide and carbon particles of various shapes and sizes, and these must be allowed to settle out before the contaminated water reaches the outflow to the tank. Typical computed particle trajectories are shown in Figs 6.14 (a, b) respectively, while Fig. 6.15 shows a computed C curve for comparison with previous tracer diagnostic tests carried out by the steelworks research personnel (Sahai, Y., and Guthrie, R. I. L., 1982, unpublished research report). This example is included to demonstrate that reactor analyses of the type presented in

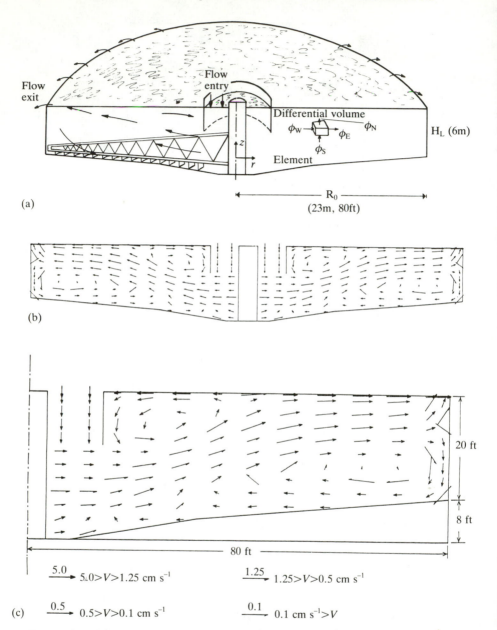

FIG. 6.13.(a) Schematic of an industrial clarifier for the treatment of blast furnace steel plant waste water. Shown are the coordinate system, a differential volume element, flow patterns, rotating vanes for sludge disposal via a central exit port, and overflow of cleaned liquid over circumferential weir. (b) Radial section through clarifier demonstrating proposed flow patterns at $18\,000$ US. gal min^{-1}. (c) Detailed flow field.

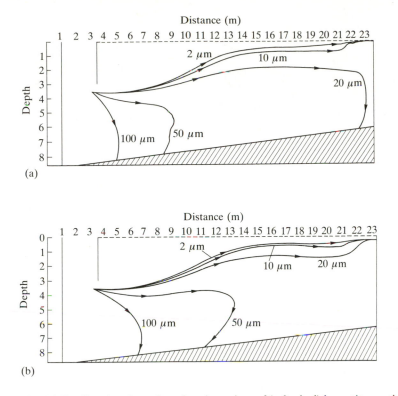

FIG. 6.14. (a) Predicted trajectories of various sizes of (spherical) hematite particles entering through clarifier feedwell, and settling during summertime. Ambient water temperature 120°F. (b) Predicted trajectories of various sizes of hematite particles entering through clarifier feedwell, and settling during wintertime. Ambient water temperature 80°F. Note the higher viscosity of water under winter conditions, leads to 20 μm particles joining the effluent.

Chapter 3 are being superseded by computational schemes such as those just demonstrated. The model was able to explain seasonal variations in effluent water quality: poorer results in the winter arising from reduced settling velocities of small particles in colder water of higher viscosity.

(d) Three-dimensional flow of steel in a rectangular tundish for a twin-slab caster

As a final example to illustrate the use of mathematical models in metallurgical systems, the isothermal flow of steel through a 60-tonne rectangular tundish into a twin-slab caster is predicted for a 0.167-scale water model. Table 6.10 provides data on physical parameters studied. For the sake of

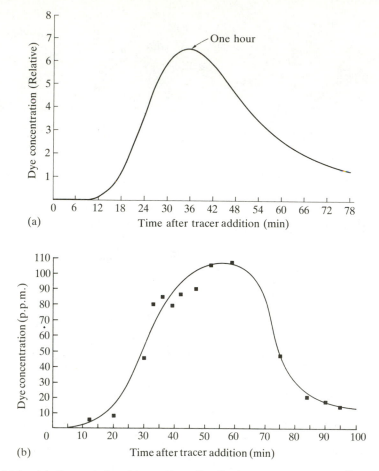

FIG. 6.15. (a) Computed residence time distribution curve corresponding to flow field in Fig. 6.13. (b) Experimental residence time distribution curve.

TABLE 6.10. *Full-scale, and model parameters, for a 60-tonne twin-slab caster tundish*

	Full Scale	Model
Tundish width, m	1.00	0.167
Tundish length, m	7.00	1.167
Tundish height, m	1.50	0.250
Filled height, m	1.28	0.214
Volumetric flow rate	11.75 tonnes min^{-1}	19. litres min^{-1}
Froude number based on inflow orifice	15.9	15.9
Reynolds number based on inflow orifice	2.52×10^4	3.71×10^5

brevity, only a few sections are shown in Fig. 6.16 (b) (i) near the vertical jet entry plane; (ii) the vertical jet exit plane; (iii) near the vertical longitudinal plane of symmetry; (iv) near the bottom surface of the tundish; (v) near the top free surface and (vi) close to the tundish side-wall. Owing to symmetry, only one quadrant of the flow field required solution. The model shows that the steel jet creates a weakly recirculating flow in plane (i), a weak vertical downflow to the jet exit in planes (ii) and (iii), a flow up the sidewalls (vi), a weakly returning surface flow and separating outflow (v) and a reversing inflow back along the lower surface which opposes the spreading flow from the jet (v).

The beauty of such numerical simulations is that once they are experimentally validated using basic physical models (as in this case), they can then be used in the design of full-scale high-temperature systems. For example, the effect of thermal natural convection, of the removal of inclusions, or the intermixing of steel grades during sequence casting, etc., are then open to analysis. (Tanaka and Guthrie 1985; Guthrie, Joo, and Nakajima 1988).

6.6. CONCLUSIONS

This chapter has demonstrated the use and value of mathematical modelling and its application to metallurgical processing operations. It is meant to serve as an introduction to the many exciting possibilities open to the process design engineer thanks to the advent of computer technology. General three-dimensional programs with two phase capabilities are now available for computational modelling of fluid flow systems. Such programs provide yet another set of tools that are potentially the most powerful and useful to the

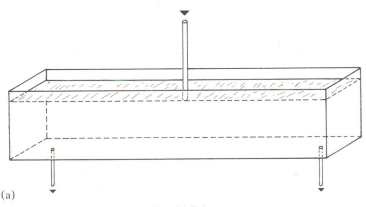

(a)

FIG 6.16(a)

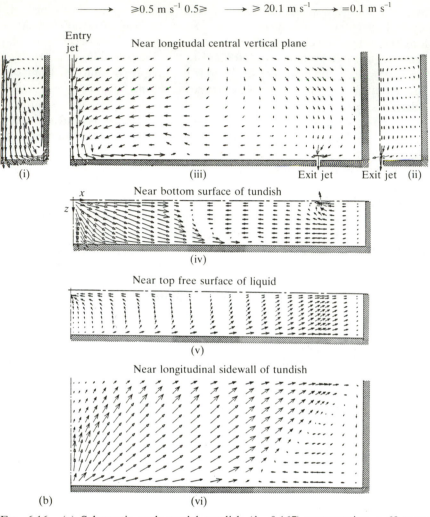

FIG. 6.16. (a) Schematic scale-model tundish ($\lambda = 0.167$) representing a 60-tonne tundish for a twin slab caster (Model length = 1.167 m width = 0.167 m, height = 0.25 m filled height = 0.214 m) submerged entry nozzle diameter = 16 mm, flow rate = 19 litres min^{-1}. (b) Predicted flow patterns:

(i) $x/L = 0.004$, $0 \leq y \leq H$,

(ii) $x/L = 0.428$, $0 \leq y \leq H$,

(iii) $z/W = 0.015$, $0 \leq y \leq H$, $0 \leq x \leq H$,

(iv) $y/H = 0.95$, $0 \leq x \leq L$, $0 \leq z \leq W$,

(v) $y/H = 0.05$, $0 \leq x \leq L$, $0 \leq z \leq W$,

(vi) $z/W = 0.473$, $0 \leq x \leq L$, $0 \leq y \leq H$.

practitioner of metallurgical engineering. As with all computer modelling, the key question is often not so much 'how good is the mathematical description of real physical phenomena' but 'how well are these phenomena understood?' The problem often boils down to choosing the right boundary conditions, knowing which factors can be ignored and which cannot, rather than mathematical complexity, slow algorithms, or insufficient storage capacity. Indeed, the use of dynamic storage allocation used by some programs (e.g., Markatos and Spalding 1983), renders the solution of even three-dimensional problems feasible on minicomputers or sophisticated microprocessors equipped with Definicon boards and the like.

Alternatively, very large programs can be run on supercomputers such as the CRAY. XMPs and the parallel processing machines now entering the market. Also exciting is the development of cellular-automata supercomputers for fluid-dynamics modelling. This is a discrete, dynamical system of cells arranged on a finite grid in which the motion and collisions of individual particles, or elements of fluid, are modelled. Such models can give rise to the Navier–Stokes equation in the macroscopic limit. Margolus *et al.* (1986) suggest that a cellular-automata machine having 10^{12} cell sites and an update cycle of 100 ps for the *whole array* will be feasible in one decade and within easy reach by two. A machine using 10^{16} sites (Avagadro's number in two dimensions) is a possibility, so that in the future we can anticipate computational tools that can be used directly to span the gap between the microscopic world of colliding particles and our macroscopic world of heat, mass, and momentum transport.

FURTHER READING

1. Carnahan, B., Luther, H. A., and Wilkes, J. O. (1969) *Applied numerical methods.* Wiley, New York.
2. Patankar, S. V. (1980). *Numerical heat transfer and fluid flow.* McGraw-Hill, New York.
3. Spalding, D. B. (1981). *Mathematics and computers in simulation,* Vol. 23, pp. 267–276. North-Holland, Amsterdam.
4. Sahai, Y. and Guthrie, R. I. L. (1982). *Advances in transport processes* (Ed. A. S. Mujumdar). Wiley Eastern.
5. Markatos, N., Rhodes, N., and Tatchell, D. G. (1982). 'A general purpose program for the analysis of fluid flow problems' In *Numerical methods for fluid dynamics* (ed. K. W. Marton and M. J. Baines). Academic Press, London. pp. 463–80.
6. Szekely, J., Evans, J. W., and Brimacombe, J. K. (1988). *The mathematical and physical modelling of primary metals processing operations,* John Wiley, New York.

APPENDIX I

NOMENCLATURE

Symbol	Meaning	SI Units
a	radius	m
a	coefficient in discretization equations	—
A	area	m^2
b	coefficient in discretization equations	—
c	velocity of molecules	$m\,s^{-1}$
c	cocfficient in discretization equations	—
c	speed of light (2.998×10^8)	$m\,s^{-1}$
C	general constants	—
C_c	molar concentrations	$kmol\,m^{-3}$
C_D	drag coefficient in steady translatory motion	—
C_P	heat capacity of material at constant pressure	$J\,kg^{-1}\,K^{-1}$
C_v	heat capacity of materials at constant volume	$J\,kg^{-1}\,K^{-1}$
d	diameter	m
d_c	collision diameter of molecules	m
d_m	equilibrium/mean separation distance between atoms	m
d_h	hydraulic diameter	m
D	diffusion coefficient	$m^2\,s^{-1}$
D_{AB}	binary diffusion coefficient of solute A in solvent B	$m^2\,s^{-1}$
D	diameter	m
e	charge of electron $(1.602178 \times 10^{-19})$	C
e	kinetic energy/unit mass of fluid	$J\,kg^{-1}$
E	kinetic energy of fluid	J
E	activation energy in Arrhenius equation	$J\,kmol^{-1}$
E	emmissive power	W
E	electric potential	V
f	Fanning friction factor for flow through a pipe or duct	
F	force	N
F	Faraday's constant (96485)	$C\,mol^{-1}$
F_{12}	geometrical shape factor for radiation from one body to another	—
G	free energy	$kJ\,mol^{-1}$
g	acceleration of gravity (Earth, sea level = 9.81)	$m\,s^{-2}$
$g(r)$	radial distribution function	
h	Planck constant (6.626×10^{-34})	J s
h	heat transfer coefficient	$W\,m^{-2}\,K^{-1}$
H	heat transfer modulus $(h\Delta t/\rho C\Delta x)$	
H	enthalpy	$J\,kmol^{-1}$

Symbol	Meaning	SI Units
I	electric current flow rate	A
I	intensity of radiation	$\mathrm{W\,m^{-2}}$
k	general constants	$\mathrm{W\,m^{-1}\,K^{-1}}$
k	thermal conductivity	$\mathrm{W\,m^{-1}\,K^{-1}}$
k	kinetic energy of turbulence/mass	$\mathrm{J\,kg^{-1}}$ or $\mathrm{m^2\,s^{-2}}$
k_L	liquid phase mass transfer coefficient boundary	$\mathrm{m\,s^{-1}}$
k_G	gas phase mass transfer coefficient at boundary	$\mathrm{m\,s^{-1}}$
K	Boltzmann constant (1.38×10^{-23})	$\mathrm{J\,K^{-1}\,kmol^{-1}}$
K	total kinetic energy	J
log	logarithm to base 10	—
ln	logarithm to base e	—
l	length, general	m
L	length dimension of body, system, etc.	m
L	latent heat of fusion	J
m	mass	kg
m_e	mass of electron (9.11×10^{-31})	kg
M	atomic or molecular weight	$\mathrm{kg\,kmol^{-1}}$
M	Modulus (Fourier $\alpha\Delta t/\Delta x^2$)	
n	number of atoms per unit volume (atoms)	$\mathrm{m^{-3}}$
N	Avagadro's number (6.023×10^{26}) (atoms)	$\mathrm{kmol^{-1}}$
$\dot{N}$	moles/unit time	$\mathrm{kmol\,s^{-1}}$
$\dot{N}''_A$	molar flux of solute A	$\mathrm{kmol\,m^{-2}\,s^{-1}}$
P	pressure	Pa
q	Sievert's constant	$\mathrm{kmol\,m^{-3}\,atm^{-1/2}}$
$\dot{q}$	rate of heat flow	W
$\dot{q}''$	heat flux	$\mathrm{W\,m^{-2}}$
$\dot{q}'''$	rate of heat generation per unit volume	$\mathrm{W\,m^{-3}}$
q'	instantaneous r.m.s. velocity in turbulent flow	$\mathrm{m\,s^{-1}}$
Q	amount of heat	J
r	radius, r_H = hydraulic radius	m
r_m	mean distance between reference atom and nearest neighbours	m
R	universal gas constant (8314)	$\mathrm{J\,kmol^{-1}\,K^{-1}}$
R	electrical resistance	Ω
R	radius	m
t	time	s
T	absolute temperature	K
T	total time	s
$\tilde{u}$	velocity (r.m.s. fluctuating component)	$\mathrm{m\,s^{-1}}$
u'	instantaneous velocity fluctuation in turbulent flow	$\mathrm{m\,s^{-1}}$
u	internal energy/unit volume	$\mathrm{J\,m^{-3}}$
U	internal energy (molar)	$\mathrm{J\,kmol^{-1}}$
U	velocity (x component)	$\mathrm{m\,s^{-1}}$
$\tilde{v}$	velocity (r.m.s. fluctuating component)	$\mathrm{m\,s^{-1}}$

Symbol	Meaning	SI Units
v'	instantaneous velocity in turbulent flow	$\mathrm{m\,s^{-1}}$
V	volume	$\mathrm{m^3}$
V	velocity (z or y component)	$\mathrm{m\,s^{-1}}$
V_b	molar volume at boiling point	$\mathrm{m^3\,kmol^{-1}}$
V_c	molar volume at critical point	$\mathrm{m^3\,kmol^{-1}}$
V_d	dead volume of a reactor	$\mathrm{m^3}$
V_m	mixed volume of a reactor	$\mathrm{m^3}$
V_p	plug flow volume of a reactor	$\mathrm{m^3}$
V_m	molar volume at melting point	$\mathrm{m^3\,kmol^{-1}}$
$\tilde{w}$	velocity (r.m.s. fluctuating component)	$\mathrm{m\,s^{-1}}$
w'	instantaneous velocity fluctuation	$\mathrm{m\,s^{-1}}$
w	width	m
W	width	m
W	velocity (z or θ component)	$\mathrm{m\,s^{-1}}$
X	mole fraction	
x	distance coordinate (normally horizontal)	m
y	distance coordinate (normally normal to flow)	m
z	distance coordinate (normally vertical)	m
z	charge number on ions	—

Greek Characters

Symbol	Meaning	SI Units
α	thermal diffusivity ($k/\rho C_P$)	$\mathrm{m^2\,s^{-1}}$
α	absorptance for radiation (fraction of energy absorbed)	—
β	coefficient in TDMA solving routine	—
β_θ	temperature coefficient of volume expansion	$\mathrm{K^{-1}}$
β_M	composition coefficient of volume expansion	$\mathrm{K^{-1}}$
γ	specific heat ratio	—
γ	Raoultian activity coefficient	—
δ	boundary layer thickness δ_m momentum, δ_θ temperature, δ_c concentration	m
δ	small thickness or distance relative to bulk dimensions;	m
Δ	difference between values	—
ε	emittance for radiation (emissivity)	—
ε	potential energy between molecules or atoms	J
ε	Lennard–Jones parameter	
$\dot{\varepsilon}$	rate of dissipation of turbulence kinetic energy/unit mass of fluid	$\mathrm{J\,kg^{-1}\,s^{-1}}$ or $\mathrm{m^2\,s^{-3}}$
η	efficiency	—

Symbol	Meaning	SI Units
η	mathematical function in error function	—
θ	temperature	K or °C
θ	angle in spherical coordinates	rad
λ	wavelength	m
λ	mean distance between molecular, atomic collisions	m
μ	absolute viscosity	Pa s
v	kinematic viscosity	$\text{m}^2\,\text{s}^{-1}$
v	frequency	s^{-1}
ρ	mass density, ρ_L = density of liquid, ρ_g = density of gas	
ρ	reflectance for radiation (fraction of energy reflected)	—
τ	shearing stress; τ_{yx} = shear stress in x direction exerted by fluid at lesser y on fluid at y	$\text{N}\,\text{m}^{-2}$
τ	transmittance for radiation (fraction of energy transmitted)	—
σ	Stefan–Boltzmann constant (5.67×10^{-8})	$\text{W}\,\text{m}^{-2}\,\text{K}^{-4}$
σ	surface tension	$\text{N}\,\text{m}^{-1}$
σ	scale of turbulence (fraction)	—
σ_t	turbulent Schmidt and Prandtl numbers	—
ϕ	angle in spherical coordinates	rad
ϕ	mathematical symbol to represent $(\theta, C, U, V, W, P, k, \varepsilon)$ in generalized form in transport equations	—
Φ	velocity potential	$\text{m}^2\,\text{s}^{-1}$
Ψ	stream function	$\text{m}^2\,\text{s}^{-1}$
ω	angular velocity	$\text{rad}\,\text{s}^{-1}$
ζ	vorticity $(= 2\omega)$	$\text{rad}\,\text{s}^{-1}$
Ω	parameter in Chapman–Enskog equation for μ	

Subscripts

b	boiling point temperature	
B	bulk	
C	cold	
c	critical temperature	
e	east (vector)	
E	East (scalar nodes)	
G	gas	
H	hot	
i	diffusing/convecting species i	
i	tensor-unit vector	

Symbol	Meaning	SI Units
j	tensor-unit vector	
k	tensor-unit vector	
L	liquid	
n	north (vector)	
nb	neighbouring nodes	
N	North (scalar nodes)	
N	number of nodes	
m	melting point temperature	
M	number (of nodes)	
M	metal	
P	central node	
r	radial	
R	radius	
s	south (vector)	
S	South (scalar nodes)	
S	surface	
t	turbulent	
w	west (vector)	
W	West (scalar nodes)	
v	vertical	
o	at origin; initial value	

Superscripts

*	interface
+	value of variable at new time step
−	value of variable at previous time interval
B	bulk
′	(dash) instantaneous value
~	r.m.s value
—	(overbar) mean value

Emboldened symbols: Vector notation.

APPENDIX II

UNITS, DIMENSIONS, AND CONVERSION FACTORS

Numerical calculations dealing with heat, mass, and momentum transfer require a consistent set of units. It is ironic that the British system of units is still used by North American industry, while Britain and most countries of the world have now adopted the Système Internationale (SI) based on the kilogram, metre, second, kelvin, joule, and kg-mole set of primary units.

TABLE A2.1. *Basic SI units*

Quantity	Name of unit	Unit symbol
Length	metre	m
Mass	kilogram	kg
Time	second	s
Electric current	ampere	A
Temperature (thermo-dynamic)	kelvin	K
Amount of substance	mole	mol
Angle	radian	rad

The mole is the molecular weight expressed in grams. Working in the SI system it will normally be necessary to convert moles to kilogram-moles (kg moles).

TABLE A2.2. *Some derived SI units having special symbols*

Physical quantity	SI unit	Unit symbol
Pressure, stress	pascal	$Pa = N\, m^{-2}$
Force	newton	$N = kg\, m\, s^{-2}$
Work, energy, quantity of heat	joule	$J = N\, m$
Power	watt	$W = J\, s^{-1}$
Electric charge	coulomb	$C = A\, s$
Electric potential	volt	$V = W\, A^{-1}$
Electrical resistance	ohm	$\Omega = V\, A^{-1}$
Frequency	hertz	$Hz = s^{-1}$

TABLE A2.3. *Common multiples and sub-multiples of SI units*

Multiplication factor	Prefix	Symbol
10^{12}	tera	T
10^9	giga	G
10^6	mega	M
10^3	kilo	k
10^2	hecto	h
10^{-1}	deci	d
10^{-2}	centi	c
10^{-3}	milli	m
10^{-6}	micro	μ
10^{-9}	nano	n
10^{-12}	pico	p

TABLE A2.4. *Units in use with the International System*

Unit	Symbol	Value in SI units
Minute	min	60 s
Hour	h	3600 s
Day	d	86400 s
Degree (of arc)	°	$2\pi/360$
Litre	l	$10^{-3}\,m^3$
Tonne	t	$10^3\,kg$
Celsius degree	°C	1 K

TABLE A2.5. *Some common units and conversions*

Quantity	Conversion to SI units
Length	$1 \text{ in} = 2.54 \times 10^{-2} \text{ m}$
	$1 \text{ ft} = 0.3048 \text{ m}$
Area	$1 \text{ ft}^2 = 9.29 \times 10^{-2} \text{ m}^2$
	$1 \text{ in}^2 = 6.45 \times 10^{-4} \text{ m}^2$
Volume	$1 \text{ in}^3 = 1.639 \times 10^{-5} \text{ m}^3$
	$1 \text{ ft}^3 = 2.832 \times 10^{-2} \text{ m}^3$
	1 Imperial (British) gallon $= 4.545 \times 10^{-3} \text{ m}^3$
	$1 \text{ US gallon} = 3.785 \times 10^{-3} \text{ m}^3$
	$1 \text{ l} = 10^{-3} \text{ m}^3$
	$1 \text{ cm}^3 (\text{cc}) = 10^{-6} \text{ m}^3$
Mass	$1 \text{ lb} = 0.454 \text{ kg}$
	1 ton (Imperial) $= 1016 \text{ kg}$
	1 ton (U.S.) $= 907 \text{ kg}$
	$1 \text{ tonne} = 10^3 \text{ kg}$
Density	$1 \text{ lb ft}^{-3} = 16.018 \text{ kg m}^{-3}$
Gravity	$g = 32.174 \text{ ft sec}^{-2} = 9.80665 \text{ m s}^{-2}$
Force	$1 \text{ lb}_f = 4.4482 \text{ N}$
	$1 \text{ poundal} = 0.13826 \text{ N}$
	$1 \text{ dyne} = 10^{-5} \text{ N}$
Pressure	$1 \text{ bar} = 10^5 \text{ Pa}$
	$1 \text{ atm} = 1.01325 \times 10^5 \text{ Pa}$
	$1 \text{ psi} = 6.895 \times 10^3 \text{ Pa}$
	$1 \text{ torr} (1 \text{ mm Hg}) = 133.3 \text{ Pa}$
	$1 \text{ μm Hg} = 0.1333 \text{ Pa}$
	$1 \text{ lb}_f \text{ ft}^{-2} = 47.88 \text{ Pa}$
Power	$1 \text{ hp} = 746.7 \text{ W}$
	$1 \text{ Btu hr}^{-1} = 0.29307 \text{ W}$
Viscosity	$1 \text{ cP} = 10^{-3} \text{ Pa s (or kg m}^{-1} \text{ s}^{-1})$
	$1 \text{ lb ft}^{-1} \text{ s}^{-1} = 1.4882 \text{ Pa s}$
Diffusivity	$1 \text{ ft}^2 \text{ h}^{-1} = 2.5807 \times 10^{-5} \text{ m}^2 \text{ s}^{-1}$
Kinematic viscosity	$1 \text{ ft}^2 \text{ h}^{-1} = 2.5807 \times 10^{-5} \text{ m}^2 \text{ s}^{-1}$
Thermal diffusivity	$1 \text{ ft}^2 \text{ h}^{-1} = 2.5807 \times 10^{-5} \text{ m}^2 \text{ s}^{-1}$
Heat, energy, work	$1 \text{ BTU} = 1.05506 \times 10^{-3} \text{ J}$
	$1 \text{ erg} = 10^{-7} \text{ J}$
	$1 \text{ cal} = 4.184 \text{ J}$
Thermal conductivity	$1 \text{ BTU h}^{-1} \text{ ft}^{-1} \text{ °F}^{-1} = 1.7307 \text{ W m}^{-1} \text{ K}^{-1}$
Heat transfer coefficient	$1 \text{ BTU h}^{-1} \text{ ft}^{-2} \text{ °F}^{-1} = 5.6783 \text{ W m}^{-2} \text{ K}^{-1}$
Specific heat	$1 \text{ BTU lb}^{-1} \text{ °F}^{-1} = 4.1869 \times 10 \times 10^3 \text{ J kg}^{-1} \text{ K}^{-1}$
Heat flux	$1 \text{ BTU h}^{-1} \text{ ft}^{-2} = 3.1546 \text{ W m}^{-2}$
Heat flow	$1 \text{ BTU h}^{-1} = 0.293 \text{ W}$

APPENDIX III

The following tables and figures have been compiled to facilitate calculations of the type presented in this text. While they cannot substitute for specialist data to be found in appropriate handbooks (listed at the end of Chapter 1), the data has been selected so as to provide a broad range of basic information on simple systems. It is hoped these will prove to be helpful as a quick reference for order of magnitude calculations and first tries.

Similarly, a few examples and exercises have been included to illustrate how the thermodynamic data provided in this Appendix can be applied to process metallurgy applications of the type considered in this book.

Note that additional data specific to transport coefficients i.e. values for k thermal conductivity; μ, viscosity; D, diffusivity; and ε, emissivities are presented in the later sections of Chapter 1.

TABLE A3.1. *The error function*

η	$\mathrm{erf}(\eta)$	η	$\mathrm{erf}(\eta)$	η	$\mathrm{erf}(\eta)$
0.00	0.00000	0.76	0.71754	1.52	0.96841
0.02	0.02256	0.78	0.73001	1.54	0.97059
0.04	0.04511	0.80	0.74210	1.56	0.97263
0.06	0.06762	0.82	0.75381	1.58	0.97455
0.08	0.09008	0.84	0.76514	1.60	0.97635
0.10	0.11246	0.86	0.77610	1.62	0.97804
0.12	0.13476	0.88	0.78669	1.64	0.97962
0.14	0.15695	0.90	0.79691	1.66	0.98110
0.16	0.17901	0.92	0.80677	1.68	0.98249
0.18	0.20094	0.94	0.81627	1.70	0.98379
0.20	0.22270	0.96	0.82542	1.72	0.98500
0.22	0.24430	0.98	0.83423	1.74	0.98613
0.24	0.26570	1.00	0.84270	1.76	0.98719
0.26	0.28690	1.02	0.85084	1.78	0.98817
0.28	0.30788	1.04	0.85865	1.80	0.98909
0.30	0.32863	1.06	0.86614	1.82	0.98994
0.32	0.34913	1.08	0.87333	1.84	0.99074
0.34	0.36936	1.10	0.88020	1.86	0.99147
0.36	0.38933	1.12	0.88679	1.88	0.99216
0.38	0.40901	1.14	0.89308	1.90	0.99279
0.40	0.42839	1.16	0.89910	1.92	0.99338
0.42	0.44749	1.18	0.90484	1.94	0.99392
0.44	0.46622	1.20	0.910331	1.96	0.99443
0.46	0.48466	1.22	0.91553	1.98	0.99489
0.48	0.50275	1.24	0.92050	2.00	0.99532
0.50	0.52050	1.26	0.92524	2.10	0.997020
0.52	0.53790	1.28	0.92973	2.20	0.998137
0.54	0.55494	1.30	0.93401	2.30	0.998857

η	erf(η)	η	erf(η)	η	erf(η)
0.56	0.57162	1.32	0.93806	2.40	0.999311
0.58	0.58792	1.34	0.94191	2.50	0.999593
0.60	0.60386	1.36	0.94556	2.60	0.999764
0.62	0.61941	1.38	0.94902	2.70	0.999866
0.64	0.63459	1.40	0.95228	2.80	0.999925
0.66	0.64938	1.42	0.90538	2.90	0.999959
0.68	0.66378	1.44	0.95830	3.00	0.999978
0.70	0.67780	1.46	0.96105	3.20	0.999994
0.72	0.69143	1.48	0.96365	3.40	0.999998
0.74	0.70468	1.50	0.96610	3.60	1.00000

Tabulated values of the error function, defined as

$$\text{erf}(\eta_1) = \frac{2}{\sqrt{\pi}} \int_0^{\eta_1} \exp(-\eta^2)\,d\eta \quad \text{where } \eta = \frac{x}{\sqrt{4\alpha t}}$$

for heat conduction, or $\dfrac{x}{\sqrt{4Dt}}$ for mass diffusion.

TABLE A3.2. *Vapour pressures of liquid metals between their melting and boiling points*

Liquid metal	A	B	C
Ag	-14400	-0.85	13.825
Al	-16946	-1.313	15.587
Cr	-20400	-1.82	18.355
Cu	-17427	-1.474	16.218
Fe	-19710	-1.27	15.395
Hg	-3305	-0.795	12.480
Li	-8320	-1.026	13.418
Mg	-7550	-1.41	14.915
Mn	-14520	-3.02	21.365
Na	-5634	-1.175	13.409
Ni	-22400	-2.01	19.075
Pb	-10093	-1.075	13.541
Si	-20900	-0.565	12.905
Sn	-15500	0	10.355
Ti	-25.229	-2.657	21.379
Zn	-6620	-1.255	14.465

The vapour pressure P in pascals is related to temperature θ in kelvin, by

$$\log P = \frac{A}{\theta} + B\log\theta + C.$$

Data derived from Kubaschewski and Alcock (1983), Pehlke. (1983), Alcock *et al.* (1984).

TABLE A3.3. *Activity coefficients of solutes in liquid metals*

Solvent	Temperature (K)	Solute	γ^0
Iron	1873	Al	0.063
	„	C	0.573
	„	Co	1.0
	„	Cr	0.874
	„	Cu	8.5
	„	Mn	1.3
	„	Ni	0.62
	„	Si	0.0013
	„	Ti	0.04
Copper	1373	Al	0.002
	1550	Au	0.155
	1200	Bi	0.47
	1823	Fe	10.57
	1500	Mn	0.511
	1823	Ni	1.91
	1473	Pb	5.27
	1190	Sb	0.045
	1833	Si	0.016
	1400	Sn	0.007
	1200	Zn	0.398
Aluminium	973	Ag	0.38
	1173	Bi	21.03
	950	Cd	56.2
	1273	Cu	0.042
	1873	Fe	0.027
	1200	Pb	78.5
	1100	Si	0.04
	973	Sn	6.64
	1000	Zn	2.16
Tin	973	Al	2.7
	873	Au	0.006
	1400	Cu	0.317
	773	Cd	1.8
	773	Pb	2.3
	905	Sb	0.46
	773	Zn	1.85
Zinc	773	Cd	3.3
	1200	Cu	0.235
	923	Mg	0.065
	923	Pb	34.6

(*Continued*)

TABLE A3.3. (*continued*)

Solvent	Temperature (K)	Solute	γ^0
	823	Sb	0.3
	773	Sn	4.6
Lead	1273	Ag	2.03
	1200	Al	22.1
	773	Cd	3.38
	1473	Cu	4.87
	1050	Sn	6.82
	923	Zn	7.94

The activity coefficient of a solute in a solution with respect to the pure form of the solute, is known as the Raoultian activity coefficient. It can be defined as

$$\gamma = \frac{P_A}{X_A P_A^0},$$

where P_A is the actual partial pressure of A in equilibrium with the solution, A; X_A is the mole fraction of the solute, and P_A^0 is the corresponding pressure of A in equilibrium with pure liquid A at that temperature.

The table lists activity coefficients γ^0, for (infinitely) dilute solutions (i.e., $X_A \to 0$), for metallic solutes in some common liquid metals. Data derived from Pehlke (1983), and Hultgren (1973).

TABLE A3.4. *Standard Gibbs energy of solution of selected elements in liquid iron, 1 wt.%, standard state (taken from Herbick and Held 1970)*

Element	ΔG^0 cal mole^{-1}
$Al(l) = Al(\%)$	$-15000 - 6.67T$
$C(gr) = C(\%)$	$5400 - 10.10T$
$Cr(s) = Cr(\%)$	$4600 - 11.20T$
$\frac{1}{2}H_2(g) = H(\%)$	$8270 + 7.28T$
$Mn(l) = Mn(\%)$	$976 - 9.11T$
$\frac{1}{2}N_2(g) = N(\%)$	$860 + 5.71T$
$\frac{1}{2}O_2(g) = O(\%)$	$-28000 - 0.69T$
$\frac{1}{2}P_2(g) = P(\%)$	$-29200 - 4.6T$
$Si(l) = Si(\%)$	$-31430 - 4.12T$
$\frac{1}{2}S_2(g) = S(\%)$	$32280 + 5.60T$

TABLE A3.5. *Calculated values of the product* $\{(wt.\%C) \times (wt.\%O)\}$ *for iron at various temperatures and carbon concentrations at an equilibrium pressure of* 1 atm

(wt.%C)	1500 °C	1600 °C	1700 °C	1800 °C	1900 °C
0.02–0.20	1.86	2.00	2.18	2.32	2.45
0.5	1.77	1.90	2.08	2.20	2.35
1.0	1.68	1.81	1.96	2.08	2.25
2.0	1.55	1.70	1.84	1.95	2.10

Tabulated values of the product (wt.% C) (wt.%O) $\times 10^3$ in liquid iron at a pressure of CO 1 atm ($CO + CO_2$) according to Fuwa and Chipman (1960) from the expression:

$$CO\,(gas) = C\,(wt.\%, \text{ iron}) + O\,(wt.\%, \text{ iron})$$

where

$$K^{eq} = \frac{[\%C]\,[\%O]}{P_{CO}} = -\frac{1168}{T} - 2.07.$$

TABLE A3.6.   *Standard heats of combustion at* 298 K (25 °C) *and* 101.325 kPa (1 atm absolute:) (g) = *gas*, (l) = *liquid*, (s) = *solid*

Compound	Combustion reaction	$\Delta H°(10^{-3}$ kJ mole$^{-1})$
Carbon	$C(s) + \frac{1}{2}O_2(g) \rightarrow CO(g)$	−110.5
Carbon monoxide	$CO(g) + \frac{1}{2}O_2(g) \rightarrow CO_2(g)$	−283.0
Carbon	$C(s) + O_2(g) \rightarrow CO_2(g)$	−393.5
Hydrogen	$H_2(g) + \frac{1}{2}O_2(g) \rightarrow H_2O(l)$	−285.
Hydrogen	$H_2(g) + \frac{1}{2}O_2(g) \rightarrow H_2O(g)$	−241.7
Methane	$CH_4(g) + 2O_2(g) \rightarrow CO_2(g)$ $+ 2H_2O(l)$	−890.4
Ethane	$C_2H_6(g) + \frac{7}{2}O_2(g) \rightarrow 2CO_2(g)$ $+ 3H_2O(l)$	−1559.9
Propane	$C_3H_8(g) + 5O_2(g) \rightarrow 3CO_2(g)$ $+ 4H_2O(l)$	−2220.1

From: Perry and Chilton (1973).

TABLE A3.7. *Normal total emissivities of various materials*

Normal total emissivity	ε
1. Metals and their oxides	
(a) Polished metal surfaces	
Aluminium, brass, copper	
Electrolytic iron, nickel	0.015–0.087
Platinum, tungsten	
(b) Oxidized surfaces	
Aluminium (600 °C)	0.11–0.19
Rolled sheet steel	0.66
Cast iron (600 °C)	0.64–0.78
Steel (600 °C)	0.79
Ingot iron	0.87–0.95
Cast sheet/plate steel	0.8
Copper plate (600 °C)	0.57
Nickel plate (600 °C)	0.37–0.48
18-8 Stainless steel (600 °C)	0.36–0.44
Aged tungsten filament	0.39
(c) Molten metals (clean)	
Copper	0.13–0.16
Cast iron	0.29
Mild steel	0.28
2. Refractories at high temperature (1000 °C)	
Alumina silicates	0.6–0.8
Fireclay	0.75
Magnesite	0.38
Zirconium silicate	0.5

Note: The emissivities of metals tend to increase with temperature. The ε of non-conductors tend to decrease with temperature. Most non-metallic materials at room temperature have ε values in the range 0.8–0.95.

TABLE A3.8. *Thermal conductivities of refractories and insulating materials*

Material	Temperature (°C)	$k(W\ m^{-1}\ K^{-1})$
Bricks		
Fireclay	1000	1.3–1.7
High alumina	1000	1.4–2.8
Basic brick	1000	3.5–4.9
Silica	1000	0.8–1.7
Silicon carbide	1000	14
Zircon	1000	1.4–2.5
Chrome	1000	1.7
Insulating materials		
Vermiculite	1000	0.2–0.3
Asbestos	100–400	0.2
Mineral wool	30	0.04

Data derived from Perry (1983), *North American Combustion Handbook* (1965), Geiger and Poirier (1980).

TABLE A3.9. *Diffusivities in solid metals*

Metal (B)	Solute (A)	$D_{AB} \times 10^{14} (m^2\ s^{-1})$
Aluminium (500 °C)	Zn	20
	Mg	8.5
	Si	7.5
	Cu	3.3
Copper (700 °C)	Zn	8
	Sn	2.3
	Al	1.8
	Ni	0.1

Data derived from Geiger and Poirier (1980).

TABLE A3.10.   *Diffusivities of gases (20 °C)*

Gases (A/B)	$(D_{AB} \times 10^6)^{\dagger}\,(\text{m}^2\,\text{s}^{-1})$
CO_2/air	16
H_2/air	14
O_2/air	21
H_2O/air	27
H_2O/CO_2	17
Ar/N_2	21
CO_2/H_2	70
H_2O/H_2	95
O_2/H_2	81

†For gases $D_{AB} \propto (\theta)^{1.75}$ Data derived from Geiger and Poirier (1980).

TABLE A3.11.   *Physical and thermal properties of gases, liquids and solids of metallurgical interest*

(a) Gases

	Temperature (°C)	ρ (kg m^{-3})	C_P (kJ kg^{-1} K^{-1})	k (W m^{-1} K^{-1})	$\mu \times 10^5$(Pa s)
Air	20	1.16	1.00	0.025	1.82
	500	0.46	1.09	0.054	3.41
	1000	0.27	1.10	0.083	4.87
	1500	0.19	1.28	0.110	6.01
Helium	20	0.17	5.2	0.144	1.98
	500	0.067	5.2	0.269	3.72
Hydrogen	20	0.08	14.3	0.18	0.88
	500	0.031	14.7	0.38	1.70
Oxygen	20	1.36	0.92	0.026	2.02
	277	0.71	0.99	0.045	3.20
Water	0	1000	4.23	0.57	1.79
	20	998	4.18	0.60	1.00
	100	964	4.21	0.68	0.29
Mercury	10	13600	0.138	8.1	1.59

TABLE A3.11 (*Continued*)

(b) Liquid metals[†]

	Temperature (°C)	ρ (kg m^{-3})	C_P (kJ kg^{-1} K^{-1})	k (W m^{-1} K^{-1})	$\mu \times 10^3$(Pa s)	Latent heat (kJ kg^{-1})	Melting point (°C)
Aluminium	m.p.	2382	1.09	104	1.5	387	660
Bismuth	316	10010	0.144	16.4	1.6	52	271
Lead	371	10500	0.159	16.1	2.4	23	327
Copper	m.p.	8000	0.494	134	4.3	203	1083
Gold	m.p.	17400	0.142	(105)	5.2	065	1063
Iron	m.p.	7030	0.627	30.5	7.0	247	1536
Mercury	10	13600	0.138	8.1	1.59	11.5	−29
Nickel	m.p.	7900	0.556	(50)	4.70	292	1453
Silver	m.p.	9330	0.293	(176)	4.0	103	961
Tin	m.p.	6980	0.266	33.5	1.8	60	232
Zinc	m.p.	6580	0.610	63	4.1	111	420

( ): Estimated via Lorenz number and electrical resistivities.

Table A3.11 (*Continued*)
(c) Solid metals (20°C)

	ρ (kg m^{-3})	C_{P} $(\text{kJ kg}^{-1} \text{K}^{-1})$	k $(\text{W m}^{-1} \text{K}^{-1})$
Aluminium	2 700	0.902	273
Bismuth	9 800	0.122	16
Chromium	7 100	0.450	90
Copper	8 920	0.386	398
Gold	19 300	0.127	315
Iron	7 860	0.450	80
Lead	11 337	1.13	35
Magnesium	1 740	1.0	156
Nickel	8 900	0.444	91
Silver	10 500	0.236	424
Tin	5 750	0.222	67
Zinc	7 140	0.389	120

[†] Data derived from Perry (1983), *North American Combustion Handbook* (1965), Geiger and Poirier (1980), Kubaschewski and Alcock (1983), Iida and Guthrie (1988), Biswas and Davenport (1980), Kreith and Black (1980).

Table A3.12.   *Electrical resistivities of metals*

Metal	Temperature (°C)	Resistivity ($\mu\Omega$ m)
Aluminium	18	0.032
	700	0.25
Bismuth	271	1.29
Lead	18	0.20
	340	0.98
Copper	1083	0.22
Gallium	30	0.26
Gold	18	0.024
Iron(steel)	18	0.12
	1536	1.39
Nickel	18	0.11
	1450	0.85
Silver	18	0.016
Tin	18	0.11
	232	0.45
Zinc	413	0.35
Magnesium	650	0.27
Mercury	100	1.03

Data from ASM *Metals Handbook* (1979).

TABLE A3.13 *Surface tensions of liquid metals*

Metal	$\sigma(\text{N m}^{-1})$
Cu	1.3
Zn	0.78
Pb	0.44
Al	0.91
Ag	1.0
Ni	1.78
Cr	1.7
Ti	1.65
Hg	0.5
(H_2O)	(0.07)
Fe	1.9
Fe(0.1%P)	1.7
Fe(0.1%N)	1.5
Fe(0.1%S)	1.2
Fe(0.1%O)	1.1

Data derived from Iida and Guthrie (1988).

TABLE A3.14. *Surface tensions of some molten oxide systems*

System	Temperature (°C)	γ mN/m
$CaO–SiO_2$	1600	485
FeO	1600	570
$CaO–SiO_2Al_2O_3$	1550	500
$CaO–Al_2O_3$	1550	600
CaF_2	1550	280
$CaO–CaF_2$ 20\80	1550	310
$CaO–Al_2O_3–CaF_2$ 40\40\20	1550	400
$CaO–Al_2O_3–CaF_2$ 20\20\60	1550	300

Data from Boni and Derge (1956) and Ogino and Hara (1977).

TABLE A3.15. *Contact angles of liquid slags in contact with liquid steel*

Slag type		Temperature (°C)	Contact angle with liquid steel
CaO–Al$_2$O$_3$	36\64	1600	65
CaO–Al$_2$O$_3$	50\50	1600	58
CaO–Al$_2$O$_3$	58\42	1600	54
CaO–Al$_2$O$_3$–SiO$_2$	44\45\11	1600	43
CaO–Al$_2$O$_3$–SiO$_2$	40\40\20	1600	40
CaO–Al$_2$O$_3$–SiO$_2$	33\33\33	1600	36
CaO–Al$_2$O$_3$–SiO$_2$	26\26\49	1600	13
CaO–SiO$_2$	58\42	1600	29
CaO–SiO$_2$	50\50	1600	31
CaO–CaF$_2$	5\95	1600	47
CaO–CaF$_2$–Al$_2$O$_3$	11\87\2	1600	36
CaO–CaF$_2$–Al$_2$O$_3$	14\71\15	1600	28
CaO–CaF$_2$–Al$_2$O$_3$	15\56\30	1600	34
CaO–CaF$_2$–Al$_2$O$_3$	45\8\47	1600	41

Data from Cramb and Jimbo (1988).

TABLE A3.16. *Contact angles of various refractories in contact with liquid steel*

Refractory	Temperature (°C)	Contact angle	Steel type
Al_2O_3 98%	1520	105	0.4%C
Al_2O_3	1600	144	pure
Al_2O_3	1600	112	3.4%C
SiO_2	1600	115	pure
SiO_2	1550	110	pure
CaO	1600	132	pure
CaO	1550	121	pure
TiO_2	1600	84	pure
Cr_2O_3	1600	88	pure
ZrO_2 89%	1520	108	0.16%C
ZrO_2	1550	119	pure
ZrO_2	1550	122	pure
MgO 95%	1520	111	0.16%C
MgO	1600	125	pure
MgO	1550	128	pure
MnO	1550	113	pure
SiC 91%	1500	60	0.16%C
Graphite	1460	120	0.16%C
TiN	1550	132	pure
BN	1550	< 50	pure
CaS	1550	87	pure
$CaO–MgO–SiO_2$	1450	104–120	Ni–Cr steel
$CaO–SiO_2–Al_2O_3$	1450	96–114	Ni–Cr steel

Data from Cramb and Jimbo (1988).

TABLE A3.17. *Contact angles of liquid slags on solid surfaces*

Slag type	Refractory type	Temperature (°C)	Contact angle
$FeO–SiO_2$ 70\30	Alumina	1250	< 10
$FeO–MnO–CaO–SiO_2–Al_2O_3$	Zirconia	1500	5–20
$FeO–MnO–CaO–SiO_2–Al_2O_3$	Alumina	1500	< 10
$FeO–MnO–CaO–SiO_2–Al_2O_3$	SiC	1500	104–122
$FeO–MnO–CaO–SiO_2–Al_2O_3$	Graphite	1500	110–132
$CaO–SiO_2 Al_2O_3$ 40\40\20	MgO	1400	9–32
$CaO–SiO_2–MgO$ 10\56\34	Si_3N_4	1550	50
$Na_2O–SiO_2–MgO$	Si_3N_4	1550	20
Mold slag	Steel	1400	0 to 30
$CaO–SiO_2–Al_2O_3$ 40\40\20	Iron	1450	30–60
$CaO–SiO_2–Al_2O_3–FeO$	Iron	1450	0–60

Data from Cramb and Jimbo (1988).

SOME WORKED EXAMPLES USING THERMODYNAMIC DATA IN APPENDIX

EXAMPLE 1

What is the vapour pressure of copper above a steel melt at 1600°C containing 1 wt.% copper?

Solution

From Table A3.2:

$$\log P_{Cu} = \frac{-17427}{1873} - 1.474 \log 1873 + 16.218.$$

Therefore,

$$P_{Cu}^0 = 107 \text{ Pa.}$$

Hence, from Table A.3.3.

$$P_{1\%Cu,Fe} = \gamma_{Cu/Fe} X_{Cu} P_{Cu}^0 \simeq \gamma_{Cu/Fe} \frac{(1/M_{Cu})}{(99/M_{Fe})} P_{Cu}^0$$

$$\simeq 8.5 \times \left(\frac{1/63.5}{99/55.8} \right) 107 \text{ Pa}$$

$$\simeq 8 \text{ Pa}$$

EXAMPLE 2

Calculate the solubility of oxygen (wt.% O) in pure liquid iron at 1600°C, given

$$Fe_xO(l) \rightleftharpoons xFe(l) + O(\%), \qquad \Delta G^0 = 28\,957 - 12.509T \text{ cal mole}^{-1} \qquad (1)$$

where x can be approximated as unity.

Solution:

For any chemical reaction, the change in Gibbs energy, ΔG, is related to the standard Gibbs energy change, ΔG^0, according to:

$$\Delta G_T = \Delta G_T^0 + RT \ln(Q), \qquad (2)$$

where Q is the reaction quotient and represents the actual conditions imposed upon the system. In eqn (1):

$$Q = \left(\frac{a_{Fe} \cdot h_0}{a_{FeO}} \right). \tag{3}$$

The activities of Fe and FeO are given with respect to the pure liquid standard states and are related to the activity coefficients (γ) and mole fraction (X);

$$a_{FeO} = \gamma_{FeO} X_{FeO}, \quad a_{Fe} = \gamma_{Fe} X_{Fe}.$$

The Henrian activity of oxygen, h_0, is taken with respect to a hypothetical 1 wt.% standard state of oxygen, where $h_0 = f_0(\%O)$. In dilute solutions, we often approximate the Henrian activity coefficient, f_0, as unity, so that wt.% O replaces h_0 in eqn (3). The use of wt.% standard states is a common industrial practice when treating dilute multicomponent systems such as modern steels.

ΔG^0 represents the change in Gibbs energy when the reactants and products are in their standard states. In eqn (1), this corresponds to pure FeO (l) dissociating into pure Fe(l) containing 1 wt.% dissolved oxygen. At 1600°C, from eqn (1), $\Delta G^0 = 5528$ cal mole^{-1}.

Equation (2) can be used as the criterion for both spontaneity and equilibrium. If one considers liquid iron containing 1 wt.% O as an actual condition, then $Q = (1 \times 1/1) = 1$, and from eqn (2), $\Delta G = 5528 + RT \ln(1) = 5528$.

Since $\Delta G > 0$, the reaction does not proceed as written, but rather from right to left. That is, some of the dissolved oxygen will react with the Fe to produce more FeO. If we consider a situation in which the liquid Fe contains only 0.01 wt.% O, from eqn (2), one calculates, $\Delta G = 5528 - 17139 = -11.61$ kcal mole^{-1}.

As the Gibbs energy change is now negative, this means that FeO will dissociate so as to increase the (%O) content. The point of equilibrium occurs when $\Delta G = 0$. Substituting $\Delta G = 0$ into eqn (2),

$$\Delta G^0 = -RT \ln(K) \tag{4}$$

where $K = (Q)^{eq}$. This is the well-known thermodynamic relation between the standard Gibbs energy change and the equilibrium constant for a reaction. At 1600°C, $\Delta G^0 = 5528$ cal, and it follows from eqn (4) that

$$K = 0.226 = \left(\frac{a_{Fe}(\%O)}{a_{FeO}} \right)^{eq}. \tag{5}$$

There is an infinite set of values of a_{Fe}, a_{FeO} and (%O) which satisfy eqn (5). For the case in which essentially pure Fe (l) is in equilibrium with pure FeO(l), then at equilibrium, it follows that the solubility of oxygen in liquid iron is 0.226 wt.% O.

It follows from eqn (5) that it is possible to reduce the residual oxygen dissolved in the Fe by increasing a_{Fe} and/or decreasing a_{FeO}. The maximum value of a_{Fe} is unity (pure Fe); a_{FeO} can be reduced to as low a value as desired by simply introducing other metallic oxides into the system (CaO, MgO, SiO_2, etc.) which will dissolve FeO in the form of a slag.

EXAMPLE 3

In pneumatic steel-making, various impurities within the iron are oxidized into an upper slag phase by the action of a pure oxygen jet at 1.0 atmosphere pressure. Estimate the lowest residual (wt.%) levels of carbon, silicon, manganese, and oxygen, that might be reached at 1600°C. Assume the slag contains 10 mole% FeO, 56 mole% CaO, 20 mole% MnO and 14 mole% SiO_2; $\gamma_{SiO_2} = 0.05$; assume the other oxides are ideal.

$$Mn(l) + \tfrac{1}{2}O_2(g) = MnO(l), \qquad \Delta G^0_{1600°C} = -56.8 \text{ kcal}; \qquad (6)$$

$$Si(l) + O_2(g) = SiO_2(l), \qquad \Delta G^0_{1600°C} = -137.6 \text{ kcal}. \qquad (7)$$

Use the standard data for the other thermodynamic values.

If an oxide is ideal (Raoultian), the activity coefficient (γ) is unity, and its activity is numerically equal to its mole fraction (X). Thus, in the slag phase,

$$a_{FeO} = X_{FeO} = 0.1,$$
$$a_{CaO} = X_{CaO} = 0.56,$$
$$a_{MnO} = X_{MnO} = 0.2,$$
$$a_{SiO_2} = \gamma_{SiO_2} X_{SiO_2} = 0.007.$$

(i) Weight per cent oxygen

In Example 2, the solubility of oxygen in equilibrium with the pure oxide ($a_{FeO} = 1.0$) was calculated as 0.226 wt.% O. Assuming the final levels of metalloids to be controlled by slag–metal equilibrium, we can first deduce the final dissolved oxygen content in the steel, covered by a layer of such a slag. Thus, substituting $a_{FeO} = 0.1$ into eqn (3) and assuming the steel is essentially pure iron ($a_{Fe} = 1$), the resulting oxygen content is 0.0226 wt.%O, or 226 p.p.m. O. In reality, kinetic factors will drive the dissolved oxygen to higher levels before a balance point between input oxygen from the jet, and oxygen equilibrium with the slag, is reached.

(ii) Weight per cent carbon

The equilibrium constant for the carbon–oxygen reaction in liquid iron (Table

A3.5) is represented by:

$$CO(g) = C(\%) + O(\%), \qquad K = \frac{h_C h_O}{P_{CO}}, \tag{8}$$

$$\log(K) = -\frac{1168}{T} - 2.07. \tag{9}$$

Assuming that the Henrian activity coefficients (f_C, f_O) can be approximated as unity at 1600°C:

$$K_{1600°C} = \frac{(\text{wt.}\%C)(\text{wt.}\%O)}{P_{CO}} = 2.02 \times 10^{-3}. \tag{10}$$

In the case of pure oxygen, the decarburization product is essentially pure CO(g) at 1 atm. Substituting $P_{CO} = 1$ atm and wt.% O = 0.0226 into eqn (10), the residual carbon content is 0.09%C.

(iii)   Weight per cent manganese and silicon

Using data on the standard Gibbs energy of solution in Table A3.4, one calculates:

$$\text{Mn(l)} = \text{Mn}(\%) \qquad \Delta G^0_{1600°C} = -16.09 \text{ kcal}, \tag{11}$$

$$\text{Si(l)} = \text{Si}(\%), \qquad \Delta G^0_{1600°C} = -39.15 \text{ kcal}, \tag{12}$$

$$\tfrac{1}{2}O_2(g) = O(\%), \qquad \Delta G^0_{1600°C} = -29.29 \text{ kcal}. \tag{13}$$

Hence for the reactions:

$$\text{Mn}(\%) + O(\%) = \text{MnO(l)}, \quad \Delta G^0_{1600°C} = -11.42 \text{ kcal mole}^{-1},$$

$$K \approx \frac{a_{MnO}}{(\text{wt.}\%\text{Mn})(\text{wt.}\%\text{O})} = 21.5. \tag{14}$$

$$\text{Si}(\%) + 2O(\%) = \text{SiO}_2(l), \quad \Delta G^0_{1600°C} = 39.87 \text{ kcal mole}^{-1},$$

$$K \approx \frac{a_{SiO_2}}{(\text{wt.}\%\text{Si})(\text{wt.}\%\text{O}^2)} = 4.5 \times 10^4. \tag{15}$$

Substituting wt.% O = 0.226, $a_{MnO} = 0.2$, and $a_{SiO_2} = 0.007$ in eqns (14) and (15), one obtains 0.41 wt.% Mn and 0.003 wt.% Si. It is noted that the standard Gibbs energy of formation of SiO$_2$ (eqn. 15) is more negative than that for MnO (eqn 14). This implies that SiO$_2$ is more stable than MnO, which results in a much lower residual level of silicon in the steel than residual manganese. To conclude, the 'turndown' composition of a steel in equilibrium with the typical steel-making slag cited, would contain 0.0226 wt.% O (or 226 ppm), 0.09 wt.% C, 0.4 wt.% manganese and negligible silicon, 0.003 wt.% (or 226 ppm).

EXAMPLE 4

Consider the possibility of blowing an argon–oxygen mixture of $9:1$ at 1 atm pressure through a steel bath in an argon–oxygen decarburization vessel with an overlying slag of the composition cited in Example 3. How would this alter the turndown carbon content of the steel?

Solution

CO(g) is the product of decarburization by oxygen:

$$C + \tfrac{1}{2}O_2(g) = CO(g) \tag{16}$$

Each mole of gas entering the reactor vessel contains 0.1 mole oxygen and 0.9 mole argon. The gaseous product will essentially contain 0.2 mole CO and 0.9 mole Ar. The equilibrium partial pressure of CO(g) is $(0.2/1.1) \times P(\text{total})$ $= 0.181$ atm.

This lowering in P_{CO}, makes it possible to obtain a lower residual wt.%C and/or wt.%O. Taking the activity of FeO in the slag to be 0.1, the oxygen content in the steel is 0.0226%. The new value of dissolved carbon (eqn 10) is then 0.016 wt.%C. The same result can be obtained by the action of pure oxygen, but at a reduced total pressure (0.181 atm) (e.g., the vacuum oxygen decarburization process, VOD).

In practice, the activities of the oxides will be reduced and the amount of oxygen dissolved in the steel will also decrease: a lower oxygen content in the Fe(l) will result in a higher residual Mn (among others).

The use of argon in the AOD (argon–oxygen decarburization) process is well established as a commercial process for the production of stainless steels. Thus, in stainless-steel-making, it is desirable to maintain the dissolved chromium in the iron phase and avoid its oxidation, whilst removing the carbon by oxidation.

In conclusion, the turndown carbon using a $9:1$ argon–oxygen mixture would be 0.016 wt.%C versus 0.09 wt.%C in the previous question. Again, kinetic factors would modify these levels.

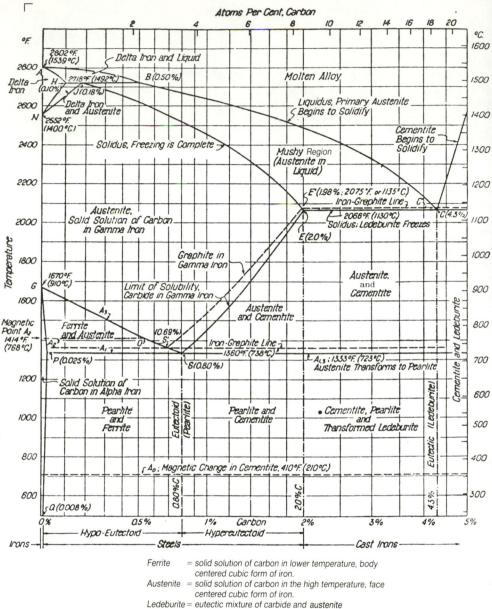

FIG. A3.1. The iron–carbon phase diagram. *Ferrite* = solid solution of carbon in the lower-temperature, body-centred cubic form of iron. *Austenite* = solid solution of carbon in the high-temperature, face-centred cubic form of iron. *Ledeburite* = eutectic mixture of carbide and austenite. *Cementite* = iron carbide, Fe_3C. *Pearlite* = mixture of ferrite and iron carbide. *Delta iron* = solid solution of carbon in the very-high-temperature body-centred cubic form of iron.

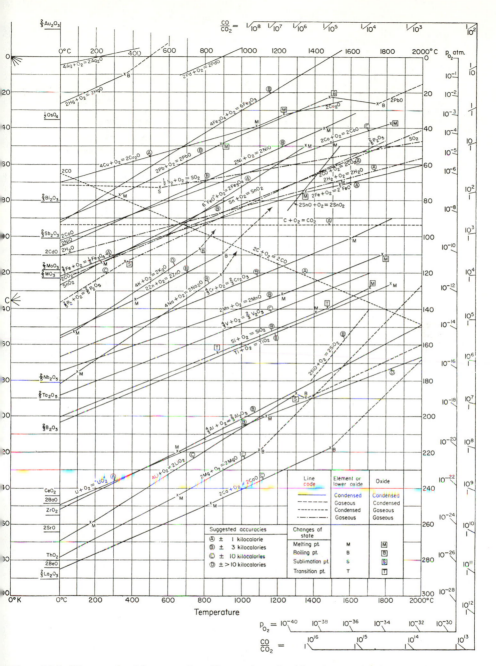

FIG. A3.2. The standard free energies of formation of oxides, the standard states being the pure condensed phases and gases at 1 atm pressure. Grids for P_{O_2}, H_2/H_2O and CO/CO_2 values are indicated by scales around the right margin, and radiate from foci marked O, H and C on the vertical axis at absolute zero temperature. (Based on the diagrams of Richardson and Jeffes (1948), modified by Darken and Gurry (1953).)

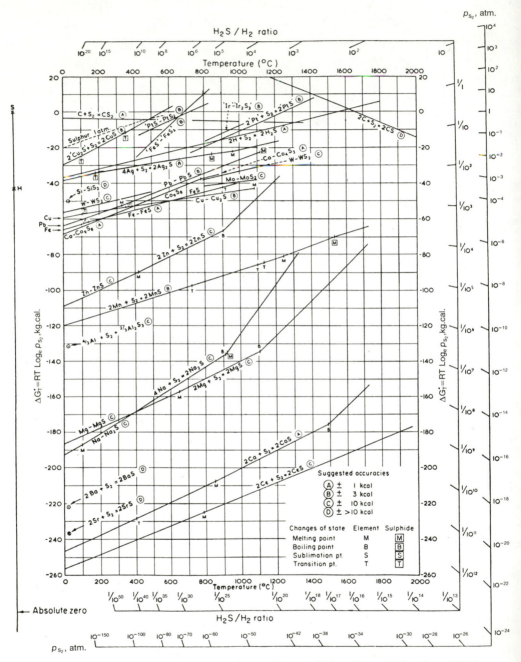

FIG. A3.3. A standard free energy diagram for sulphides. (From Richardson 1974.)

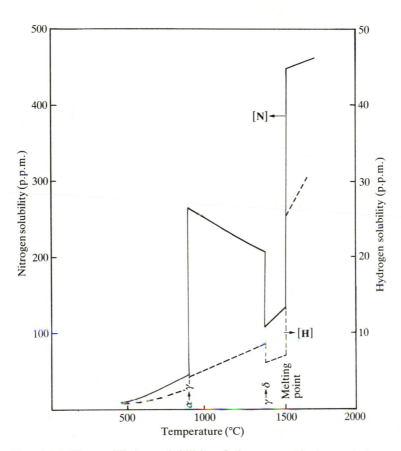

F$_{\text{IG}}$. A3.4. The equilibrium solubilities of nitrogen and hydrogen in iron.

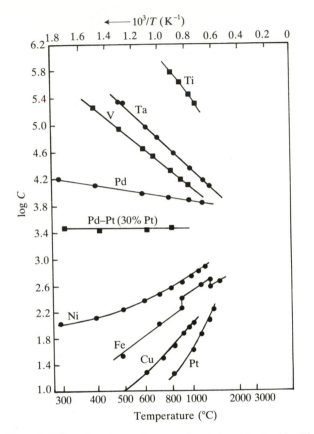

FIG. A3.5. The solubility of hydrogen in several metals, at 1 atm H_2. The concentration C is in atoms of hydrogen per million metal atoms. To convert to p.p.m. by weight divide C by the metal molecular weight. The data on which the figure is based have mostly been derived from A. Sieverts and co-workers (cf. G. Borelius 1929).

1A	2A	3B	4B	5B	6B	7B	8B	8B	8B	1B	2B	3A	4A	5A	6A	7A	0
1 H 1.008																1 H 1.008	2 He 4.003
3 Li 6.941	4 Be 9.012											5 B 10.81	6 C 12.01	7 N 14.01	8 O 16.00	9 F 19.00	10 Ne 20.18
11 Na 22.99	12 Mg 24.31											13 Al 26.98	14 Si 28.09	15 P 30.97	16 S 32.06	17 Cl 35.45	18 Ar 39.95
19 K 39.10	20 Ca 40.08	21 Sc 44.96	22 Ti 47.90	23 V 50.94	24 Cr 52.00	25 Mn 54.94	26 Fe 55.85	27 Co 58.93	28 Ni 58.70	29 Cu 63.55	30 Zn 65.38	31 Ga 69.72	32 Ge 72.59	33 As 74.92	34 Se 78.96	35 Br 79.90	36 Kr 83.80
37 Rb 85.47	38 Sr 87.62	39 Y 88.91	40 Zr 91.22	41 Nb 92.91	42 Mo 95.94	43 Tc (98)	44 Ru 101.1	45 Rh 102.9	46 Pd 106.4	47 Ag 107.9	48 Cd 112.4	49 In 114.8	50 Sn 118.7	51 Sb 121.8	52 Te 127.6	53 I 126.9	54 Xe 131.3
55 Cs 132.9	56 Ba 137.3	57 *La 138.9	72 Hf 178.5	73 Ta 180.9	74 W 183.9	75 Re 186.2	76 Os 190.2	77 Ir 192.2	78 Pt 195.1	79 Au 197.0	80 Hg 200.6	81 Tl 204.4	82 Pb 207.2	83 Bi 209.0	84 Po (209)	85 At (210)	86 Rn (222)
87 Fr (223)	88 Ra 226.0	89 †Ac 227.0	104 φ (261)	105 φ (262)	106 φ (263)												

SOME PHYSICAL CONSTANTS

c speed of light	2.9979246×10^8 m s^{-1}
e electron charge	$1.60217733 \times 10^{-19}$ C
F Faraday's constant	96485.3 C mol^{-1}
c speed of light	2.9979246×10^8 m s^{-1}
e electron charge	$1.60217733 \times 10^{-19}$ C
F Faraday's constant	96485.3 C mol^{-1}
c speed of light	2.9979246×10^8 m s^{-1}
e electron charge	$1.60217733 \times 10^{-19}$ C
F Faraday's constant	96485.3 C mol^{-1}
g gravitational constant (Earth sea-level)	9.81 m s^{-2}
h Planck's constant	6.626×10^{-34} Js
K Boltzmann constant (molar)	1.38×10^{-23} J K^{-1} kmole^{-1}
N Avagadro's number	$6.023 \times 10^{+26}$ molecules kmole^{-1}
P_A atmospheric pressure	1.01×10^5 Pa
R universal gas constant	8314 J kmole^{-1} K^{-1}
	8.31451 J mol^{-1} K^{-1}
	0.082058 litre atm K^{-1} mol^{-1}
	1.9872 cal K^{-1} mol^{-1}
σ Stefan–Boltzmann constant	5.67×10^{-8} W m^{-2} K^{-4}

REFERENCES

Alcock, C. B., Itkin, V., and Harrigan, M. K. (1984). *Can. Met. Quart.* **23**, 309–13.

American Society for Metals (1979). *Metals handbook* (9th edn). American Society for Metals, Metals Park, Ohio.

Amsden, A. A. and Harlow, F. H. (1970). *Los Alamos Scientific Laboratory Report*, LA, 4370.

Baliga, B. R. and Patankar, S. V. (1983). *Numerical Heat Transfer* **6**, 245–56.

Bell W. W. (1968). *Special functions for scientists and engineers.* D. Van Nostrand Co., London.

Bird, R. B., Stewart, W. W., Lightfoot, E. N. (1960). *Transport phenomena.* Wiley, New York.

Biswas, A. K. and Davenport, W. G. (1980). *Extractive metallurgy of copper.* Pergamon Press, Oxford.

Blasius, H. (1908). *Z. Math. u. Phys. Sci.*, **56**, 1–37.

Boni, R. E. and Derge, G. (1956). *J. Metals* **8**, 59.

Borelius, G. (1927). *Ann. Physik.* **83**, 121.

Burmeister, L. C. (1983). *Convective heat transfer*, p. 278. Wiley, New York.

Byrne, M., Cramb, A. W., Fenick, T. W. (1985). *Steelmaking Proceedings*, Detroit, ISBN 0-932897-0209, ISS of AIME, pp. 451–61.

Buckingham, E. (1914). *Phys. Rev.* **4**, 345.

Cahn (1968). *Trans. Met. Soc., AIME*, **242**, 166.

Carslaw, H. S. and Jaeger, J. C. (1959). *Conduction of heat in solids.* Oxford University Press, Oxford.

Chapman, S. and Cowling, T. G. (1960). *The mathematical theory of non-uniform gases* (2nd edn), Chs. 10 and 14. Cambridge University Press.

Chiesa, F. and Guthrie, R. I. L. (1971). *Metall. Trans.* **2**, 2833–8.

Christian, J. W. (1975), *Theory of transformation in metals and alloys, Part I. Equilibrium and general kinetic theory* (2nd edn) pp. 380–1, 390–1. Pergamon Press, Oxford.

Colburn, A. P. (1933). *Trans. AIChE* **29**, 174–210.

Compaan, K. and Haven, Y. (1956). *Trans. Faraday Soc.* **52**, 786.

Courant, R., Isaacson, E. and Rees M. (1952). *Comm. Pure Appl. Math.* **5**, 243.

Cramb, A. W. and Jimbo, I. (1988). *Proc. Phillrook Memorial Conference*, ISS of AIME, Toronto, p. 25.

Crank, J. (1975). *The mathematics of diffusion* (2nd edn). Oxford University Press, Oxford.

Croxton, C. A. (1975). *Introduction to liquid state physics.* Wiley, New York.

Danckwerts, P. V. (1951). *Ind. Eng. Chem.* **43**, (6), 1460–7.

Darken, L. S. (1948). *Trans. AIME, Int. Met. Div. Tech. Pub.*, 2443.

Darken, L. S. and Gurry, R. W. (1953). *Physical chemistry of metals*, p. 453. McGraw-Hill, New York.

Davis, J. T. (1972). *Turbulence phenomena.* Academic Press, New York.

Einstein, A. (1905). *Ann Physik* **19**, 371.

Fujii, T. and Imura, H. (1972). *Int. J. Heat Transfer* **15**, 755.

Fuwa, T. and Chipman, J. (1960). *J. Trans. Metall. Soc. AIME* **218**, 887–90.

Gebhart, B. (1970). *Heat transfer* (2nd edn). McGraw-Hill, New York.

Geiger, G. H. and Poirier, D. R. (1980). *Transport phenomena in metallurgy*. Addison-Wesley, Reading, MA.

Green, D. W. and Maloney, J. O. *Perry's chemical engineers' handbook* (6th edn). McGraw-Hill, New York.

Gosman, A. D., Pun, W. M., Runchal, A. K., Spalding, D. B., and Wolfontein, M. (1969). *Heat and mass transfer in recirculating flows*. Academic Press, New York.

Grieveson, P. and Turkdogan, E. T. (1964). *J. Phys. Chem.* **68**, 1547–52.

Grimison, E. D. (1937). *Trans. ASME* **59**, 583–94.

Guthrie, R. I. L. and Bradshaw, A. V. (1969). *Chem. Eng. Sci.* **24**, 913–17.

Guthrie, R. I. L. and Stubbs, P. (1973), *Can Metall. Quart.* **12**, (4), 465–73.

Guthrie, R. I. L., Henein, H., and Clift, R. (1975). *Metall. Trans.* **6B**(2), 320–9.

Guthrie R. I. L. (1979). *Can. Metall. Quart.* **18** (4), 361–77.

Guthrie, R. I. L. Perrin, A. R. (1982). *Proc. 3rd Technical Conference*, Pittsburgh, Pa. ISS of AIME, pp. 151–6.

Guthrie, R. I. L., Joo, S., Nakajima, H. (1988). *Int. Symp. on Direct Rolling and Direct Charging of Strand Cast Billets* (ed. J. J. Jonas), Vol. 9., Proc. Met. Soc. of C.I.M., Montreal, August. Pergamon Press (in press).

Hardy, J., Pomeau, Y., and de Pazzis, O. (1973). *J. Math. Phys.* **14**(12), 1746–59.

Harlow, F. H. and Welch, J. E. (1965). *Phys. Fluids* **8**, 2181.

Harries, D. S. and Guthrie R. I. L. (1982). *CIM Bulletin* **75**, (837), 117–21.

Herbick and Held (1970). *Making, shaping and treating of steel*. Pittsburgh, Pa.

Higbie, R. (1935). *Trans. AIChE* **31**, 365–89.

Hills, A. W. D. (1966). *Heat and mass transfer in process metallurgy*, Proc. Symp. 1966, John Percy Research Group. Institution of Mining and Metallurgy, London.

Hottel, H. C. and Sarofim, A. F. (1967). *Radiative heat transfer*. McGraw-Hill, New York.

Iida, T. and Guthrie, R. I. L. (1988). *The physical properties of liquid metals*. Oxford University Press, Oxford.

Iida, T. and Morita, Z. (1975). *Proc. Int. Conf. on the Science and Technology of Iron and Steel*, AIME, Chicago, Illinois.

Iida, T., Morita, Z., and Takeuchi, S. (1975). *J. Japan Inst. Metals* **39**, 1169.

Iida, T. Guthrie, R. I. L. and Morita, Z. (1982). *Proc. of the Int. Symp. on the Phys. Chem. of Iron and Steelmaking*, C.I.M., Annual Conf. of Metallurgists, Toronto, 25–32.

Irons, G. A. and Guthrie, R. I. L. (1978). *Metall. Trans.*, **9B**, 101–10.

Jakob, M. (1964). *Heat transfer*. Wiley, New York.

Jost, W. (1960). *Diffusion in solids, liquids and gases*, p. 76. Academic Press, New York.

Kasama, A., Iida, T., and Morita, Z. (1976). *J. Japan Inst. Metals* **40**, 1030.

Kolmogorov, A. N. (1941). *Dokl. Akad. Sci. SSSR* **30**, 301; **31**, 538; **32**, 16.

Kreith, F. and Black, W. Z. (1980). *Basic heat transfer*. Harper and Row, New York.

Kubachewski, O. and Alcock, C. B. (1983), *Metallurgical thermochemistry* (5th edn). Pergamon Press, Oxford.

Lamb, H. (1945). *Hydrodynamics* (6th edn). Dover, New York.

Launder, B. E. and Spalding, D. B. (1972). *Lectures in mathematical models of turbulence*. Academic Press, New York.

Levenspiel, O. (1972) (2nd edn). *Chemical reaction engineering*. Wiley, New York.

Lewis and Whitman (1924). *Ind. Eng. Chem.* **16**, 1215–20.

Lewis, R. D., Foreman, J. W. Watson, H. J., and Thornton, J. R. (1968). *Phys. Fluids* **11**, 433.

Lorenz, L. (1872). *Ann. Phys. Chem.* **147**, 429.

Margolus, N., Toffoli, T., and Vichniac, G. (1986). *Phys. Rev. Lett.* **16**, 1694–6.

Markatos, N., Rhodes, N., and Tatchell, D. G. (1982). A general purpose program for the analysis of fluid flow problems. In *Numerical methods for fluid dynamics* (ed. K. W. Morton and M. J. Baines), pp. 463–80. Academic Press, London.

Mazumdar, D. and Guthrie, R. I. L. (1987). An analysis of numerical methods for solving the particle trajectory equation. *J. Appl. Math. Modelling* **12** (7), pp. 398–402.

Nikuradse, J. (1932). *Ver Deutsch. Ing. Forsch* **356**, 21.

North American Manufacturing Company (1965). *North American combustion handbook*. Cleveland, Ohio.

Nusselt, W. (1931). *Forsch. Geb. Ingenieurwes* **2**, 309.

Ogino, K. and Hara, S. (1977). *Tetsu to Hagane* **63**, No. 13, 2141–51.

Osida, I. (1939). *Proc. Phys. Math. Soc. Japan* **21**, 353.

Patera, A. T. (1984). *J. Comp. Phys* **54**, 468–88.

Peaceman, D. W. and Rachford, H. H. (1955). *J. Soc. Ind. Appl. Math* **3**, 28.

Pehlke, R. D. (1983). *Unit processes of extractive metallurgy*. Elsevier, New York.

Perrin, R., Guthrie, R. I. L., and Stonehouse B. C. (1987). The process technology of batch annealing, *Proc. 1987 Mechanical Working and Steel Processing Proceedings*, ISS of AIME, pp. 261–8.

Perry, R. H. and Chilton, C. H. (1973). *Chemical engineers' handbook* (5th edn). McGraw-Hill, New York.

Perry, R. H. and Green, D. (1984). *Chemical engineers' handbook*. McGraw-Hill, New York.

Prandtl L. (1945). *Nachr. Akad. Wiss. Göttingen*, cited by Launder and Spalding (1972).

Prandtl, L. and Tietjens O. G. (1957). *Applied hydro- and aeromechanics*. Dover Publications, New York.

Protopapas, P. and Parlee, N. A. D. (1976). *High Temp. Sci.* **8**, 141.

Rayleigh (1915). *Nature* **15**, 66.

Richardson, F. D. (1974). *Physical chemistry of metals in metallurgy* 2, p. 109. Academic Press, New York.

Richardson, F. D. and Jeffes, J. (1948). *J. Iron Steel Inst.* **160**, 261.

Rodi, W. (1980). *Turbulence models and their application in hydraulics—a state of the art review*. Int. Assoc. for Hydraulic Research, 1980.

Sahai, Y. and Guthrie, R. I. L. (1982). Viscosity models for gas stirred ladles. *Metall. Trans.* **13B**, 125.

Salcudean, M. and Guthrie, R. I. L. (1979). *Metall. Trans. B.* **10B**, 423–8.

Scarborough, J. B. (1958), *Numerical mathematical analysis* (4th edn). Johns Hopkins Press, Baltimore.

Schack, A. (1965). *Industrial heat transfer*. Wiley, New York.

Schlichting, H. (1955). *Boundary layer theory*. Ch. 5, §10. McGraw-Hill, New York.

Schlichting, H. (1979). *Boundary layer theory* (7th edn.). McGraw-Hill, New York.

Seban, R. A. and Shimazaki, T. T. (1951). *Trans. ASME* **73**, 803.

Singh, N., Birkekak, R. C., and Drake, R. M. (1969). *Prog. Heat Mass Transfer* **12**, 87.

Smigelskas, A. D. and Kirkendall, E. O. (1947). *Trans. AIME* **171**, 130.

Solar, M. Y. and Guthrie, R. I. L. (1972). *Metall. Trans.* **3**, 2007.

Spalding, D. B. and Launder, B. E. (1974). *Comp. Meth. in Mech. Eng.* **3**, 269.

Steinberger, R . L. and Treybal, R. E. (1960). *AIChEJ* **6**, 327.

Sutherland, W. (1905), *Phil. Mag.* **9**, 781.

Szekely, J. (1964), *J. Iron Steel Inst.* **6**, 505–8.

Szekely, J., Evans, J. W., and Brimacombe, J. K. (1988). *The mathematical and physical modeling of primary metals processing operations.* Wiley, New York.

Tanaka, S. and Guthrie, R. I. L. (1985). Proc. Japan–Canada Seminar on Secondary Steelmaking Refining, Casting, Physical Metallurgy and Properties. Can. Steel Ind. Res. Assoc. & Iron Steel Inst. Japan, C-3-1 to C-3-19.

Welty, J. R., Wicks, C. E., and Wilson, R. E. (1976). *Fundamentals of momentum, heat, and mass transfer* (2nd edn). Wiley, New York.

Whitaker, S. (1972). *AIChEJ* **18**, 361.

White, F. M. (1979). *Fluid mechanics.* McGraw-Hill, New York.

Williams, E. J. and Burton, E. J. (1955). *J. Iron Steel Inst.* **179**, 17–22.

Witte, L. C. (1968). *J. Heat Transfer* **90**, 9.

Yuge, T. (1960). *J. Heat Transfer, Sec. C.* **82**, 214–38.

AUTHOR INDEX

Alcock 431
Amsden 411
Andrade 50, 52
Arkharov 29
Asle, M. J. 70

Baliga 411
Batchelor, G. K. 150
Bernoulli 123, 348
Bird, R. B. 44, 69, 359
Black 295, 296, 359
Blasius 367
Boldt, J. R. 69
Boltzmann, L. 20
Bolz, R. E. 70
Boni 440
Borelius 454
Bradshaw, P. 149
Bridgman, P. W. 211
Brimacombe, J. K. 421
Buckingham 152, 155
Burton 183
Byrne 127

Cahn 227
Carnahan 385, 421
Carslaw, M. S. 264, 285
Chapman 40, 43, 44
Chiesa 299
Chilton, C. H. 70, 434
Chipman 434
Christian 227
Clift, R. 150
Cohen 56
Colburn 296
Cole, G. H. A. 69
Compaan 255
Courant 406
Cowling 43
Crank, J. 264, 285
Croxton, C. A. 45, 49, 51, 69

Daily, J. W. 150
Danckwerts
Darcy 139
Darken, L. S. 226, 285, 359
Darken, L. S. 359
Davies, J. T. 149
Derge 440
D'Alembert 123

Einstein 56, 57, 66
Emley, E. F. 69
Enskog 40, 43, 44
Euler 123
Evans, J. W. 211, 421
Eyring 50, 56

Fick 212
Fourier 18, 212
Frenkel 50
Fuwa 434

Geiger, G. H. 35, 69, 285
Gourtsoyannis 30, 64
Grace, J. R. 150
Grieveson 29
Grimison 297
Gurry, R. W. 285, 359
Guthrie 30, 31, 47, 50, 51, 56, 59, 61, 69, 70, 150, 175, 295, 305, 335, 396, 411, 412, 413, 415, 421

Hara 440
Harleman, D. R. F. 150
Harlow 408, 411
Harries 396
Hatch, J. E. 69
Haven 255
Higbie 318
Hills 342
Holman, J. P. 359
Hottel 277
Hultgren 433

Iida 31, 47, 50, 51, 56, 59, 61, 64, 69, 70
Irons 335, 412

Jaeger 264, 285
Jakob, M. 285, 298
John, H. M. 211
Johnstone, R. E. 211
Joo 419

Kasama 52
Kirkendall 254
Kirkwood 45
Kolmogorov 114

SUBJECT INDEX